茂兰研究·4

中国茂兰两栖爬行动物

Amphibians & Reptiles of Maolan, China

邓怀庆　姚正明　周 江　**主编**

贵　州　师　范　大　学
贵州茂兰国家级自然保护区管理局

科学出版社
北　京

内 容 简 介

本书介绍了中国茂兰地区两栖爬行动物94种，其中两栖动物2目8科18属32种（含1未定种），爬行动物2目12科39属62种，包含新种1种：荔波异角蟾；茂兰地区特有种3种：荔波壁虎、茂兰瘰螈、荔波睑虎。多数物种都配有彩色照片和在茂兰地区的分布图，以及详细的形态、生态环境描述和可量性状的测量数据。

本书是迄今对茂兰地区两栖爬行动物最系统的考察总结，可供从事动物研究的科研人员、大中专院校师生、野生动物保护组织及爱好者识别野外两栖爬行动物时参考，可作为农林等相关部门及保护区进行管理执法、宣传教育等的材料，也可作为中小学生了解大自然的科普读物。

图书在版编目（CIP）数据

中国茂兰两栖爬行动物/邓怀庆，姚正明，周江主编. — 北京：科学出版社，2019.11
ISBN 978-7-03-061541-1

Ⅰ.①中… Ⅱ.①邓…②姚…③周… Ⅲ.①两栖动物–爬行纲–荔波县 Ⅳ.①Q959.608

中国版本图书馆CIP数据核字（2019）第109279号

责任编辑：张会格　付　聪　赵小林/责任校对：郑金红/责任印制：肖　兴
封面设计：北京图阅盛世文化传媒有限公司
装帧设计：北京美光设计制版有限公司

科学出版社 出版
北京东黄城根北街16号
邮政编码：100717
http://www.sciencep.com

北京汇瑞嘉合文化发展有限公司 印刷
科学出版社发行　各地新华书店经销

*

2019年11月第　一　版　开本：889×1194　1/16
2019年11月第一次印刷　印张：16 1/2
字数：535 000

定价：248.00元

（如有印装质量问题，我社负责调换）

《中国茂兰两栖爬行动物》编委会

主编单位

贵州师范大学

贵州茂兰国家级自然保护区管理局

主　　编

邓怀庆　姚正明　周　江

副 主 编

覃龙江　谭成江　李　刚　龙　健　王万海　罗　涛

编　　委

（以姓氏笔画为序）

王万海　邓怀庆　玉　萍　龙　健　兰洪波　刘　弢　刘绍飞
李　刚　杨　旭　杨亮亮　吴尚川　宋　波　陈　健　范祥迪
罗　涛　周　江　周　俊　洪之国　洪文青　费仕鹏　姚　芊
姚正明　莫家伟　高　凯　常开奎　覃龙江　曾　祥　蒙建国
谭成江　熊志斌

序 | Foreword

茂兰，一个充满诗意的名字，一个生命力旺盛的家园，拥有极其丰富的物种多样性和独特性。

目前，贵州省内已知两栖爬行动物不足200种，而在茂兰地区有94种，茂兰物种之丰富，种类之独特，在我国乃至世界也是不多见的。能孕育如此丰富的物种多样性的地方其生态也必然是独特的。

贵州喀斯特地貌极其发育，拥有各个发育阶段的喀斯特地貌类型，尤为著名的是荔波的"中国南方喀斯特"世界自然遗产地。历史上对于贵州脊椎动物物种多样性的调查和动物地理学的研究未能系统深入地开展，尤其是对于不同类型与发育阶段的喀斯特生态系统中两栖和爬行动物的物种多样性及动物地理学的研究缺乏系统性的成果。该书的出版可以弥补这方面的遗憾。

回顾贵州两栖动物的研究历史，自1945年Boring在《中国两栖类》一书中描述了贵州的7种两栖动物以后，20世纪50～60年代，中国科学院成都生物研究所的刘承钊、胡淑琴、赵尔宓、费梁、叶昌媛等先生在贵州毕节、梵净山、雷公山、安龙、兴义、罗甸等地进行了一系列的调查，将贵州分布的两栖动物记录增加至47种；1974～1979年，遵义医学院的伍律、董谦、须润华、李德俊、刘积琛等对贵州两栖动物做了进一步的调查，使贵州的两栖动物记录增加至62种；其后，六盘水师范学院的田应洲和贵州师范大学的谷晓明等对贵州两栖纲有尾目进行了深入的调查研究，发现了瘰螈属尾斑瘰螈种组的龙里瘰螈、织金瘰螈、武陵瘰螈和茂兰瘰螈4个新种；贵州师范大学的郑建洲及贵阳学院的魏刚、徐宁等也对贵州的两栖动物进行了大量的调查及研究。迄今，贵州记录的两栖动物物种数量增加至80种。

贵州爬行动物的研究始于1935年，Pope在《中国爬行动物》一书中总结了贵州9种爬行动物。其后，不断有爬行动物省级新纪录被报道，1962年刘承钊先生在贵州毕节、威宁及水城调查时发现爬行动物新纪录4种；1963年胡淑琴等在贵州印江、梵净山、雷山、安龙、兴义及罗甸等地调查时发现47种省级新纪录，将贵州爬行动物记录增加至85种。1974～1980年，遵义医学院的伍律及李德俊等对贵州爬行动物进行了广泛深入的调查，记录了爬行动物105种。此后，又不断有新纪录被报道，当前贵州有爬行动物110种。

2015～2017年，贵州茂兰国家级自然保护区管理局与贵州师范大学共同开展了茂兰地区两栖爬行动物资源调查，调查共发现94种两栖爬行动物，其中发现了新种：荔波异角蟾，茂兰地区新纪录：凹顶泛树蛙、无声囊泛树蛙、南草蜥、蟒蛇、颈棱蛇等。这些珍贵的野生动物丰富了茂兰地区的生物多样性，维持了茂兰地区的生态系统稳定。

2018年，由贵州师范大学邓怀庆博士和周江教授、贵州茂兰国家级自然保护区管理局姚正明先生为主要编写人员编写的《中国茂兰两栖爬行动物》一书，图文并茂，既有物种的学名、详细的形态描述、鉴别特征描述、生活环境描述，也有物种的彩色图片，同时配备了物种在茂兰地区的分布图，不仅可以帮助读者了解茂兰地区丰富的两栖爬行动物资源，为茂兰地区两栖爬行动物的保护及监测提供基础资料，也可以作为科研人员及动物爱好者识别野生两栖爬行动物的图鉴，还可以作为中小学生了解大自然的科普读物。

《中国茂兰两栖爬行动物》不仅仅是对茂兰地区生物多样性物种资源的介绍，更是反映中国南方喀斯特自然遗产地保护成效的专著。相信此书将成为人们了解和探究中国茂兰地区及喀斯特环境中的两栖爬行动物的极佳载体，也将成为人们了解和探究中国南方喀斯特的一扇窗户。

中国科学院成都生物研究所 研究员
亚洲两栖爬行动物研究杂志 主编
2018年12月5日于成都

前 言 | Preface

茂兰地区地处贵州黔南布依族苗族自治州荔波县。这里山清水秀，安静祥和。这里的动物就像一个个精灵，与山、水、森林、岩洞、瀑布相互交融，那样和谐，那样自然，处处彰显着喀斯特森林生态环境的神奇与完美。

茂兰自然保护区始建于1986年，1988年被国务院批准为国家级自然保护区，1996年加入联合国教科文组织"人与生物圈"保护区网络。2007年被收入联合国教科文组织的《世界遗产名录》，2011年荔波县茂兰地区被列入中国十大美丽乡镇。茂兰国家级自然保护区（以下简称：茂兰保护区）位于云贵高原南缘，面积约2万hm^2，地理位置为北纬$25°09'20''\sim 25°20'50''$，东经$107°52'10''\sim 108°45'40''$，属中亚热带季风湿润气候，具有世界上保存最完好的喀斯特森林资源，孕育了众多的生物资源，主要保护对象为亚热带喀斯特森林生态系统及其珍稀野生动植物资源。两栖爬行动物是原始的陆生脊椎动物，在动物由水生向陆生进化过程中占据着非常重要的地位，是研究动物系统发育的良好材料，也是生态环境保护成效的重要指示物种，具有较高的生态和经济价值。

2015年，贵州茂兰国家级自然保护区管理局与贵州师范大学合作，共同开展茂兰保护区两栖爬行动物本底资源调查。2015年6月至2017年8月对两栖爬行动物进行了4次系统调查，探访到94种生活在此地的两栖爬行动物。其中，两栖动物2目8科18属32种，占贵州两栖动物种数的40.0%，比早期调查的两栖动物记录（2目6科19种）多了13种（贵州省林业厅，1987）；爬行动物2目12科39属62种，占贵州爬行动物种数的56.4%，比《贵州茂兰国家级自然保护区生物多样性保护及发展对策》（2011年）上记载的爬行动物（3目10科47种）多了15种。本次调查发现两栖动物新种1种，即荔波异角蟾；发现树蛙属未定种1种。茂兰地区两栖爬行动物中有国家Ⅰ级重点保护野生动物蛇类1种，即蟒蛇；国家Ⅱ级重点保护野生动物两栖类3种，即虎纹蛙、文县疣螈、细痣疣螈。荔波县茂兰地区特有种3种：荔波壁虎、茂兰瘰螈、荔波睑虎。茂兰地区丰富的生物多样性资源有效地维持了该地区生态系统的稳定。此次调查丰富了中国茂兰地区的两栖爬行动物物种名录。

《中国茂兰两栖爬行动物》一书将有助于人们了解茂兰地区丰富的野生动物资源，可为茂兰地区两栖爬行动物物种多样性的长期监测及保护提供基础资料。本书依据野外实地调查数据，参考国内外相关两栖爬行动物鉴定资料，详细描述了本次采集的两栖爬行动物形态特征，并对前人文献中存疑种类进行了分析。需要说明的是，本书科、属的世界分布仅记录大片区的分布地。本书既可以作为相关科研人员、野生动物保护组织及爱好者识别野外两栖爬行动物的手册，也可作为中小学生了解大自然的科普读物。

感谢贵州茂兰国家级自然保护区管理局姚正明、谭成江、覃龙江先生及茂兰保护区各管理站工作人员给予本次野外调查工作的支持与帮助。本次工作主要由贵州师范大学生命科学学院邓怀庆博士和周江教授及贵州茂兰国家级自然保护区管理局姚正明先生负责。模式标本核对得到了中国科学院成都生物研究所王跃招、戴强老师的帮助；野外调查主要由李刚、罗涛、洪之国、龙健、刘殿、周俊、曾祥、王万海、蒙建国、费仕鹏、吴尚川、刘绍飞、莫家伟、姚芊、玉萍、兰洪波、熊志斌、常开奎、范祥迪、杨旭、陈健、高凯、宋波、吴贵才、蒋奉昌、陈斌、杨东妹、陈应宽、陶冬、张然等完成，同时得到安徽大学张保卫教授团队的帮助（参与了野外调查及物种鉴定工作）；标本数据的测量由常开奎、范祥迪、王作波、杨旭、孙雯、张天燕、韩兰、沈鑫鑫、黄娜婷、熊远林等完成；文本材料收集及标本照片拍摄主要由罗涛、李刚完成；文本编辑及校对主要由洪文青、杨亮亮完成。在此表示诚挚的谢意！

由于作者经验不足，水平有限，在编写过程中难免有所疏漏，敬请读者批评指正。

2018年9月于贵阳相宝山

目 录 | Contents

序
前言
第一章　两栖爬行动物概述 ·· 1
　　一、茂兰地区两栖爬行动物研究简史 ·· 2
　　　　（一）两栖动物研究简史 ··· 2
　　　　（二）爬行动物研究简史 ··· 2
　　二、茂兰地区自然概况 ·· 2
　　　　（一）地理位置 ··· 2
　　　　（二）地质地貌 ··· 3
　　　　（三）气候特征 ··· 3
　　　　（四）水系 ·· 3
　　　　（五）植被概况 ··· 4
　　三、茂兰地区两栖爬行动物调查方法 ·· 4
　　　　（一）两栖动物的调查方法 ·· 4
　　　　1. 样线法 ··· 4
　　　　2. 捕尽法 ··· 4
　　　　（二）爬行动物的调查方法 ·· 4
　　　　1. 编目法 ··· 4
　　　　2. 样带法 ··· 5
　　四、茂兰地区两栖爬行动物物种多样性 ·· 5
　　　　（一）两栖动物多样性 ·· 5
　　　　（二）两栖动物分布地人为干扰情况 ·· 5
　　　　（三）爬行动物多样性 ·· 6
　　五、茂兰地区两栖爬行动物区系及其地理区划 ··· 6
第二章　两栖动物术语及分类描述 ··· 11
　　一、两栖纲分类检索名词术语 ·· 12
　　　　（一）有尾目特征及分类检索常用术语 ·· 12

1. 成体形态结构及说明 ·· 12
2. 形态量度 ··· 12
3. 外部形态特征常用术语 ··· 13
（二）无尾目特征及分类检索常用术语 ··· 13
1. 成体形态结构及说明 ·· 13
2. 成体外形量度 ·· 13
3. 外部形态特征常用术语 ··· 14
4. 皮肤表面结构 ·· 18
5. 骨骼系统 ··· 19
    1）肩带与胸骨组合的类型 ·· 19
    2）椎体类型 ·· 20

二、茂兰地区两栖动物物种分类描述 ··· 21

（一）有尾目 CAUDATA ··· 21

1. 蝾螈科 Salamandridae Goldfuss, 1820 ·· 21
    1）瘰螈属 *Paramesotriton* Chang, 1935 ·· 22
    2）疣螈属 *Tylototriton* Anderson, 1871 ··· 24

（二）无尾目 ANURA ·· 29

2. 角蟾科 Megophryidae Bonaparte, 1850 ·· 29
    3）异角蟾属 *Xenophrys* Günther, 1864 ·· 30
3. 蟾蜍科 Bufonidae Gray, 1825 ·· 32
    4）蟾蜍属 *Bufo* Garsault, 1764 ·· 32
    5）头棱蟾属 *Duttaphrynus* Frost et al., 2006 ·· 38
4. 雨蛙科 Hylidae Rafinesque, 1815 ·· 40
    6）雨蛙属 *Hyla* Laurenti, 1768 ··· 41
5. 姬蛙科 Microhylidae Günther, 1858 ··· 47
    7）姬蛙属 *Microhyla* Tschudi, 1838 ··· 47
6. 叉舌蛙科 Dicroglossidae Anderson, 1871 ··· 55
    8）陆蛙属 *Fejervarya* Bolkay, 1915 ·· 55
    9）虎纹蛙属 *Hoplobatrachus* Peters, 1863 ·· 58
    10）双团棘胸蛙属 *Gynandropaa* Dubois, 1992 ······································ 61
    11）棘胸蛙属 *Quasipaa* Dubois, 1992 ··· 65
7. 蛙科 Ranidae Batsch, 1796 ·· 71
    12）琴蛙属 *Nidirana* Dubois, 1992 ·· 72
    13）水蛙属 *Hylarana* Tschudi, 1838 ··· 75
    14）臭蛙属 *Odorrana* Fei, Ye and Huang, 1990 ····································· 80

15）侧褶蛙属 *Pelophylax* Fitzinger, 1843 ········· 90
　8. 树蛙科 Rhacophoridae Hoffman, 1932 (1858) ········· 93
　　　16）原指树蛙属 *Kurixalus* Ye, Fei and Dubois, 1999 ········· 94
　　　17）泛树蛙属 *Polypedates* Tschudi, 1838 ········· 97
　　　18）树蛙属 *Rhacophorus* Kuhl and van Hasselt, 1822 ········· 101

第三章　爬行动物术语及分类描述 ········· 109
一、爬行纲分类检索名词术语 ········· 110
　（一）龟鳖目特征及分类检索常用术语 ········· 110
　　1. 龟壳背甲的骨板 ········· 110
　　2. 龟壳背甲的盾片 ········· 110
　　3. 龟壳腹甲的骨板 ········· 110
　　4. 龟壳腹甲的盾片 ········· 110
　　5. 龟壳的甲桥 ········· 110
　（二）有鳞目特征及分类检索常用术语 ········· 111
　　Ⅰ. 蜥蜴亚目 ········· 111
　　1. 头部的鳞片及相关结构 ········· 111
　　　1）头背的鳞片及相关结构 ········· 111
　　　2）头侧的鳞片及相关结构 ········· 111
　　　3）头腹的鳞片及相关结构 ········· 112
　　2. 躯干部的鳞片 ········· 112
　　3. 四肢的鳞片及相关结构 ········· 112
　　4. 尾部的鳞片 ········· 113
　　5. 蜥蜴各部分的量度 ········· 113
　　Ⅱ. 蛇亚目 ········· 114
　　1. 头部的鳞片 ········· 114
　　　1）头背的鳞片 ········· 114
　　　2）头侧的鳞片 ········· 114
　　　3）头腹的鳞片 ········· 116
　　2. 躯干部的鳞片 ········· 116
　　3. 尾部的鳞片 ········· 117
　　4. 鳞片的排列方式及其结构 ········· 117
　　5. 牙齿的类型 ········· 117
　　6. 其他与分类检索有关的名词术语 ········· 118
　　7. 斑纹的描述 ········· 118

二、茂兰地区爬行动物物种分类描述 ·················· 118
 （一）龟鳖目 TESTUDINES ·················· 118
  1. 鳖科 Trionychidae Bell, 1828 ·················· 118
   1）鳖属 *Pelodiscus* Fitzinger, 1835 ·················· 119
 （二）有鳞目 SQUAMATA ·················· 120
  Ⅰ. 蜥蜴亚目 Lacertilia ·················· 120
  2. 睑虎科 Eublepharidae Boulenger, 1883 ·················· 121
   2）睑虎属 *Goniurosaurus* Barbour, 1908 ·················· 121
  3. 壁虎科 Gekkonidae Gary, 1825 ·················· 124
   3）壁虎属 *Gekko* Laurenti, 1768 ·················· 124
  4. 石龙子科 Scincidae Gary, 1825 ·················· 127
   4）蜓蜥属 *Sphenomorphus* Fitzinger, 1843 ·················· 127
   5）石龙子属 *Plestiodon* Wiegmann, 1834 ·················· 129
  5. 蜥蜴科 Lacertidae Gray, 1825 ·················· 134
   6）草蜥属 *Takydromus* Daudin, 1802 ·················· 134
  6. 蛇蜥科 Anguidae Gray, 1825 ·················· 138
   7）脆蛇蜥属 *Dopasia* Daudin, 1803 ·················· 139
  7. 鬣蜥科 Agamidae Gray, 1827 ·················· 141
   8）棘蜥属 *Acanthosaura* Gray, 1831 ·················· 141
   9）拟树蜥属 *Pseudocalotes* Fitzinger, 1843 ·················· 143
  Ⅱ. 蛇亚目 Serpentes ·················· 145
  8. 蚺科 Pythonidae Fitzinger, 1826 ·················· 146
   10）蚺属 *Python* Daudin, 1803 ·················· 146
  9. 钝头蛇科 Pareatidae Romer, 1956 ·················· 147
   11）钝头蛇属 *Pareas* Wagler, 1830 ·················· 148
  10. 蝰科 Viperidae Bonaparte, 1840 ·················· 151
   12）原矛头蝮属 *Protobothrops* Hoge and Romano-Hoge, 1983 ·················· 151
   13）尖吻蝮属 *Deinagkistrodon* Gloyd, 1979 ·················· 155
   14）绿蝮属 *Viridovipera* Malhotra and Thorpe, 2004 ·················· 157
   15）竹叶青属 *Trimeresurus* Lacepède, 1804 ·················· 159
   16）亚洲蝮属 *Gloydius* Hoge and Romano-Hoge, 1981 ·················· 161
  11. 眼镜蛇科 Elapidae Boie, 1827 ·················· 163
   17）中华珊瑚蛇属 *Sinomicrurus* Slowinski, Boundy and Lawson, 2001 ·················· 163
   18）眼镜蛇属 *Naja* Laurenti, 1768 ·················· 166
   19）环蛇属 *Bungarus* Daudin, 1803 ·················· 168

12. 游蛇科 Colubridae Cope, 1893 ··170
　20）两头蛇属 *Calamaria* Boie, 1826 ···171
　21）斜鳞蛇属 *Pseudoxenodon* Boulenger, 1890 ···173
　22）剑蛇属 *Sibynophis* Fitzinger, 1843 ···177
　23）林蛇属 *Boiga* Fitzinger, 1826 ···179
　24）小头蛇属 *Oligodon* Boie, 1827 ···182
　25）三索蛇属 *Coelognathus* Fitzinger, 1843 ···188
　26）翠青蛇属 *Cyclophiops* Boulenger, 1888 ···189
　27）鼠蛇属 *Ptyas* Fitzinger, 1843 ···191
　28）绿蛇属 *Rhadinophis* Vogt, 1922 ···196
　29）链蛇属 *Lycodon* Boie, 1826 ···198
　30）玉斑蛇属 *Euprepiophis* Fitzinger, 1843 ···205
　31）紫灰蛇属 *Oreocryptophis* Cantor, 1839 ···206
　32）晨蛇属 *Orthriophis* Utiger et al., 2002 ···208
　33）腹链蛇属 *Amphiesma* Duméril, Bibron and Duméril, 1854 ·································213
　34）东亚腹链蛇属 *Hebius* Thompson, 1913 ···215
　35）颈槽蛇属 *Rhabdophis* Fitzinger, 1843 ···221
　36）异色蛇属 *Xenochrophis* Günther, 1864 ···225
　37）后棱蛇属 *Opisthotropis* Günther, 1872 ···228
　38）华游蛇属 *Sinonatrix* Rossman and Eberle, 1977 ···230
　39）颈棱蛇属 *Macropisthodon* Boulenger, 1893 ···234

主要参考文献 ··236
中文名索引 ···246
拉丁名索引 ···248

第一章
两栖爬行动物概述

一、茂兰地区两栖爬行动物研究简史

（一）两栖动物研究简史

贵州茂兰地区地处中亚热带季风湿润气候区，为典型的喀斯特峰丛漏斗和峰丛洼地地貌。保护区内地貌复杂多样，野生动植物种类繁多，是名副其实的科研圣地，一直吸引着国内外研究团队前来科考调查。近年来，随着科研技术的不断进步，保护区内不断地有新种被发现。在两栖动物研究方面，1984年，李德俊、魏刚等对贵州荔波县茂兰地区进行了首次两栖动物调查，调查历时32天，采集到两栖动物2目6科6属19种，其中锯腿原指树蛙（锯腿水树蛙）、阔褶水蛙为贵州省内新纪录。2007年4月至2009年7月，熊洪林、刘燕等对荔波县茂兰保护区内泽陆蛙的繁殖和生态进行了研究。2006年，贵州师范大学的谷晓明教授等在茂兰保护区采集到两栖动物有尾目蝾螈科瘰螈属1新种，命名为茂兰瘰螈 Paramesotriton maolanensis，该新种的文章于2012年发表在 Zootaxa 上。2015年6月至2017年8月，贵州师范大学生命科学学院的邓怀庆博士及周江教授的科研团队（30多人）先后在荔波县茂兰地区对两栖动物进行了4次调查，总计调查时间达3个月，调查到两栖动物32种，隶属于2目8科18属。调查期间研究生李刚、陈健、许铁龙等在茂兰保护区周边一洞穴内采集到两栖动物无尾目角蟾科异角蟾属1新种，命名为荔波异角蟾 Xenophrys liboensis，该新种的文章于2017年在SCI期刊 Asian Herpetological Research 上发表。另外，调查期间还采集到树蛙属1未定种，该种与侏树蛙和白线树蛙比较相似，但又有一定的区别，可能为1新种。

（二）爬行动物研究简史

1984年，李德俊、魏刚等在对荔波县茂兰保护区进行两栖动物调查的同时也做了爬行动物的调查，采集到爬行动物3目9科27属44种，其中贵州省内新纪录5种，分别为四川龙蜥 Japalura szechwanensis、丽棘蜥 Acanthosaura lepidogaster、细鳞拟树蜥 Pseudocalotes microlepis、里氏睑虎 Goniurosaurus lichtenfelderi、山烙铁头蛇指名亚种 Ovophis monticola monticola。2010年，杨剑焕等在茂兰地区采集到蝰科原矛头蝮属1新种，命名为茂兰原矛头蝮 Protobothrops maolanensis。2012年，贵州师范大学生命科学学院周江教授应荔波县世界自然遗产管理局之邀，带领团队成员对荔波县进行了一次两栖爬行动物的调查，记录了爬行动物46种，隶属于3目10科。2015年6月至2017年8月，贵州师范大学生命科学学院的邓怀庆博士及周江教授的科研团队在前人调查结果的基础上，对荔波县茂兰地区的爬行动物进行了系统调查，历时100多天，确定茂兰地区共有爬行动物62种，隶属于2目12科39属，其中采集标本45种，隶属于2目9科32属。

二、茂兰地区自然概况

（一）地理位置

茂兰地区位于贵州黔南州荔波县，毗邻广西木论国家级自然保护区，与广西河池环江毛南族自治县（以下简称：环江）接壤。地理位置为北纬25°09′20″～25°20′50″，东经107°52′10″～108°45′40″。这一区域地处中亚热带季风湿润气候区，为典型的喀斯特峰丛漏斗和峰丛洼地地貌，主要保护对象为亚热带喀斯特森林生态系统及其珍稀野生动植物资源。

茂兰地区以茂兰保护区为核心地带。保护区海拔430～1078.6m，总面积21 285hm^2。其中，核心区8305hm^2，缓冲区8130hm^2，实验区4850hm^2。保护区内开发成为生态旅游区的面积为1915hm²。保护区森林覆盖率为88.61%，核心区达92%。

在世界动物地理区划上属于东洋界，中国动物地理区划上属于西南区，处于华南区和华中区动物相互交汇地带。保护区内地貌类型多样，森林植被类型丰富，为动物的栖息提供了多样化的生境条件。

（二）地质地貌

茂兰地区具有全球最大的原始喀斯特森林生态系统，是中国西南地区生物多样性最丰富的地区之一，是地球同纬度地区残存下来的一片面积最大、分布集中、原生性强、相对稳定的喀斯特森林生态系统。中国茂兰地区处于贵州高原向广西丘陵平原过渡的斜坡地带，地势西北高东南低，最高海拔1078.6m，最低海拔430m，平均海拔880m，山峰与洼地相对高差为150～300m。喀斯特地貌发育十分完全，形态多样，主要有落水洞、漏斗、洼地、盆地、峰林、峰丛等。

茂兰地区的土壤类型分为红壤、黄壤、红色石灰土、黑色石灰土、紫色土、潮土、水稻土7个土类。其中以黑色石灰土、黄壤、红壤为主。在古生代，荔波县茂兰地区为海侵区，沉积物以浅海相碳酸盐类为主，于中生代末褶皱隆起，脱离海侵，上升为陆地。新生代造山运动后持续上升，形成了强烈溶蚀的喀斯特峰林、峰丛低山山地。该区喀斯特地貌十分发育，喀斯特形态多种多样，锥峰高耸而密集，洼地深陷而陡峭，锥峰洼地层层叠叠，呈现出峰峦叠嶂的喀斯特峰丛景观。地貌类型以峰丛漏斗和峰丛洼地为主，仅东部有小面积的峰丛谷地分布。在地貌上，该区处于宽缓褶皱构造控制的宽阔河间地带，东濒三岔河，西临樟江。中国西南部的大部分喀斯特地区已遭到石漠化的侵扰。而茂兰地区的洞塘、翁昂、永康喀斯特形态和类型（如峰丛洼地与峰丛坡立谷）极度发育，森林覆盖率极高。在碳酸盐岩石构成的裸岩山间和嶙峋岩石上竟异乎寻常地滋生出连片乔木林，乔木林面积近20 000hm²，覆盖率达88.61%，是世界上同纬度地区残存下来的仅有的一片分布集中、原生性强、相对稳定的喀斯特森林生态系统。

（三）气候特征

根据气候带划分，茂兰地区属于中亚热带季风湿润气候区。由于地理纬度较低，区内山峦起伏，河谷较深，地形地貌复杂，气候类型多样，具有亚热带高原山地季风湿润气候特征。茂兰地区总的气候特征是：气候温热，四季分明，冬无严寒，夏无酷暑，夏长冬短，无霜期长，雨量充沛，日照尚足，雨热同季，灾害性天气少。

茂兰地区平均气温为18.3℃，气温分布的总趋势是南高北低。保护区内地形复杂，气候垂直变化和地形小气候明显。最热月为7月，最冷月为1月。7月平均气温为27.0℃左右，极端最高气温不超过40.0℃；1月平均气温在5.5℃左右，极端最低气温可达-10.0℃以下。保护区内霜期短，无霜期在270天以上。初霜期和终霜期分别约在12月中旬和2月上旬。降水主要集中在夏季。6～8月各月雨量均在2000mm以上，占全年总雨量的50%左右；冬季（12月至次年2月）仅占全年总雨量的5%左右；秋季（9～11月）占全年总雨量的15%左右；春季（3～5月）降水占全年雨量的30%左右。4月下旬海洋季风逐渐增强，进入雨季，4～10月雨量占全年总雨量的81%。10月以后，海洋季风减弱，逐渐被南下的大陆季风取代，降水量显著减少。

（四）水系

茂兰地区地表水系不发育，区内瑶兰河、瑶所河、板寨河及洞腮河分属区内与之相应的4条地下河系，是地下河明流的局部段落。河流流量较小，枯流量一般为50～100L/s。板寨河、瑶兰河坡降小于1.8%，瑶所河、洞腮河注入山谷时形成总落差60～80m的跌水。该地地下水存在着二元结构，即枯枝落叶垫积层充填的上层喀斯特裂隙水与下层喀斯特水同时并存，上层水流小且动态稳定，下层水流大，动态变化也相对较大，这是茂兰地区喀斯特森林所具有的独特的水文地质现象。喀斯特水来源于碳酸盐含水岩组，以常见的下降泉、上升泉、多湖泉、喀斯特潭及地下河出入口和地下河天窗等形式出露。上层喀斯特裂隙水是森林植被对喀斯特作用的具体表现形式，主要是由喀斯特森林的持水作用形成，凡喀斯特森林茂密的山麓坡下、漏斗洼地中、山坳上均有分布，使茂兰地区内到处是泉水淙淙、流水潺潺。茂兰地区水质类型为 HCO_3-$Ca+Mg$ 水及 HCO_3-Ca 水，硬度 $8°\sim11°$，属微硬水；pH $7.2\sim7.7$，属中性水；固体物含量 $140\sim190mg/L$，是很好的饮用水。

（五）植被概况

茂兰地区的植被区系在中国植被分区上处于亚热带常绿阔叶林区东部（湿润）常绿阔叶林亚区中亚热带常绿阔叶林带。自然植被以发育在喀斯特地貌上的原生性的常绿落叶阔叶混交林为主，是一种非地带性的植被。中国茂兰地区内的自然植被分为以下几种类型：①常绿落叶阔叶混交林；②针阔叶混交林；③竹林；④灌木林；⑤藤刺灌丛；⑥灌草丛。

茂兰地区的植被中，灌木林、藤刺灌丛和灌草丛是喀斯特森林顶极群落次生演替中不同阶段的表现，主要受人为干扰而形成，其分布基本是以村寨为中心的 $1\sim2km$ 半径范围内，多在坡下部和洼地中。灌木林主要是因为长期樵采退化形成的，藤刺灌丛和灌草丛是经常性火烧和长期放牧形成的。

地貌类型和地形坡位的不同导致生境的差异，在长期协同进化过程中，形成了多样性的植物分布格局。在无明显垂直带差异的情况下，山顶、山脊两侧分布着以华南五针松 *Pinus kwangtungensis*、短叶黄杉 *Pseudotsuga brevifolia*、翠柏 *Calocedrus macrolepis*、黄枝油杉 *Keteleeria davidiana* var. *calcarea* 等为优势的针阔叶混交林。局部地形平缓、具有土壤堆积条件的地形部位，如鞍部、丫口，常有以丝栗栲 *Castanopsis fargesii*、山杜英 *Elaeocarpus sylvestris* 等为主的较小面积的常绿阔叶树占优势，绝大部分地段（包括不同坡位、洼地、漏斗底部等）发育的是常绿落叶阔叶混交林（玉屏等，2011）。

三、茂兰地区两栖爬行动物调查方法

（一）两栖动物的调查方法

在茂兰地区两栖动物调查过程中，针对不同的季节及不同的生境，采取了不同的调查方式。主要调查方法如下。

1. 样线法

在两栖动物的非繁殖期，对中国茂兰地区的板寨、瑶山、捞村、洞塘、永康、翁昂等地的调查主要采取样线统计法。在调查区域内选择适合两栖动物生存的生境，用全球定位系统（GPS）测定样线的长度，设置长 $3\sim5km$、宽 $20m$ 的样线。以一定的速度尽量沿直线行进（每条样线最少两人同时行走调查，遇到河流、溪流等两人各走一边，以增加调查精细度），仔细观察记录两边蛙类的种类和数量。以调查路线的长度和宽度确定调查面积，计算单位面积内蛙的数量。

2. 捕尽法

对于茂兰地区内草滩、农田等生境的调查，一般采取捕尽法。将一定面积样方内的蛙类全部捕尽，计算密度。调查面积依情况而定。确定样方大小后，数人同时从不同方向开始捕捉，尽量在较短时间内完成，以确保调查结果的准确性。为了保护两栖动物，在统计后将其放回原生境。部分物种采集少数个体制作标本，采用乙醇或福尔马林进行保存，并存放于贵州师范大学花溪校区生态实验室。物种根据《贵州两栖类志》（伍律等，1986）及《中国两栖动物及其分布彩色图鉴》（费梁等，2012）进行鉴定，根据图鉴无法识别的请相关专家进行识别，并通过测序进行分子鉴定。

（二）爬行动物的调查方法

在茂兰地区爬行动物的调查过程中，采用编目法和样带法分别对爬行动物资源进行调查。

1. 编目法

在茂兰地区的板寨、瑶山、捞村、洞塘、永康、翁昂等地，选择适宜的采集地随机行走，对适宜爬行动物生存的生境仔细搜索，对遇见的物种进行 GPS 定位，记录生境状况，并用数码相机拍摄照片。该方法主要用于短时间的重要地点物种多样性的调查，可以提供物种丰富度信息及各个采集地的物种详细信息。

2. 样带法

在茂兰地区的板寨、瑶山、捞村、洞塘、永康、翁昂等地，采用以样带法为主、样带调查与定点观察相结合的调查方法。白天以样带法为主，辅以样方法并熟悉地形，夜晚以样线法为主，20:00～23:00 利用电筒照明，沿水沟、湖边、溪流或附近农田区采用自下而上的方法进行直接调查统计，记录物种种类、数量及生境，并采集少量标本。所采集的物种，每个物种采集 1～4 号标本，用 75% 乙醇固定 12 小时后转 99% 无水乙醇保存。物种鉴定根据《贵州爬行类志》（伍律等，1985）及《中国爬行动物图鉴》（中国野生动物保护协会，2002）进行鉴定，根据图鉴无法识别的请相关专家进行识别，并通过测序进行分子鉴定。

四、茂兰地区两栖爬行动物物种多样性

（一）两栖动物多样性

茂兰地区目前已记录的两栖动物有 32 种，隶属于 2 目 8 科 18 属，占贵州两栖动物的 40.00%。茂兰地区共发现有尾目蝾螈科 3 个物种，即茂兰瘰螈、细痣疣螈和文县疣螈，约占茂兰地区两栖动物的 9.38%。无尾目 29 种，其中以蛙科种类最多，有 8 种，约占茂兰地区两栖动物的 25.00%；角蟾科种类最少，仅 1 种，约占茂兰地区两栖动物的 3.13%；其余分别为蟾蜍科 2 种（约占茂兰地区两栖动物的 6.25%），雨蛙科 2 种（约占茂兰地区两栖动物的 6.25%），姬蛙科 4 种（约占茂兰地区两栖动物的 12.50%），叉舌蛙科 6 种（约占茂兰地区两栖动物的 18.75%），树蛙科 6 种（约占茂兰地区两栖动物的 18.75%）。

（二）两栖动物分布地人为干扰情况

良好的栖息环境可以孕育较多的物种，物种种类的多少与区域内的生境类型、人为影响及海拔紧密相关。茂兰保护区内存在的人为干扰主要有人为捕食、过度放牧、薪材砍伐等。根据距道路及居民地的远近对干扰进行等级划分，分为强、中、弱三级。核心区白鹇山拥有良好的森林植被，水源充足，人为干扰程度低，调查到的两栖动物有 18 种，而同为核心区的干排和洞多（面积较小）分别调查到 8 种和 13 种；核心区和缓冲区（8 种、9 种和 11 种）的两栖动物物种数低于实验区（10 种、16 种和 23 种），有些珍稀的国家Ⅱ级重点保护野生动物也生活于实验区（表 1-1）。茂兰保护区两栖动物在海拔 500～700m 区域的分布多于海拔 700～900m 的区域。

表 1-1 物种分布地、生境及人为干扰程度

分布地	物种数（国家级保护动物）	分布区域	生境	海拔（m）	干扰程度
甲乙	16 种（虎纹蛙）	实验区	森林、河流、农田、灌丛、沼泽	508～535	**
板寨	23 种（细痣疣螈）	实验区	森林、盆地、沼泽、农田+河流	557～598	**
白鹇山	18 种（细痣疣螈）	核心区	原始森林、森林+沼泽、盆地+沼泽	598～778	**
长衫瑶寨	11 种（无）	缓冲区	农田+河流、沼泽（农田、森林过渡带、溪流）	557～859	**
干排	8 种（无）	核心区	原始森林（常绿阔叶林、林间溪流）	798～821	*
洞多	13 种（无）	核心区	盆地、农田、原始森林、山间溪流	685～757	*

续表

分布地	物种数（国家级保护动物）	分布区域	生境	海拔（m）	干扰程度
肯地山	8种（无）	缓冲区	针叶林、针叶阔叶混交林、林间溪流	801～859	**
垃牙	9种（无）	缓冲区	农田+水沟、沼泽、林间公路、森林和农田的过渡带	700～785	**
瑶兰	10种（无）	实验区	农田+河流、农田、农田+沼泽、农田和森林的过渡带	632～690	***

*代表人为干扰程度弱；**表示人为干扰程度中等；***表示人为干扰程度强

（三）爬行动物多样性

茂兰地区目前已记录的爬行动物共62种，隶属于2目12科39属，约占贵州爬行动物种数的56.4%。其中，龟鳖目最少，仅1科1属1种，即中华鳖，约占茂兰地区爬行动物的1.6%。有鳞目蜥蜴亚目有6科8属12种，约占茂兰地区爬行动物的19.4%；蛇亚目种类最丰富，有5科30属49种，约占茂兰地区爬行动物的79.0%，其中游蛇科种类最多，有20属37种，约占茂兰地区爬行动物的59.7%。

五、茂兰地区两栖爬行动物区系及其地理区划

贵州在世界动物地理区划上属于东洋界，在中国动物地理区划上属于西南区，处于华南区和华中区动物相互交汇地带。中国茂兰地区位于贵州南部，与华南区的广西毗邻，保护区内两栖爬行动物多为华中区和华南区物种。

在茂兰地区最终确定有94种两栖爬行动物，其中，两栖动物32种（含树蛙属未定种1种），除黑斑侧褶蛙、斑腿泛树蛙和锯腿原指树蛙在东洋界和古北界均有分布，以及树蛙属未定种未统计在内之外，其余28种皆为东洋界物种，约占茂兰地区两栖动物的87.50%。仅在华中区分布的有4种，分别是茂兰瘰螈、荔波异角蟾、棘侧蛙和务川臭蛙，约占茂兰地区两栖动物的12.50%；在华中区和华南区分布的有26种，分别为细痣疣螈、文县疣螈、中华蟾蜍指名亚种、中华蟾蜍华西亚种、黑眶蟾蜍、三港雨蛙、粗皮姬蛙、饰纹姬蛙、小弧斑姬蛙、花姬蛙、泽陆蛙、虎纹蛙、棘腹蛙、双团棘胸蛙、棘胸蛙、滇蛙、沼水蛙、阔褶水蛙、大绿臭蛙、绿臭蛙、花臭蛙、黑斑侧褶蛙、锯腿原指树蛙、斑腿泛树蛙、无声囊泛树蛙、大树蛙、峨眉树蛙，约占茂兰地区两栖动物的81.25%；仅在华中区和西南区分布的有1种，即华西雨蛙武陵亚种，约占茂兰地区两栖动物的3.13%；仅在华中区、华南区和西南区同时有分布的有9种，分别是粗皮姬蛙、小弧斑姬蛙、虎纹蛙、双团棘胸蛙、棘胸蛙、滇蛙、沼水蛙、阔褶水蛙和无声囊泛树蛙，约占茂兰地区两栖动物的28.13%，具体见表1-2（不含树蛙属中的未定种）。

爬行动物62种，东洋界和古北界均有分布的有20种，约占茂兰地区爬行动物的32.3%，其余42种均为东洋界物种，约占茂兰地区爬行动物的67.7%，其中仅华中区分布的有2种，约占茂兰地区爬行动物的3.2%；仅在华中区和华南区分布的有12种，约占茂兰地区爬行动物的19.4%；仅在华中区和西南区分布的有2种，约占茂兰地区爬行动物的3.2%；华中区、华南区和西南均有分布的有38种，约占茂兰地区爬行动物的61.3%；分布界线向北仅延伸进入华北区的有2种，即北草蜥和草腹链蛇；进入蒙新区的有原矛头蝮、平鳞钝头蛇、乌华游蛇、锈链腹链蛇、黑线乌梢蛇、绞花林蛇和黑头剑蛇等17种；进入东北区的有福建蝮、短尾蝮、黑眉晨蛇、虎斑颈槽蛇和赤链蛇5种；分布最广的是虎斑颈槽蛇和黑眉晨蛇，在全国七大区都有分布，具体见表1-3。

从两栖动物和爬行动物的区系特征对比来看，茂兰地区两栖爬行动物以东洋界物种为主，两栖动物中华中区和华南区区系成分约占81.25%，而爬行动物中华中区和华南区区系成分约占80.65%，向北扩散到古北界的物种，两栖动物仅有3种（约占茂兰地区两栖动物的9.38%），而爬行动物则多达20种（约占茂兰地区爬行动物的32.26%），是两栖动物古北界物种数的6倍多，反映了爬行动物的迁徙和扩散能力大于两栖动物，两栖动物对环境变化的适应强于爬行动物。

表 1-2 茂兰地区两栖动物物种地理区系

物种名	拉丁学名	保护级别*	IUCN 濒危等级	东洋界	古北界	主要分布范围
茂兰瘰螈	Paramesotriton maolanensis		未予评估（NE）	√		华中区
细痣疣螈	Tylototriton asperrimus	Ⅱ级	近危（NT）	√		华中区、华南区
文县疣螈	Tylototriton wenxianensis	Ⅱ级	易危（VU）	√		华中区、华南区
荔波异角蟾	Xenophrys liboensis		数据缺乏（DD）	√		华中区
中华蟾蜍华西亚种	Bufo gargarizans andrewsi		无危（LC）	√		西南区、蒙新区、华中区、华南区
中华蟾蜍指名亚种	Bufo gargarizans gargarizans		无危（LC）	√		华中区、华南区、蒙新区、华北区、东北区
黑眶蟾蜍	Duttaphrynus melanostictus		无危（LC）	√		华中区、华南区、西南区、华北区、蒙新区
华西雨蛙武陵亚种	Hyla gongshanensis wulingensis		无危（LC）	√		华中区、西南区
三港雨蛙	Hyla sanchiangensis		无危（LC）	√		华中区、华南区
粗皮姬蛙	Microhyla butleri		无危（LC）	√		华中区、华南区、西南区
饰纹姬蛙	Microhyla fissipes		无危（LC）	√		华中区、蒙新区、华北区、西南区、华南区
小弧斑姬蛙	Microhyla heymonsi		无危（LC）	√		华中区、华南区、西南区
花姬蛙	Microhyla pulchra		无危（LC）	√		华中区、华南区、西南区、蒙新区
泽陆蛙	Fejervarya multistriata		无危（LC）	√		华中区、华南区、华北区、西南区、蒙新区
虎纹蛙	Hoplobatrachus chinensis	Ⅱ级	濒危（EN）	√		华中区、华南区、西南区
双团棘胸蛙	Gynandropaa phrynoides		濒危（EN）	√		西南区、华南区、华中区
棘腹蛙	Quasipaa boulengeri		易危（VU）	√		华中区、西南区、蒙新区、华南区
棘侧蛙	Quasipaa shini		易危（VU）	√		华中区
棘胸蛙	Quasipaa spinosa		易危（VU）	√		华中区、华南区、西南区
滇蛙	Nidirana pleuraden		无危（LC）	√		华中区、华南区、西南区
沼水蛙	Hylarana guentheri		无危（LC）	√		西南区、华南区、华中区
阔褶水蛙	Hylarana latouchii		无危（LC）	√		华中区、华南区、西南区
大绿臭蛙	Odorrana graminea		无危（LC）	√		华中区、华南区、西南区、华北区
绿臭蛙	Odorrana margaretae		无危（LC）	√		华中区、华南区、蒙新区、华北区
花臭蛙	Odorrana schmackeri		无危（LC）	√		华中区、华南区
务川臭蛙	Odorrana wuchuanensis		极危（CR）	√		华中区
黑斑侧褶蛙	Pelophylax nigromaculatus		近危（NT）	√	√	华中区、华南区、蒙新区
锯腿原指树蛙	Kurixalus odontotarsus		无危（LC）	√	√	华中区、华南区、青藏区、西南区
斑腿泛树蛙	Polypedates megacephalus		无危（LC）	√	√	华中区、华南区、蒙新区、西南区
无声囊泛树蛙	Polypedates mutus		无危（LC）	√		华中区、华南区、西南区
大树蛙	Rhacophorus dennysi		无危（LC）	√		华中区、华南区
峨眉树蛙	Rhacophorus omeimontis		无危（LC）	√		华中区、华南区

*保护级别此处仅列国家重点野生动物保护等级

表 1-3　茂兰地区爬行动物物种地理区系

物种名	拉丁学名	保护区内分布	IUCN 濒危等级	东洋界	古北界	主要分布范围
中华鳖	Pelodiscus sinensis	洞塘瑶所（青龙潭）	濒危（EN）	√		华中区、西南区
里氏睑虎	Goniurosaurus lichtenfelderi	茂兰	易危（VU）	√		华南区
荔波睑虎	Goniurosaurus liboensis	板寨、瑶兰	濒危（EN）	√		华中区
多疣壁虎	Gekko japonicus	茂兰	无危（LC）	√		西南区、华南区、华中区
荔波壁虎	Gekko liboensis	洞腮	无危（LC）	√		西南区
铜蜓蜥	Sphenomorphus indicus	板寨、瑶兰	无危（LC）	√		西南区、华南区、华中区、华北区、蒙新区、青藏区
蓝尾石龙子	Plestiodon elegans	板寨	无危（LC）	√		西南区、华南区、华中区
中华石龙子	Plestiodon chinensis	洞长	无危（LC）	√		华中区、华南区
北草蜥	Takydromus septentrionalis	黎明关（板寨）、洞长、瑶兰	无危（LC）	√	√	华中区、华北区
南草蜥	Takydromus sexlineatus meridionalis	洞腮、瑶兰	无危（LC）	√		华中区、西南区
脆蛇蜥	Dopasia harti	茂兰	濒危（EN）	√		华中区、华南区
丽棘蜥	Acanthosaura lepidogaster	黎明关（板寨）	无危（LC）	√		华中区、华南区、西南区
细鳞拟树蜥	Pseudocalotes microlepis	翁昂（干排）	无危（LC）	√		华中区、华南区
蟒蛇	Python bivittatus	翁昂（干排）	极危（CR）	√	√	华中区、华南区、西南区、青藏区
平鳞钝头蛇	Pareas boulengeri	甲乙	无危（LC）	√	√	蒙新区、西南区、华中区
福建钝头蛇	Pareas stanleyi	瑶所	无危（LC）	√		华中区、华南区
原矛头蝮	Protobothrops mucrosquamatus	翁昂、干排、茂兰镇	无危（LC）	√		西南区、华南区、华中区、蒙新区
茂兰原矛头蝮	Protobothrops maolanensis	翁昂、干排、茂兰镇	数据缺乏（DD）	√		华南区、华中区
尖吻蝮	Deinagkistrodon acutus	黎明关（板寨）	濒危（EN）	√		西南区、华中区、华南区
福建绿蝮	Viridovipera stejnegeri	黎明关（板寨）	无危（LC）	√		西南区、华南区、华中区、东北区、华北区、蒙新区
白唇竹叶青蛇	Trimeresurus albolabris	甲乙	无危（LC）	√		西南区、华南区、华中区
短尾蝮	Gloydius brevicaudus	茂兰	近危（NT）	√		西南区、华南区、华中区、东北区、蒙新区、华北区
中华珊瑚蛇	Sinomicrurus macclellandi	板寨（翁根坡）	易危（VU）	√	√	西南区、华南区、华中区、蒙新区、青藏区、华北区
孟加拉眼镜蛇	Naja kaouthia	瑶兰	无危（LC）	√		华南区、华中区、西南区
银环蛇	Bungarus multicinctus	板寨	无危（LC）	√		西南区、华南区、华中区
钝尾两头蛇	Calamaria septentrionalis	板寨	无危（LC）	√		华中区、华南区
大眼斜鳞蛇	Pseudoxenodon macrops	板寨、干排	无危（LC）	√	√	西南区、华南区、华中区、华北区、蒙新区、青藏区
横纹斜鳞蛇	Pseudoxenodon bambusicola	板寨、干排	无危（LC）	√		华中区、华南区
黑头剑蛇	Sibynophis chinensis	洞长	无危（LC）	√		西南区、华南区、华中区、蒙新区
繁花林蛇	Boiga multomaculata	瑶所	无危（LC）	√		西南区、华南区、华中区
绞花林蛇	Boiga kraepelini	甲乙	无危（LC）	√	√	蒙新区、华南区、华中区

续表

物种名	拉丁学名	保护区内分布	IUCN 濒危等级	东洋界	古北界	主要分布范围
台湾小头蛇	Oligodon formosanus	洞腮	无危（LC）	√		西南区、华南区、华中区
中国小头蛇	Oligodon chinensis	洞长、瑶兰、甲乙	无危（LC）	√		西南区、华南区、华中区
紫棕小头蛇	Oligodon cinereus	板寨	无危（LC）	√		华南区、华中区、西南区
三索蛇	Coelognathus radiatus	瑶兰	无危（LC）	√		华南区、华中区、西南区
翠青蛇	Cyclophiops major	板寨（白鹇山）、翁昂	无危（LC）	√		西南区、华南区
滑鼠蛇	Ptyas mucosa	板寨（长衫瑶寨）	濒危（EN）	√	√	华南区、华中区、西南区、青藏区
灰鼠蛇	Ptyas korros	板寨	易危（VU）	√		华南区、华中区、西南区
黑线乌梢蛇	Ptyas nigromarginata	瑶兰	易危（VU）	√	√	西南区、华南区、华中区、蒙新区
灰腹绿蛇	Rhadinophis frenatum	翁昂	无危（LC）	√		华南区、华中区
赤链蛇	Lycodon rufozonatum	板寨（长衫瑶寨）	无危（LC）	√		西南区、华南区、华中区、东北区、华北区、蒙新区
北链蛇	Lycodon septentrionalis	干排、翁昂	无危（LC）	√		西南区、华南区
黑背链蛇	Lycodon ruhstrati	洞长	无危（LC）	√		华南区、华中区
黄链蛇	Lycodon flavozonatum	板寨	无危（LC）	√		西南区、华南区、华中区
玉斑蛇	Euprepiophis mandarinus	洞长	易危（VU）	√	√	华中区、华北区、蒙新区、华南区、西南区、青藏区
紫灰蛇	Oreocryptophis porphyraceus	翁昂	近危（NT）	√	√	西南区、华南区、华中区、华北区、蒙新区、青藏区
百花晨蛇	Orthriophis moellendorffi	板寨、瑶山、翁昂、洞塘	濒危（EN）	√		华中区
黑眉晨蛇	Orthriophis taeniurus	瑶兰	濒危（EN）	√	√	东北区、华北区、蒙新区、青藏区、西南区、华南区、华中区
草腹链蛇	Amphiesma stolatum	板寨	无危（LC）	√	√	华中区、华南区、西南区、华北区
锈链腹链蛇	Hebius craspedogaster	板寨	无危（LC）	√	√	华中区、蒙新区、华北区
无颞鳞腹链蛇	Hebius atemporalis	板寨、翁昂、洞长	近危（NT）	√		西南区、华南区、华中区
坡普腹链蛇	Hebius popei	洞塘（五眼桥）	无危（LC）	√		西南区、华南区、华中区
丽纹腹链蛇	Hebius optatum	板寨、翁昂、瑶兰	近危（NT）	√		西南区、华南区、华中区
棕黑腹链蛇	Hebius sauteri	瑶所	无危（LC）	√		华中区、华南区、西南区
虎斑颈槽蛇	Rhabdophis tigrinus	板寨、翁昂	无危（LC）	√	√	西南区、华南区、华中区、东北区、华北区、蒙新区、青藏区
红脖颈槽蛇	Rhabdophis subminiatus	翁昂	无危（LC）	√		华南区
异色蛇	Xenochrophis piscator	瑶兰	无危（LC）	√		华中区、华南区
黄斑异色蛇	Xenochrophis flavipunctatus	板寨	无危（LC）	√		华南区、华中区
山溪后棱蛇	Opisthotropis latouchii	板寨	无危（LC）	√		华中区、华南区
乌华游蛇	Sinonatrix percarinata	板寨、翁昂、瑶兰	易危（VU）	√	√	西南区、华南区、华中区、蒙新区
环纹华游蛇	Sinonatrix aequifasciata	瑶兰	易危（VU）	√		华中区、华南区、西南区
颈棱蛇	Macropisthodon rudis	板寨	无危（LC）	√		华中区、华南区、西南区

第二章

两栖动物术语及分类描述

一、两栖纲分类检索名词术语

（一）有尾目特征及分类检索常用术语

1. 成体形态结构及说明

有尾目 Caudata 是两栖动物中最不特化的一目，终生有尾，多数有四肢，幼体与成体比较相似。体长形，分为头、躯干和尾三部分，颈部较明显，四肢匀称。皮肤光滑湿润，紧贴皮下肌肉，富含皮肤腺，全无小鳞。耳无鼓膜和鼓室。幼体用鳃呼吸，成体用肺呼吸，也有些种类终生具鳃，肺很不发达或无肺，皮肤呼吸占主要地位。椎体双凹型或后凹型，有肋骨。体皮肤无鳞，多数具有四肢，少数只有前肢而无后肢。变态不显著。有的具有外鳃，终生生活于水中，有的在变态后移到陆上湿地生活。雄性无交接器。多卵生。幼体先出前肢再出后肢。

有尾目成体的外部形态参见图 2-1。现就该目分类检索常用的形态特征及其术语与量度说明如下。

2. 形态量度

在分类上常用的量度有下列各项。

体全长（total length = TOL）：吻端至尾末端的长度。
头体长（snout-vent length = SVL）：吻端至肛孔后缘的长度。
头长（head length = HL）：吻端至颈褶或口角（无颈褶者）的长度。
头宽（head width = HW）：头或颈褶左右两侧之间的最大距离。
吻长（snout length = SL）：吻端至眼前角之间的距离。
躯干长（trunk length = TRL）：颈褶至肛孔后缘的长度。
眼间距（interorbital space = IOS）：左右上眼睑内侧缘之间的最小距离。
眼径（diameter of eye = ED）：与体轴平行的眼的直径。
尾长（tail length = TL）：肛孔后缘至尾末端的长度。
尾高（tail height = TH）：尾上、下缘之间的最大距离。
尾宽（tail width = TW）：尾基部（即肛孔）两侧之间的最大宽度。
前肢长（length of foreleg = FLL）：前肢基部至最长指末端的长度。
后肢长（length of hind leg = HLL）：后肢基部至最长趾末端的长度。
腋至胯距（space between axilla and groin = AGS）：前肢基部后缘至后肢基部前缘之间的距离。

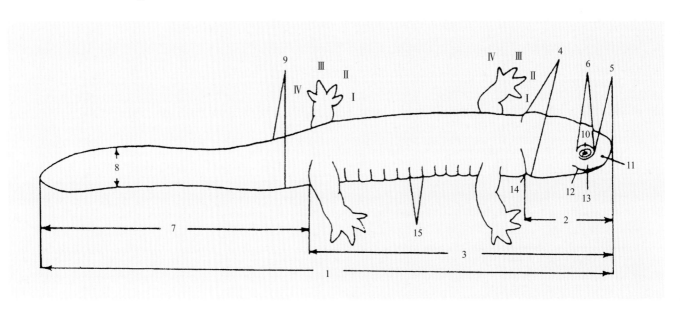

图2-1 有尾目（山溪鲵属 *Batrachuperus*）成体（引自费梁等，2005）
1.体全长；2.头长；3.头体长；4.头宽；5.吻长；6.眼径；7.尾长；8.尾高；9.尾宽；10.上眼睑；11.鼻孔；12.口裂；13.唇褶；14.颈褶；15.肋沟。Ⅰ、Ⅱ、Ⅲ、Ⅳ分别表示指和趾的顺序

3. 外部形态特征常用术语

犁骨齿（vomerine tooth）：着生在犁腭骨上的细齿，其齿列的位置、形状和长短均具有分类学意义（图2-2）。

囟门（fontanel）：颅骨背壁未完全骨化所留下的孔隙。位于前颌骨与鼻骨之间者称前颌囟；位于左右额骨与顶骨之中缝者称额顶囟。

唇褶（labial fold）：颌缘皮肤肌肉组织的帘状褶。通常在上唇侧缘后半部，掩盖着对应的下唇缘，如山溪鲵属、北鲵属等物种。

颈褶（jugular fold）：颈部两侧及其腹面的皮肤皱褶；通常作为头部与躯干部的分界线。

肋沟（costal groove）：躯干部两侧、位于两肋骨之间形成的体表凹沟。

尾鳍褶（tail fin fold）：位于尾上（背）、下（腹）方的皮肤肌肉褶襞称为尾鳍褶；在尾上方者称为尾背鳍褶，反之为尾腹鳍褶。不同于无尾目蝌蚪的膜状尾鳍。

角质鞘（horny cover）：一般指四肢掌、蹠及指、趾底面皮肤的角质化表层，呈棕黑色。

卵胶袋或卵鞘袋（egg sack or egg sac）：成熟卵在输卵管内向后移动时，管壁分泌的蛋白质将卵粒包裹后产出，蛋白层吸水膨胀形成袋状物，卵粒在袋内呈单行或多行交错排列。

童体型或幼态性熟（neoteny）：指性腺成熟能进行繁殖，但又保留有幼体形态特征（如具外鳃或鳃孔）的现象。

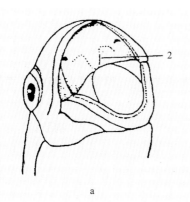 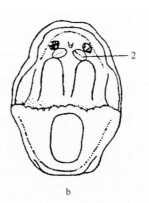 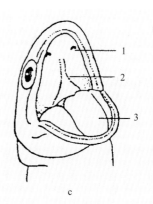

图2-2 有尾目头部（引自费梁等，2005）
a. 小鲵属 *Hynobius*；b. 山溪鲵属 *Batrachuperus*；c. 蝾螈科 Salamandridae。图中，1. 内鼻孔；2. 犁骨齿；3. 舌

（二）无尾目特征及分类检索常用术语

1. 成体形态结构及说明

无尾目 Anura 成体的外部形态参见图2-3。该目分类学上除常用的外形量度外，其形态特征还包括外部形态特征、皮肤表面结构和骨骼系统，分别简述如下。

2. 成体外形量度

分类学上常用的外形量度有下列各项。

头体长（snout-vent length = SVL）：吻端至体后端的长度。

头长（head length = HL）：吻端至上、下颌关节后缘的长度。

头宽（head width = HW）：头两侧之间的最大距离。

吻长（snout length = SL）：吻端至眼前角的长度。

鼻间距（internasal space = INS）：左右鼻孔内缘之间的距离。

眼间距（interorbital space = IOS）：左右上眼睑内侧缘之间的最小距离。

上眼睑宽（width of upper eyelid = UEW）：上眼睑的最大宽度。

眼径（diameter of eye = ED）：与体轴平行的眼的直径。

鼓膜径（diameter of tympanum = TD）：鼓膜最大的直径。

前臂及手长（length of lower arm and hand = LAHL）：肘关节至第三指末端的长度。

前臂宽（diameter of lower arm = LAD）：前臂最粗处

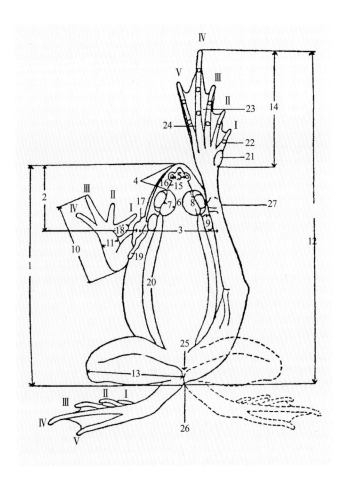

图2-3 黑斑侧褶蛙成体（引自费梁等，2005）
1.头体长；2.头长；3.头宽；4.吻长；5.鼻间距；6.眼间距；7.上眼睑宽；8.眼径；9.鼓膜；10.前臂及手长；11.前臂宽；12.后肢或腿全长；13.胫长；14.足长；15.吻棱；16.颊部；17.咽侧外声囊；18.婚垫；19.颞褶；20.背侧褶；21.内蹠突；22.关节下瘤；23.蹼；24.外侧蹠间蹼；25.肛；26.示左右跟部相遇；27.示胫跗关节前达眼部。手上的Ⅰ、Ⅱ、Ⅲ、Ⅳ表示指的顺序；足上的Ⅰ、Ⅱ、Ⅲ、Ⅳ、Ⅴ表示趾的顺序

的直径。

后肢或腿全长（hindlimb or leg length = HLL）：体后端正中部位至第四趾末端的长度。

胫长（tibia length = TL）：胫部两端之间的长度。

胫宽（tibia width = TW）：胫部最粗处的直径。

跗足长（length of foot and tarsus = TFL）：胫跗关节至第四趾末端的长度。

足长（foot length = FL）：内蹠突的近端至第四趾末端的长度。

3. 外部形态特征常用术语

吻及吻棱（snout and canthus rostralis）：眼前角至上颌前端称为吻或吻部；吻背面两侧的线状棱称为吻棱。吻部的形状及吻棱的明显与否随属、种的不同而异。

颊部（genal region）：鼻眼之间的吻棱下方至上颌上方部位。其垂直或倾斜程度随属、种不同而异。

内鼻孔（internal nares or choanae）：位于口腔顶壁前端1对与外鼻孔相通的小孔（图2-4）。

咽鼓管孔（pores of eustachian tube）：位于口腔顶壁近两口角的1对小孔，与内耳相通，又称欧氏管孔（图2-4）。

上颌齿（maxillary teeth）：着生于上颌骨和前颌骨上的细齿（图2-4）。

犁骨棱与犁骨齿（vomerine ridge and vomerine teeth）：犁骨向腹面凸起而隐于口腔上皮的脊棱称犁骨棱。犁骨齿是着生在犁骨或犁骨棱上的1排或1团细齿，位于内鼻孔内侧或后缘（图2-4）。犁骨齿的有无及位置、形状、大小可作为分类特征之一。

齿状突（tooth-like projections）：在下颌前方近联合处有1对明显高出颌缘的齿状骨质突起。

声囊（vocal sac）：大多数种类的雄性，在咽喉部由咽部皮肤或肌肉扩展形成的囊状突起，称为声囊（图2-5～图2-8）。在外表能观察到者为外声囊（external vocal sac），反之即为内声囊（internal vocal sac）。

外声囊又可分为3种。①单咽下外声囊（external single subgular vocal sac）：指咽喉部腹面的皮肤皱褶形成一松弛的泡状突囊，表面颜色一般较深（如华西雨蛙武陵亚种 Hyla gongshanensis wulingensis、泽陆蛙 Fejervarya multistriata 等）（图2-5）。②咽侧下外声囊（external subgular vocal sac）：指两口角腹面的皮肤皱褶形成的1对袋状突囊（如滇蛙 Nidirana pleuraden、沼水蛙 Hylarana guentheri、虎纹蛙 Hoplobatrachus chinensis 等）（图2-6）。③咽侧外声囊（external lateral vocal sac）：指紧靠两口角下缘后侧之皮肤

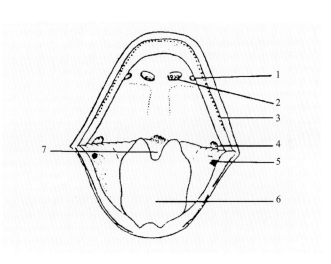

图2-4 无尾目口腔（引自费梁等，2005）
1.内鼻孔；2.犁骨齿；3.上颌齿；4.咽鼓管孔；5.声囊孔；6.舌；7.舌后端缺刻

皱褶所形成的 1 对袋状突囊，亦称颈侧外声囊（如黑斑侧褶蛙 Pelophylax nigromaculatus 等）（图 2-7）。

内声囊系由肌肉褶襞形成的且被皮肤所掩盖的突囊，可分为 2 种。①单咽下内声囊（internal single subgular vocal sac）：指咽喉部腹面的肌肉褶襞形成的一弧状突囊（如峨眉异角蟾 Xenophrys omeimontis、花背蟾蜍 Bufo raddei 等）。②咽侧下内声囊（internal subgular vocal sac）：指两口角腹面的肌肉褶襞形成的 1 对袋状突囊（如中国林蛙 Rana chensinensis、峨眉树蛙 Rhacophorus omeimontis 等）（图 2-8）。

声囊孔（opening of vocal sac）：在舌两侧或近口角处各有 1 圆形或裂隙状的孔，称为声囊孔，声囊与口腔之间以此孔相通。

指、趾长顺序（digital formula）：用阿拉伯数字表示指、趾长短的顺序，如 3、4、2、1，即表示第三指（趾）最长，依次递减，第一指（趾）最短。

指、趾吸盘（digital disc or disk）：指、趾末端扩大成圆盘状，其底部增厚成半月形肉垫，可吸附于物体上。

指、趾沟（digital groove）：指位于沿指、趾吸盘边缘和腹侧的凹沟。根据凹沟的位置又可分为 2 种。①边缘沟或环缘沟（circum margin groove）：指吸盘边缘或游离缘，且在吸盘顶端贯通的凹沟，呈马蹄形，故又称马蹄形沟（horse shoe-shaped groove），如雨蛙属、树蛙科等中的物种（图 2-9a）。②腹侧沟（lateroventral groove）：指吸盘腹面两侧接近边缘的凹沟，或长或短，且在吸盘顶端互不相通，其间距或宽或窄，有的几乎相连，如蛙科中的臭蛙属、水蛙属和趾沟蛙属等中的物种（图 2-9b）。

关节下瘤（subarticular tubercle）：指、趾底面活动关节之间的褥垫状突起（图 2-10）。

指基下瘤（supernumerary tubercle below the base of finger）：位于掌部远端（即在指基部）的瘤状突起（图 2-10）。

掌突与蹠突（metacarpal and metatarsal tubercle）：掌和蹠底面基部的明显隆起，内侧称为内掌突与内蹠突，外侧称为外掌突与外蹠突（图 2-10）。它们的形状、大小、存在

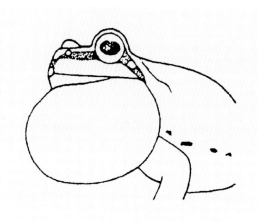

图 2-5 单咽下外声囊（中国雨蛙 Hyla chinensis）
（引自费梁等，2005）

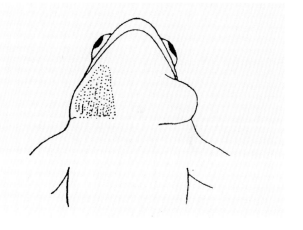

图 2-6 咽侧下外声囊（滇蛙）
（引自费梁等，2005）

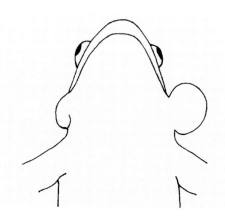

图 2-7 颈侧外声囊（黑斑侧褶蛙）
（引自费梁等，2005）

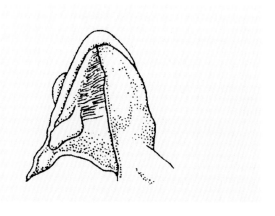

图 2-8 咽侧下内声囊（中国林蛙）
（引自费梁等，2005）

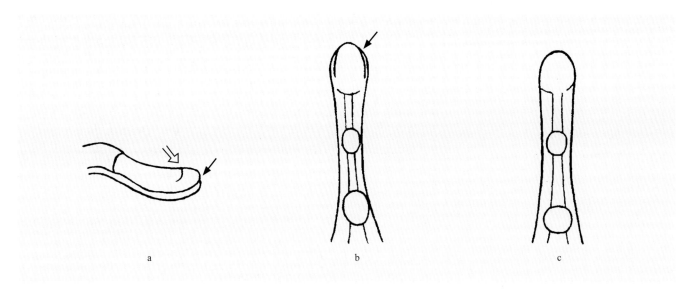

图2-9 指、趾端有沟或无沟（引自费梁等，2005）
a. 边缘沟（◀）及背面横凹痕（↙）；b. 腹侧沟（◀）；c. 无沟

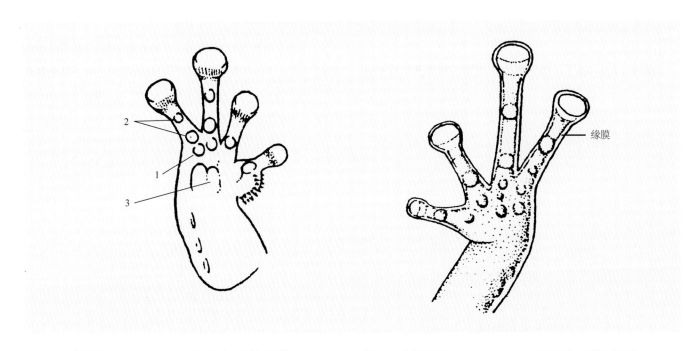

图2-10 武夷湍蛙 *Amolops wuyiensis* 手部腹面观（引自费梁等，2005）
1. 指基下瘤；2. 关节下瘤；3. 掌突

图2-11 白颊费树蛙 *Feihyla palpebralis* 手部腹面观（引自费梁等，2005）
示指间无蹼和缘膜

与否及内外两者的间距因种类不同而异，可作为分类依据。

跗瘤（tarsal tubercle）：着生在胫跗关节后端的明显隆起。

缘膜（fringe）：指、趾两侧的膜状皮肤褶（图2-11）。

蹼（web）：连接指与指或趾与趾的皮膜。指间一般无蹼，仅少数树栖种类的指间有蹼；趾间一般都具蹼。蹼的发达程度则因种类不同而异，同一种内两性之间亦可能存在差异。

（1）指间蹼形态主要以外侧2指即第三、第四指之间蹼的形态划分，大致可分为以下5个类型。

微蹼或蹼迹（rudimentary web）：指侧缘膜在指间基部相连而成的很弱的蹼，如华西雨蛙、侧条跳树蛙 *Chirixalus vittatus*（图2-12）。

1/3蹼（one third web）：指间蹼较明显，蹼缘缺刻深，最深处未达到外侧2指的第二关节或关节下瘤中央的连线，如洪佛树蛙 *Rhacophorus hungfuensis* 等（图2-13）。

半蹼（half web）：指间蹼明显，蹼缘缺刻最深处与外侧2指的第二关节下瘤的连线约相切，如大树蛙、峨眉树蛙 *Rhacophorus omeimontis* 等（图2-14）。

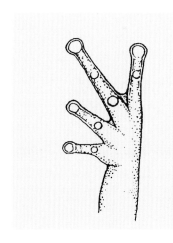

图2-12 侧条跳树蛙手部,示指间微蹼或蹼迹（引自费梁等,2005）

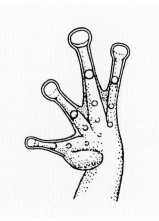

图2-13 洪佛树蛙手部,示指间1/3蹼（引自费梁等,2005）

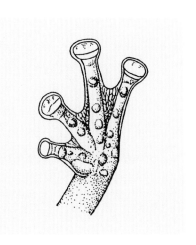

图2-14 峨眉树蛙手部,示指间半蹼（引自费梁等,2005）

全蹼（entire web）：指间蹼达指端，蹼缘略凹陷，其凹陷最深处远超过外侧2指第二关节下瘤的连线，如白颌大树蛙 Rhacophorus maximus 等（图2-15）。

满蹼（full web）：指间蹼均达指端，蹼缘凸出或平齐于指吸盘基部，如黑蹼树蛙 Rhacophorus kio 等（图2-16）。

（2）趾间蹼形态主要以外侧3趾间即第三、第四趾和第四、第五趾之间的形态划分，有的种趾间无蹼（图2-17），有蹼者大致可分为以下6个类型。

微蹼或蹼迹：趾侧缘膜在趾间基部相连接处有很弱的皮膜，如高山掌突蟾 Paramegophrys alpinus 等（图2-18）。

1/3蹼：趾间的蹼均不达趾端，蹼缘缺刻很深，其最深处未达到第三、第四趾及第四、第五趾间的第二关节或关节下瘤中央的连线，如沙坪角蟾、白颊费树蛙等（图2-19）。

半蹼：趾间的蹼均不达趾端，蹼缘缺刻较深，其最深处与两趾的第二关节下瘤连线约相切，如经甫树蛙、中国林蛙等（图2-20）。

2/3蹼（two thirds web）：趾间蹼较发达，除第四趾侧的蹼不达趾端而仅达第三关节下瘤及其附近之外，其余各趾的蹼均达趾端，但蹼缘缺刻最深处超过两趾第二关节下瘤的连线，如崇安湍蛙、花臭蛙 Odorrana schmackeri 等（图2-21）。

全蹼：各趾的蹼均达趾端，其蹼缘凹陷成弧形，凹陷最深处远超过两趾第二关节下瘤的连线，如理县湍蛙、无指盘臭蛙 Odorrana grahami 等（图2-22）。

满蹼：趾间的蹼达趾端，其蹼缘凸出或平齐于趾端的连接，如大蹼铃蟾、隆肛蛙 Nanorana quadranus 和尖舌浮蛙等（图2-23）。

雄性线（linea masculina）：雄体腹斜肌与腹直肌之间的带状结缔组织，呈白色、粉红色或红色；部分种类在背侧

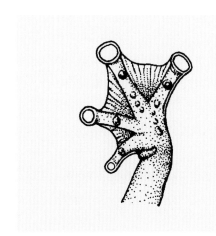

图2-15 白颌大树蛙手部,示指间全蹼（引自费梁等,2005）

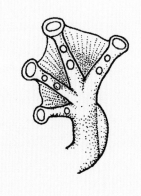

图2-16 黑蹼树蛙手部,示指间满蹼（引自费梁等,2005）

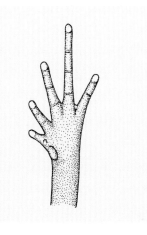

图2-17 莽山异角蟾 Xenophrys mangshanensis 足部,示无蹼（引自费梁等,2005）

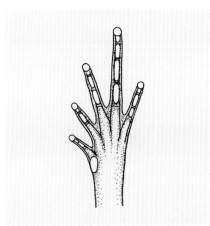
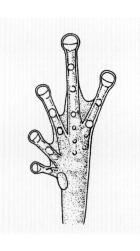
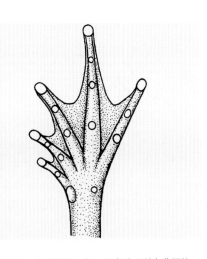

图2-18　高山掌突蟾足部，示微蹼或蹼迹（引自费梁等，2005）
图2-19　白颊费树蛙 Feihyla palpebralis 足部，示1/3蹼（引自费梁等，2005）
图2-20　中国林蛙足部，示半蹼（引自费梁等，2005）

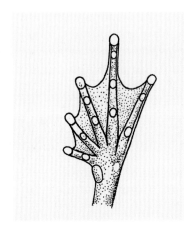
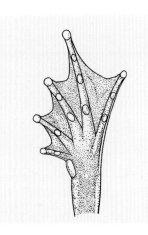
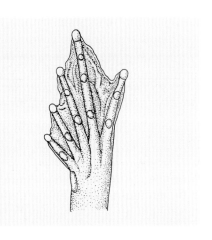

图2-21　花臭蛙足部，示2/3蹼（引自费梁等，2005）
图2-22　无指盘臭蛙足部，示全蹼（引自费梁等，2005）
图2-23　隆肛蛙足部，示满蹼（引自费梁等，2005）

亦有此线。大多存在于高等类群的种类中，低等类群很少有。

4. 皮肤表面结构

仅根据皮肤表面的隆起状态，以肉眼所能观察到的加以说明。

头棱或头侧棱（cephalic ridge）：有的种类在头部两侧（即从吻端）经眼部前内侧至鼓膜上方，由皮肤形成的非角质化、角质化或骨质化的脊棱，统称头棱或头侧棱。按其所在部位可分为：①吻上棱（canthal ridge）；②眶上棱（supraorbital ridge）；③眶前棱（preorbital ridge）；④眶后棱（postorbital ridge）；⑤鼓上棱（supratympanic ridge）；⑥顶棱（parietal ridge）（图2-24）。这些棱的形状、发达程度和存在与否均可作为物种的鉴别依据，如黑眶蟾蜍、隆枕蟾蜍和喜山蟾蜍的头棱互有区别。

颞褶（temporal fold）：自眼后经颞部背侧达肩部的皮肤增厚所形成的隆起（图2-25）。

背侧褶（dorsolateral fold）：在背部两侧，一般起自眼后，伸达胯部的一对纵走皮肤腺隆起（图2-25）。

跗褶（tarsal fold）：在后肢跗部背腹交界处的纵走皮肤腺隆起（图2-26）；内侧者为内跗褶，外侧者为外跗褶。

肤褶或肤棱（skin fold or skin ridge）：皮肤表面略微增厚而形成的分散的细褶。

耳后腺（parotoid gland）：位于眼后至枕部两侧，由皮肤增厚形成的明显腺体。其大小和形态因种而异。

颌腺（maxillary gland）或口角腺（rictal gland）：位于两口角后方的成团或窄长皮肤腺体。

肩腺（shoulder gland）：位于雄性体侧肩部后上方的扁平皮肤腺体（图2-27），如弹琴蛙 Nidirana adenopleura、滇蛙。

胸腺（pectoral gland, chest gland）：位于雄性胸部的一对扁平皮肤腺体（图2-28）。一般在繁殖季节明显，而且上面多被着生的棕褐色或黑色角质刺团所掩盖。

腋腺或胁腺（axillary gland）：位于腋部或胁内侧的一对扁平腺体（图2-28）。雌雄性均有，一般色较浅，雄性

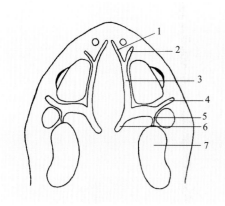

图2-24 蟾蜍 *Bufo* sp. 头部头棱（引自费梁等，2005）

1.吻上棱；2.眶前棱；3.眶上棱；4.眶后棱；5.鼓上棱；6.顶棱；7.耳后腺

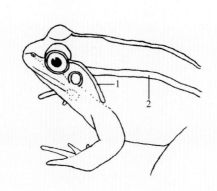

图2-25 林蛙 *Rana* sp. ♂，示侧面观（引自费梁等，2005）

1.颞褶；2.背侧褶

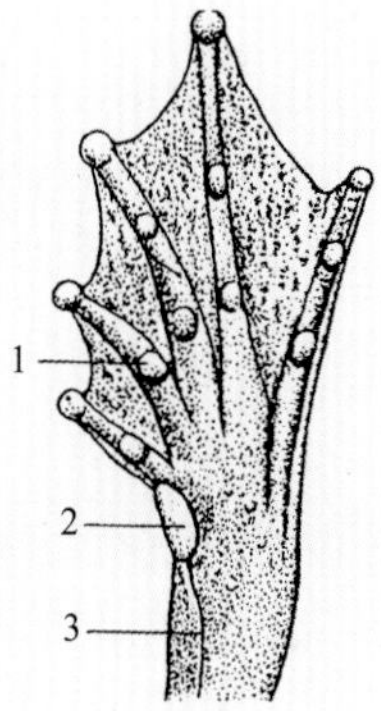

图2-26 大头蛙 *Rana kuhlii* 足部（引自费梁等，2005）

1.关节下瘤；2.蹠突；3.趾褶

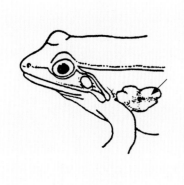

图2-27 蛙♂，示肩腺（♂）（引自费梁等，2005）

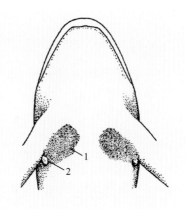

图2-28 蟾♂（引自费梁等，2005）

1.胸腺；2.腋腺或胁腺

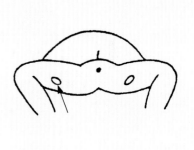

图2-29 突蟾，示股腺（♂）（引自费梁等，2005）

的腋腺在胸腺外侧，有的种类在繁殖季节时，其上还着生有深色角质刺。

肱腺或臂腺（humeral gland）：位于雄性前肢或上臂基部前方的扁平皮肤腺，如沼水蛙 *Hylarana guentheri*、黑带水蛙 *Hylarana nigrovittata*。

股腺（femoral gland）：位于股部后下方的疣状皮肤腺体（图2-29），如金顶齿突蟾 *Scutiger chintingensis*。

胫腺（tibial gland）：在胫跗部外侧的粗厚皮肤腺体，如胫腺侧褶蛙 *Pelophylax shuchinae*。

瘰粒（wart）：皮肤上排列不规则、分散或密集而表面较粗糙的大隆起，如蟾蜍属物种。

疣粒及痣粒（tubercle and granule）：较瘰粒小的光滑隆起称为疣粒；较疣粒小的隆起称为痣粒，有的呈小刺状。二者的区别是相对的，仅为描述方便而提出。

角质刺（keratinized spine，horny spine）：皮肤局部角质化的衍生物，呈刺或锥状，多为黑褐色。其大小、强弱、疏密和着生的部位因种而异。

婚垫与婚刺（nuptial pad and nuptial spine）：雄性第一指基部内侧的局部隆起称为婚垫，少数种类的第二、第三指内侧亦存在。婚垫上着生的角质刺称婚刺（图2-30，图2-31）。

5. 骨骼系统

1）肩带与胸骨组合的类型

弧胸型（arcifera）：主要特征是上喙软骨颇大且呈弧状，其外侧与前喙软骨和喙骨相连，一般是右上喙软骨重叠在左上喙软骨的腹面，肩带可通过上喙软骨在腹面左右交错活动；前胸骨与正胸骨仅部分发达或不发达（图2-32a）。我国产的铃蟾科 Bombinatoridae、角蟾科 Megophryidae、蟾蜍科 Bufonidae 和雨蛙科 Hylidae 均属于弧胸型。

固胸型（firmisternia）：主要特征是上喙软骨极小，其外侧与前喙软骨和喙骨相连，左右上喙软骨在腹中线紧密连接而不重叠，有的种类甚至合并成1条窄小的上喙骨；

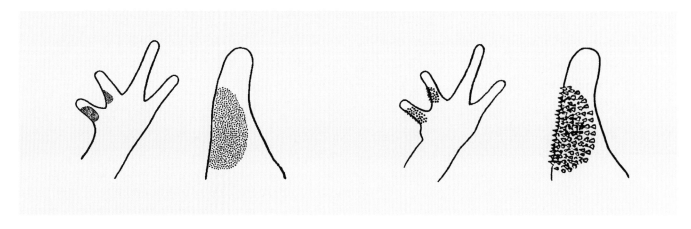

图2-30 峨眉异角蟾 Xenophrys omeimontis，示婚刺细密（引自费梁等，2005）

图2-31 棘指异角蟾 Xenophrys spinata，示婚刺粗大（引自费梁等，2005）

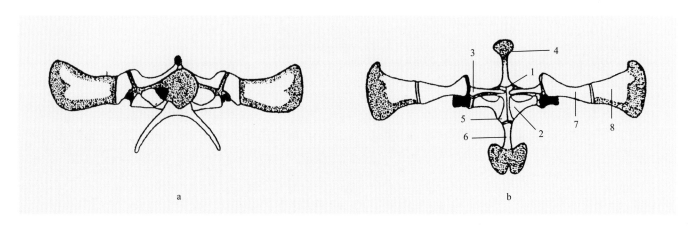

图2-32 肩带类型（腹面观）（引自费梁等，2005）

a. 弧胸型（铃蟾）；b. 固胸型（黑斑侧褶蛙）。1. 前喙软骨；2. 喙骨；3. 锁骨；4. 前胸骨（上胸骨、肩胸骨）；5. 上喙骨；6. 后胸骨（正胸骨和剑胸骨）；7. 肩胛骨；8. 上肩胛骨

肩带不能通过上喙软骨左右交错活动（图2-32b）。蛙科Ranidae、树蛙科Rhacophoridae和姬蛙科Microhylidae属于固胸型。

2）椎体类型

无尾目的脊柱有10枚椎骨，即颈椎（寰椎）1枚、躯椎7枚、荐椎（或骶椎）和尾杆骨（或尾椎）各1枚。椎骨的椎体均不发达，按前后接触面的凹凸差异，组成如下4种类型（图2-33～图2-36均为腹面观）。

后凹型脊椎（opisthocoelous）：各个椎骨的椎体都是前凸后凹的。铃蟾科即属此型，其前3枚躯椎各具1对短肋，荐椎横突宽大，尾杆骨髁1或2个；尾杆骨近端常有1或2对退化的横突（图2-33）。

前凹型脊椎（procoelous）：各个椎骨的椎体都是前凹后凸的；荐椎横突较宽大，尾杆骨髁2个。蟾蜍科和雨蛙科属于此类型（图2-34）。

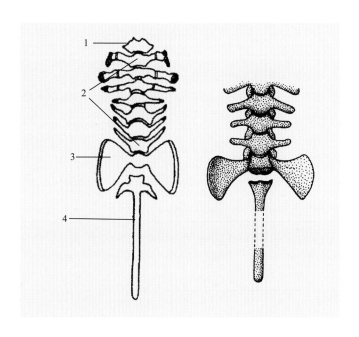

图2-33 后凹型脊椎（大蹼铃蟾）（引自费梁等，2005）　图2-34 前凹型脊椎（中华蟾蜍）（引自费梁等，2005）

1. 颈椎；2. 躯椎；3. 荐椎；4. 尾杆骨（尾椎）

变凹型脊椎（anomocoelous）：大部分或全部椎体都是前凹后凸的，间或也有若干个椎体前后都是凹的（即为双凹）；荐椎横突宽大；荐椎与尾杆骨完全愈合而无关节，或者具关节而仅有1个尾杆骨髁。角蟾科属于此类型（图2-35）。

参差型脊椎（diplasiocoelous）：第1～第7枚椎骨的椎体为前凹型；第8枚椎骨的椎体为双凹；荐椎的椎体前后都是凸的（即为双凸），其前凸面与第8枚的后凹面相关节，而其后凸面为2个尾杆骨髁与尾杆骨相关节；荐椎横突呈柱状或略宽大。蛙科、树蛙科和姬蛙科属于该类型（图2-36）。

间介软骨（intercalary cartilage）：指、趾最末两个骨节之间一小块软骨（图2-37），有的可能骨化。雨蛙科和树蛙科均有此软骨。

"Y"形骨（"Y"-shaped phalange）：指、趾最末节骨的远端分叉，呈"Y"形（图2-37），如树蛙科。

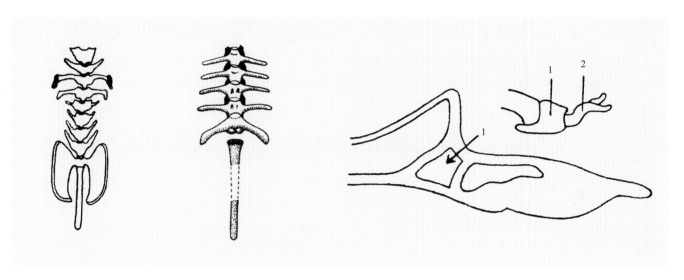

图2-35　变凹型脊椎（无蹼齿蟾）（引自费梁等，2005）　　图2-36　参差型脊椎（黑斑侧褶蛙）（引自费梁等，2005）　　图2-37　树蛙指、趾的末端骨节（引自费梁等，2005）
1. 间介软骨；2. "Y"形骨

茂兰地区目检索表

有尾，体型长；少数种类后肢退化或终生有鳃；幼体先长出前肢 ··· 有尾目 Caudata
无尾，体宽短，均具四肢，跗部长；变态后无鳃；幼体先长出后肢 ··· 无尾目 Anura

二、茂兰地区两栖动物物种分类描述

（一）有尾目CAUDATA

1. 蝾螈科 Salamandridae Goldfuss, 1820

Salamandridae Goldfuss, 1820, Handb. Zool., 2: 129. Type genus: *Salamandra* Laurenti, 1768.

Salamandridae: Gray, 1825, Ann. Philos., London, Ser. 2, 10: 215; Bonaparte, 1840, Nuovi Ann. Sci. Nat., Bologna, 4: 100.

Salamandroidea Fitzinger, 1826, Neue Class. Rept.: 41. Explicit family.

Tritonidae Boie, 1828, Isis von Oken, 21: 363. Type genus: *Triton* Laurenti, 1768.

Cercopi Wagler, 1828, Isis von Oken, 21: 859. Unavailable name formed explicitly as a family containing "Salamandra etc.".

Salamandrina Hemprich, 1829, Grundniss Naturgesch. Höhere Lehr., ed. 2: xix; Wiegmann, 1832, in Wiegmann and Ruthe (eds.),

Handbuch der Zool., Amph.: 203; Bonaparte, 1838, Nuovi Ann. Sci. Nat., Bologna, 1: 393; Bonaparte, 1840, Nuovi Ann. Sci. Nat., Bologna, 4: 101.

Tritones Tschudi, 1838, Classif. Batr.: 26. Type genus: *Triton* Laurenti, 1768.

Tritonides Tschudi, 1838, Classif. Batr.: 26. Type genus: *Triton* Laurenti, 1768.

Pleurodelina Bonaparte, 1839, Iconograph. Fauna Ital., 2 (Fasc. 26): unnumbered; Bonaparte, 1839, Mem. Soc. Sci. Nat. Neuchâtel, 2: 16; Bonaparte, 1840, Nuovi Ann. Sci. Nat., Bologna, 4: 11. Type genus: *Pleurodeles* Michahelles, 1830. Treated as a subfamily of Salamandridae.

Salamandrinae Fitzinger, 1843, Syst. Rept.: 33. Type genus: *Salamandrina* Fitzinger, 1826.

Seiranotina Gray, 1850, Cat. Spec. Amph. Coll. Brit. Mus., Batr. Grad.: 29. Type genus: *Seiranota* Barnes, 1826 (= *Salamandrina* Fitzinger, 1826). Synonymy with Pleurodelidae by Cope, 1875, Bull. U.S. Natl. Mus., 1: 11. Synonymy with Salamandridae by Boulenger, 1882, Cat. Batr. Grad. Batr. Apoda Coll. Brit. Mus., ed. 2: 2.

Pleurodelidae Bonaparte, 1850, Conspect. Syst. Herpetol. Amph.: 1 p.

Bradybatina Bonaparte, 1850, Conspect. Syst. Herpetol. Amph.: 1 p. Type genus: *Bradybates* Tschudi, 1838.

Triturinae Brame, 1957, List World's Recent Caudata: 9. Type genus: *Triturus* Rafinesque, 1815.

全长一般不超过230mm，头、躯略扁平，皮肤光滑或有瘰疣，肋沟不明显，陆栖种的尾略呈圆柱状或略侧扁；四肢较发达，指4，趾5。均有肺，个别属肺退化成残迹状。犁骨齿2长列，呈"Λ"形。睾丸分叶，肛腺3对，体内受精，雌螈将雄螈排出的精包（或精子团）纳入或植入泄殖腔壁。

本科分为3个亚科，包括肋突螈亚科Pleurodelinae、Salamandrinae、Salamandrininae。现有21属123种，中国分布5属49种，分布于秦岭以南，均隶属于肋突螈亚科。茂兰地区内分布有2属3种，即瘰螈属1种和疣螈属2种。

茂兰地区属检索表

头侧为腺质脊棱；背面瘰粒分散均匀；体腹面橘红色斑显著·· 瘰螈属 *Paramesotriton*

头侧骨质棱显著；体侧瘰疣密集，构成13～16个结节状瘰粒，或瘰疣连续隆起成行；体腹面无橘红色斑················ 疣螈属 *Tylototriton*

1）瘰螈属 *Paramesotriton* Chang, 1935

Mesotriton Bourret, 1934, Annexe Bull. Gen. Instr. Publique, Hanoi, 1934: 83. Type species: *Mesotriton deloustali* Bourret, 1934, by monotypy. Preoccupied by *Mesotriton* Bolkay, 1927.

Paramesotriton Chang, 1935, Bull. Mus. Natl. Hist. Nat., Paris, Ser. 2, 7: 95. Also published by Chang, 1935, Bull. Soc. Zool. France, 60: 425. Replacement name for *Mesotriton* Bourret, 1934.

Trituroides Chang, 1935, Bull. Soc. Zool. France, 60: 425. Type species: *Cynops chinensis* Gray, 1859, by monotypy. Synonymy (with *Paramesotriton*) by Freytag, 1962, Mitt. Zool. Mus. Berlin, 38: 451-459.

Allomesotriton Freytag, 1983, Zool. Abh. Staatl. Mus. Tierkd. Dresden, 39: 47. Type species: *Trituroides caudopunctatus* Liu and Hu, 1973, by original designation.

Allomesotriton: Pang, Jiang and Hu, 1992, in Jiang (ed.), Collect. Pap. Herpetol.: 89. Treatment as a subgenus of *Paramesotriton*.

Karstotriton Fei and Ye, 2016, Amph. China, 1: 360. Type species: *Paramesotriton zhijinensis* Li, Tian and Gu, 2008, by original designation. Coined as a subgenus of *Paramesotriton*.

头部扁平，躯干圆柱形，尾部明显侧扁；四肢较细弱而长，无蹼。皮肤粗糙，头侧有腺质脊棱；背正中脊棱明显；背面瘰粒小，分散均匀，沿体两侧的较大而明显；舌小而圆或卵圆，小于口腔底部的1/2，前后端与口腔黏膜相连，两侧游离。幼体体较细长，无平衡枝，外鳃3对，鳃丝羽状；尾背鳍褶细弱，起始于尾基部；一般营水栖生活，或不远离水域。卵产于溪流或溪边静水池。睾丸豆形，每侧分为2叶。

犁骨齿列呈"Λ"形，前端会合，后端分离；前颌骨仅单枚，鼻突长，鼻突中间有骨缝；左右鼻骨被鼻突分开；额鳞弧粗壮而明显；上颌骨较短，不达方骨；翼骨较短，不与上颌骨连接；基舌软骨上有1对指状突；2对角鳃骨均骨化，上鳃骨仅1对，硬骨质。

瘰螈属物种中除越南瘰螈 *Paramesotriton deloustali* (Bourret, 1934) 在越南北部地区有分布外，本属物种均分布于中国西南和华南地区，中国现有14种。茂兰地区分布有1种，即茂兰瘰螈 *Paramesotriton maolanensis*。

（1）茂兰瘰螈 *Paramesotriton maolanensis* Gu et al., 2012

Paramesotriton maolanensis Gu et al., 2012, Zootaxa, 3510: 41. Holotype: GZNU2006030001, by original designation. Type locality: "Wengang, 25°40' N, 107°53' E, 817m a.s.l., Libo County, Guizhou Province, P. R. China".

Paramesotriton maolanensis Fei and Ye, 2016, Amph. China, 1: 364.

[同物异名]　无

[鉴别特征]　皮肤较为光滑，头和身体无颗粒状疣；体型较大；眼睛退化，上眼睑和下眼睑闭合；头后侧角状突起大；后枕骨到尾端的背鳍褶为黄色，在背中部的肤褶较为发达；体背两侧的横沟不显；舌骨体短，第二角鳃骨为硬骨，而非软骨，上鳃骨分离，不相连。

[形态描述]　本种体型较大；雄螈体全长177.4～192.0mm，雌螈体全长197.4～207.8mm，头长明显大于头宽；吻短，吻端平截，突出在下唇前方，吻棱明显，吻长明显大于眼径；鼻孔位于吻端；眼睛相对退化，自然条件下上眼睑和下眼睑为闭合状；唇褶发达，上、下颌具细齿；舌呈椭圆形，左右两侧游离，犁骨齿列呈"Λ"形，齿列的前缘在两内鼻孔之间会合；肋沟不明显，背脊棱显著；前肢贴体前伸时，指端达到吻端；后肢略长于前肢，略比前肢粗壮，前后肢贴体相对时，互达掌、蹠部；第三趾与第四趾几乎等长；指、趾端具角质膜，无缘膜、无蹼；无掌突、蹠突。

[第二性征]　雄螈泄殖腔孔显著隆起，肛裂为1纵缝，长6～7mm，内侧有指状乳突；雌螈泄殖腔孔隆起小，肛裂小，椭圆形，呈圆锥状，其内侧无指状乳突。

生活时身体呈黑褐色；背脊棱为不连续的黄色纵纹；喉部腹面和体腹色较背部浅，并缀以不规则大型的橘红色斑块和黄色小型斑块；掌、蹠部为灰白色。

[生活习性]　本种生活于水流平缓的大水塘或有地下水流出的水塘中，水塘周围植被茂盛，水质清澈。平时通常会栖息在水塘底部，较难寻找，有洪水时会跳出水面。

[地理分布]　茂兰地区见于翁昂。

[模式标本]　正模标本（GZNU2006030001）保存于贵州师范大学

[模式产地]　贵州荔波，25°40′N，107°53′E，海拔817m

[保护级别]　无

[IUCN濒危等级]　未予评估（NE）

[讨论]　该物种为2012年发表的新种，且仅在茂兰国家级自然保护区内有发现，2016～2017年在其分布地进行重新调查时未发现该物种，表明该物种在该区域内数量稀少，分布狭窄，需重点保护。

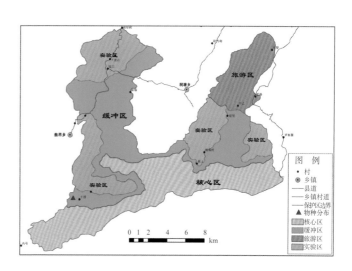

在茂兰地区的分布

生境

2）疣螈属 *Tylototriton* Anderson, 1871

Tylototriton Anderson, 1871, Proc. Zool. Soc., London, 1871: 423. Type species: *Tylototriton verrucosus* Anderson, 1871, by monotypy.

Triturus (*Tylototriton*) Boulenger, 1878, Bull. Soc. Zool. France, 3: 308.

Tylotriton Boettger, 1885, Ber. Offenbach. Ver. Naturkd., 24-25: 165. Incorrect subsequent spelling.

Yaotriton Dubois and Raffaëlli, 2009, Alytes, 26: 59. Type species: *Tylototriton asperrimus* Unterstein, 1930, by original designation. Coined as a subgenus of *Tylototriton*.

Yaotriton Fei, Ye and Jiang, 2012, Colored Atlas Chinese Amph. Distr.: 87. Treatment as a genus.

Qiantriton Fei, Ye and Jiang, 2012, Colored Atlas Chinese Amph. Distr.: 78, 594. Type species: *Tylototriton kweichowensis* Fang and Chang, 1932, by original designation. Coined as a subgenus.

Liangshantriton Fei, Ye and Jiang, 2012, Colored Atlas Chinese Amph. Distr.: 44. Type species: *Tylototriton taliangensis* Liu, 1950.

Qianotriton Fei and Ye, 2016, Amph. China, 1: 284.

本属物种皮肤粗糙，布满瘰粒和疣粒，有的体侧有13～16枚瘰粒或瘰疣连续隆起成纵列。头扁平而宽短，无唇褶，头侧骨质脊棱明显，背正中脊棱明显。舌卵圆，前后端粘连，两侧游离。指、趾无蹼，尾一般侧扁较长。

犁骨齿列于颌骨和额骨之间，呈"Λ"形，前端会合；前颌骨2，鼻突短，不达额骨；鼻骨大，介于前颌骨和额骨之间，额鳞弧粗壮；上颌骨长，与方骨相接触，翼骨前端或以韧带或直接与上颌骨连接。基舌软骨前端有2对指状突；角鳃骨2对，上鳃骨1对，骨质细小。

卵单生（仅红瘰疣螈有时连成单行）。卵产于静水中或塘边陆地潮湿环境中。睾丸豆形，每侧1～3叶。

早期幼体眼后下方有平衡枝；羽状外鳃发达；体尾肥实，尾背鳍褶起于体背部。成体一般营陆栖生活。

本属物种现有25种，中国分布有15种，国内主要分布于湖北、广东、云南、河南、四川、重庆、甘肃、安徽、湖南、广西、海南等地；国外分布于印度（东北部）、尼泊尔、缅甸、泰国及越南（北部）。贵州分布有3种，即细痣疣螈 *Tylototriton asperrimus*、贵州疣螈 *Tylototriton kweichowensis* 和文县疣螈 *Tylototriton wenxianensis*。茂兰地区分布有2种，即细痣疣螈和文县疣螈。

茂兰地区种检索表

体两侧的瘰粒几乎连成纵列，分界不清；背、腹面疣粒大小较为一致，腹面无细密横纹，体两侧无肋沟；肛裂周缘与体色相同···文县疣螈 *Tylototriton wenxianensis*

体两侧瘰粒13～16枚，圆形，各瘰粒界限明显；胸、腹部有许多细密横纹，体两侧有肋沟；肛裂周缘色浅············ 细痣疣螈 *Tylototriton asperrimus*

（2）细痣疣螈 *Tylototriton asperrimus* Unterstein, 1930

Tylototriton asperrimus Unterstein, 1930, Sitzungsber, Ges. Naturforsch. Freunde Berlin, 1930: 314. Syntypes: ZMB (2 specimens); ZMB 34089 considered holotype (lectotype designation by implication) by Bauer, Good and Günther, 1993, Mitt. Zool. Mus. Berlin, 69: 298. Type locality: "Kwangsi"; corrected to "Yaoshan [= Mt. Yaoshan], Kwangsi Prov. [= Guangxi], China" by Bauer, Good and Günther, 1993, Mitt. Zool. Mus. Berlin, 69: 298. Fan, 1931, Bull. Dept. Biol. Coll. Sci. Sun Yatsen Univ., 11: 1-154, provided evidence that the type locality is or near to the town of Loshiang.

Tylototriton (*Echinotriton*) *asperrimus* Zhao and Hu, 1984, Stud. Chinese Tailed Amph.: 19.

Tylototriton asperrimus asperrimus Fei, Ye and Yang, 1984, Acta Zool. Sinica, 30: 89.

Pleurodeles (*Tylototriton*) *asperrimus* Risch, 1985, J. Bengal Nat. Hist. Soc., N.S., 4: 142.

Pleurodeles (*Echinotriton*) *asperrimus* Dubois, 1987 "1986", Alytes, 5: 11.

Echinotriton asperrimus Zhao, 1990, in Zhao (ed.), From Water onto Land: 219; Nguyen, Ho and Nguyen, 2005, Checklist Amph. Rept. Vietnam: 9.

Echinotriton asperrimus asperrimus Zhao and Adler, 1993, Herpetol. China: 112.

Tylototriton asperrimus Nussbaum, Brodie and Yang, 1995, Herpetologica, 51: 265.

Tylototriton (*Yaotriton*) *asperrimus* Dubois and Raffaëlli, 2009, Alytes, 26: 68.

Yaotriton asperrimus Fei, Ye and Jiang, 2012, Colored Atlas Chinese Amph. Distr.: 88.

[同物异名] *Echinotriton asperrimus*（细痣棘螈）；*Pleurodeles asperrimus*（细痣肋突螈）；*Yaotriton asperrimus*（细痣瑶螈）

[鉴别特征] 体两侧瘰粒圆形，各瘰粒界限明显；胸、腹部有许多细密横纹，体两侧有肋沟；周身黑棕色，口角后上方无浅色突起，仅指、趾、肛缘及尾腹鳍褶下缘橘红色；头侧脊棱明显；尾短于头体长。

[形态描述] 依据茂兰地区标本描述：雄螈体全长110.23～134.92mm，头部扁平，吻端平截；头侧脊棱甚显著，耳后腺后部向内弯曲，鼻孔近吻端；无囟门，头顶"V"形脊棱与背部中央脊棱相连，无唇褶，舌卵圆形，前后端与口腔底部粘连，两侧游离；口角距眼后角下后方较远；犁骨齿列呈"∧"形，颈褶甚明显，其躯干呈圆柱状且略扁，体侧有肋沟，背脊棱显著；前后肢较细，指式为3＞2＞4＞1；趾式为3＞4＞2＞5＞1；指、趾无缘膜和角质鞘，掌突和蹠突扁平或不显；尾侧扁，尾末端钝尖，背鳍褶较高而薄，起始于尾基部，腹鳍褶较厚。体尾背面黑褐色，仅指、趾、肛缘和尾下缘为橘红色；体腹面黑灰色。皮肤粗糙，满布瘰粒和疣粒，体两侧各有圆形瘰粒13～16枚，排成纵列，瘰粒间界限明显，胸、腹部有细密横纹。

[第二性征] 雄螈肛孔纵长，内壁有小乳突；雌螈肛部呈丘状隆起，肛孔略呈圆形。

[生活习性] 该物种生活于海拔550～600m的山间及有腐殖质的水塘中，白天常栖息于水底，晚上成对夜出觅食。雨季来临时，成螈会爬到水源地周围附近的山林。在翁昂和板寨地区小河中的落叶上和流水缓慢的山间小溪中看见过本种的卵。2017年6月在板寨进行调查时发现有较大的种群，但未看见卵群。

[地理分布] 茂兰地区分布于洞塘、翁昂等地。国内分布于贵州、广西、广东。国外分布于越南（北部）。

[模式标本] 全模标本（ZMB, 2号标本），正模标本（ZMB 34089）保存于德国柏林洪堡大学自然历史博物馆（Museum of Natural History, Humboldt University of Berlin, Berlin, Germany, ZMB）

[模式产地] 广西金秀县大瑶山（Unterstein, 1930）

[保护级别] 国家Ⅱ级重点野生保护动物

[IUCN濒危等级] 近危（NT）

[讨论] 最近文献研究表明：疣螈属为一单系，与棘螈属 Echinotriton 互为姐妹群体关系（Weisrock et al., 2006; Zhang et al., 2008; Phimmachak et al., 2015）。疣螈属内分化为两个支系，分别为棕黑疣螈种组 Tylototriton verrucosus group 和细痣疣螈种组 Tylototriton asperrimus group（Yuan et al., 2011; Nishikawa et al., 2013），也称为瑶螈亚属 Yaotriton 和疣螈亚属 Tylototriton（Phimmachak et al., 2015）。

Unterstein（1930）以广西标本发表了新种细痣疣螈 Tylototriton asperrimus。之后，赵尔宓和胡其雄（1984）将其记载为 Tylototriton（Echinotriton）asperrimus。赵尔宓（1990）与 Zhao 和 Adle（1993）又将该种改订为细痣棘螈 Echinotriton asperrimus。但是我国学者多数不采用以上分类系统（田婉淑和江耀明，1986；叶昌媛等，1993；费梁等，1990, 2010）。费梁等（2012）采用瑶螈属 Yaotriton 将其修订为细痣瑶螈 Yaotriton asperrimus。Amphibia China（2017）基于形态和分子数据的研究仍采用疣螈属 Tylototriton 的分类系统而记录为细痣疣螈 Tylototriton asperrimus。目前，Amphibia Web（2017）共收录该属22个物种，世界两栖物种（Frost, 2017）共收录该属21个物种（将大凉疣螈归于凉螈属）。

在贵州分布的疣螈属物种有细痣疣螈 Tylototriton asperrimus、文县疣螈 Tylototriton wenxianensis 和贵州疣螈 Tylototriton kweichowensis 3种，其中贵州疣螈和细痣疣螈与文县疣螈区别明显，但是细痣疣螈与文县疣螈的形状相似，所以以往的调查研究中往往有鉴别失误的情况。费梁等（2006）在《中国动物志 两栖纲（上卷）总论 蚓螈目 有尾目》261页写道："此外贵州省绥阳、遵义、大方地区的本种标本是否属于细痣疣螈或者属于文县疣螈？还有待深入研究。"同时，费梁等（2006, 2010）将贵州绥阳、荔波、遵义、大方地区作为贵州地区细痣疣螈的分布地，将贵州雷山作为文县疣螈的分布地；费梁等（2012）和 Amphibia China（2017）将贵州荔波作为细痣疣螈在贵州的分布地，将大方、绥阳、遵义、雷山作为文县疣螈在贵州的分布地。Yuan 等（2011）基于线粒体基因对细痣疣螈种组系统发育关系进行了探讨，使用了贵州荔波和雷山的标本数据，其结果显示，贵州荔波和雷山所产的细痣疣螈与四川龙门山所采集的文县疣螈互为姐妹群。因此，Yuan 等（2011）建议将贵州荔波和雷山所产的细痣疣螈改名为文县疣螈，并作为其分布地。有的学者（袁智勇，个人通信）认为，"贵州绥阳、遵义地区分布的是文县疣螈，荔波应该也有文县疣螈和细痣疣螈，荔波一带应该是细痣疣螈和文县疣螈的交界"。

细痣疣螈是国家Ⅱ级重点保护野生动物，该物种对环境要求很高，易受人为干扰，在茂兰地区见于白鹇湖、翁根

雄性成体量度 （单位：mm）

标本编号	体全长	头长	头体长	头宽	吻长	眼径	上眼睑宽	尾高	尾宽	尾长
GZNU19980712001	135.23	25.40	67.72	12.88	4.89	4.91	4.00	7.55	5.26	67.51
GZNU19980712002	113.62	19.79	61.69	14.38	5.27	4.08	3.49	6.53	5.10	51.93
GZNU19980712003	129.04	18.85	60.20	14.09	4.02	4.49	3.37	7.3	4.65	68.84
GZNU19980712004	124.87	16.49	62.84	13.81	4.95	4.13	3.09	7.49	4.75	62.03
GZNU19980712005	158.04	20.81	87.92	15.11	5.76	5.22	4.15	8.83	6.15	70.12
GZNU19980712006	99.91	13.69	58.69	14.60	4.45	4.30	3.78	7.37	4.36	41.22

洞和佳荣拉蒋村3处，其中前两处位于远离居民居住地的茂兰保护区核心区内，人为干扰较弱，种群数量较多；后一处离村寨较近，距离最近的村寨仅600m左右，且其生活地点的山谷中还有农田和鱼塘数个，存在人为严重干扰的问题，建议加强对此处栖息地的保护。

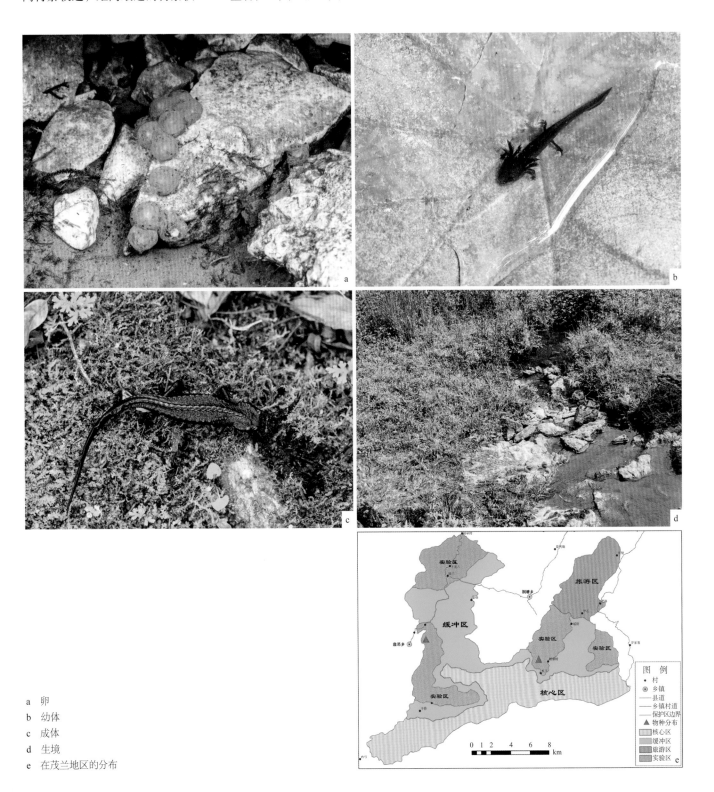

a 卵
b 幼体
c 成体
d 生境
e 在茂兰地区的分布

（3）文县疣螈 *Tylototriton wenxianensis* Fei, Ye and Yang, 1984

Tylototriton asperrimus wenxianensis Fei, Ye and Yang, 1984, Acta Zool. Sinica, 30: 89. Holotype: CIB 638164, by original designation. Type locality: "Wenxian [= Wenxian County], Gansu [Province], alt. 946m", China.

Tylototriton asperrimus pingwuensis Deng and Yu, 1984, Acta Herpetol. Sinica, Chengdu, N.S., 3 (2): 75. Holotype: KIZ 74005, by original designation. Type locality: "Duiwoliang, Pingwu, Sichuan, alt. 1400m", China. Synonymy by Ye, Fei and Hu, 1993, Rare and Economic Amph. China: 80, following Zhao, Hu, Jiang and Yang, 1988, Studies on Chinese Salamanders: 63.

Pleurodeles (*Tylototriton*) *asperrimus wenxianensis* Risch, 1985, J. Bengal Nat. Hist. Soc., N.S., 4: 142.

Echinotriton asperrimus wenxianensis Zhao and Adler, 1993, Herpetol. China: 112.

Tylototriton wenxianensis Ye, Fei and Hu, 1993, Rare and Economic Amph. China: 80.

Tylototriton (*Yaotriton*) *wenxianensis* Dubois and Raffaëlli, 2009, Alytes, 26: 68.

Tylototriton wenxianensis wenxianensis Chen, Wang and Tao, 2010, Acta Zootaxon. Sinica, 35: 666.

Yaotriton wenxianensis Fei, Ye and Jiang, 2012, Colored Atlas Chinese Amph. Distr.: 94.

[同物异名] *Yaotriton wenxianensis*（文县瑶螈）；*Echinotriton asperrimus wenxianensis*；*Echinotriton asperrimus wenxianensis*；*Pleurodeles asperrimus wenxianensis*

[鉴别特征] 形似细痣疣螈，与细痣疣螈的区别在于体两侧无肋沟，瘰粒间分界不明显，几乎形成纵带；腹面瘰粒不呈横纹状。

[形态描述] 依据茂兰地区标本描述 雄螈全长110～135mm，头部扁平，吻端平截；头侧脊棱甚显著，鼻孔近吻端；头顶"V"形脊棱与背部中央脊棱相连；犁骨齿列呈"Λ"形，颈褶明显，体侧无肋沟，体侧瘰粒间分界不明显，几乎

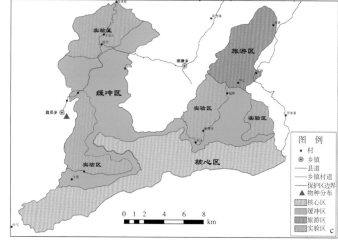

a 背面观
b 腹面观
c 在茂兰地区的分布

形成纵带；腹面疣粒显著，不呈横纹状。背脊棱显著；尾侧扁，尾末端钝尖，背鳍褶较高而薄，始于尾基部，腹鳍褶窄而厚。体尾背面黑褐色，仅指、趾和尾部下缘为橘红色；体腹面黑灰色。

[第二性征] 雄螈肛孔纵长，内壁有小乳突；雌螈肛部呈丘状隆起，肛孔较短，略呈圆形。

[生活习性] 本种生活于茂兰地区翁昂乡的一个小盆地中。该盆地四面环山，山底有一终年不断的溪流，在盆地中央形成多个大小不一的小水塘，水塘周围杂草丛生，泥地松软，4月调查时便在该地水塘内发现多只成体。

[地理分布] 茂兰地区分布于翁昂。国内见于甘肃、四川、重庆、贵州。

[模式标本] 正模标本（CIB 638164）保存于中国科学院成都生物研究所

[模式产地] 甘肃文县，海拔946m

[保护级别] 三有保护动物；国家Ⅱ级重点保护野生动物

[IUCN 濒危等级] 易危（VU）

（二）无尾目ANURA

2. 角蟾科 Megophryidae Bonaparte, 1850

Megalophreidina Bonaparte, 1850, Conspect. Syst. Herpetol. Amph.: 1. Type genus: *Megalophrys* Wagler, 1830 (= *Megophrys* Kuhl and van Hasselt, 1822).

Megalophryinae Fejérváry, 1922 "1921", Arch. Naturgesch., Abt. A, 87: 25.

Megophryinae Noble, 1931, Biol. Amph.: 492.

Megalophryninae Tamarunov, 1964, in Orlov (ed.), Osnovy Paleontologii, 12: 129.

Leptobrachiini Dubois, 1980, Bull. Mens. Soc. Linn. Lyon, 49: 471. Type genus: *Leptobrachium* Tschudi, 1838.

Megophryini Dubois, 1980, Bull. Mens. Soc. Linn. Lyon, 49: 471.

Leptobrachiinae Dubois, 1983, Bull. Mens. Soc. Linn. Lyon, 52: 272; Dubois, 1983, Alytes, 2: 147.

Oreolalaxinae Tian and Hu, 1985, Acta Herpetol. Sinica, Chengdu, N.S., 4 (3): 221. Type genus: *Oreolalax* Myers and Leviton, 1962. Considered a synonym of Leptobrachiinae by Dubois, 1987 "1986", Alytes, 5: 173-174; Fei and Ye, 2005, in Fei et al. (eds.), Illust. Key Chinese Amph.: 56.

Oreolalaginae Dubois, 1987 "1986", Alytes, 5: 173-174. Justified emendation.

Megophryidae Ford and Cannatella, 1993, Herpetol. Monogr., 7: 94-117.

肩带弧胸型。肩胸骨小，多为软骨质，中胸骨长，大多数骨化；肩胛骨长（大于锁骨的1/2），近端二叉状，前部不与锁骨重叠。椎体变凹型，角蟾亚科和拟髭蟾亚科的前凹式椎骨由骨化或钙化了的椎间体与1枚椎体愈合而成，椎间体为游离状态，其前后椎骨形成典型双凹，荐椎前椎骨8枚；无肋骨；荐椎横突显著宽大；荐椎与尾杆骨以单个骨髁与尾杆骨凹相关节；尾杆骨近端多有横突或仅有残迹。鼻骨或大或小，左右相接触或不相接触；通常无额顶囟；一般上颌有齿，个别属无齿或有齿突；有腭骨，有方轭骨，多与上颌骨重叠；鼓膜显著或隐于皮下或缺，有耳柱骨或不同程度退化；有或无犁骨齿（拟髭蟾亚科种类无犁骨齿）。舌骨前突宽大，无舌骨前角和翼状突（或与前突合并），无副舌骨或有"Y"形残迹。环状软骨在背面不连接或连接。跟、距骨仅在两端愈合；远列跗骨2~3枚。

皮肤光滑或具有大小疣粒；舌卵圆，后端游离，一般有缺刻；瞳孔大多纵置；指趾末端不呈吸盘状；指间和外侧蹼间无蹼；一般趾间无蹼或蹼不发达；关节下瘤多不明显，或趾下有肤棱。配对时，抱握胯部。

卵及蝌蚪在水中发育，蝌蚪为有唇齿左孔型（acosmanura）。口部形态除角蟾类口部呈漏斗状外，其余属、种口周围有唇乳突，上、下唇有唇齿，且外排最短；角质颌强，适于刮取藻类，甚至还可咬食小蝌蚪。

Cannatella（1985）将Bonaparte于1850年建立的角蟾亚科Megophryinae提升为角蟾科Megophryidae。本书采用

这一科名，并将原属于锄足蟾科 Pelobatidae 的产于亚洲东部、东南部和南部的属、种均归属于角蟾科。现本科共有 2 亚科 5 属 242 种。角蟾亚科 Megophryinae 共计 1 属 91 种，拟髭蟾亚科 Leptobrachiinae 共计 4 属 151 种，分布于亚洲东部、南部和东南部。中国共有 5 属 118 种，多分布于秦岭以南地区。

3）异角蟾属 *Xenophrys* Günther, 1864

Xenophrys Günther, 1864, London: Ray Society by R. Hardwicke. The reptiles of British India. Type species: *Xenophrys parva* Günther, 1864, Type locality: Khasi Hills, India.

Megalophrys Boulenger, 1908, Proc. Zool. Soc., London, 78: 407-430. A revision of the Oriental pelobatid batrachians (genus *Megalophrys*).

Xenophrys Chen et al., 2017, Molecular Phylogenetics and Evolution 106: 28-43. A novel multilocus phylogenetic estimation reveals unrecognized diversity in Asian horned toads, genus *Megophrys sensu lato* (Anura: Megophryidae).

Chen 等（2017）研究表明异角蟾属为有效属，该属分布在中国（南方、喜马拉雅山）和东南亚的内陆地区，真正的角蟾属局限分布在巽他古陆（Sundaland），两者的地理分布可能在马来半岛的南部有重叠。目前，异角蟾属记录有 47 个物种。中国南方是该属物种多样性最高的地区。我国现记录有 31 个物种（Chen et al.，2017）。茂兰地区仅有 1 种，即荔波异角蟾。

（4）荔波异角蟾 *Xenophrys liboensis* Zhang et al., 2017

Xenophrys liboensis Zhang et al., 2017, Asian Herpetol. Res., 8: 77. Holotype: GNUG20160408008, by original designation. Type locality: "Libo County, Guizhou Province, China (25.4731°N, 108.1054°E, elevation 634m a.s.l.)".

[同物异名] 无

[鉴别特征] 体型大，鼓膜明显；有犁骨齿；前臂及手长大于体长之半；左右跟部无重叠；第Ⅰ～第Ⅲ趾之间具蹼迹，但无边缘沟；趾侧具缘膜；背部可见"X"形细肤棱；上眼角状突起显著；生活时虹膜呈棕色。

[形态描述] 依据茂兰地区标本描述：头长略小于头宽，吻端突出，且显著突出于下唇，吻棱发达，瞳孔纵置，鼓膜明显，呈卵圆形，具上颌齿和犁骨齿；舌端卵圆形，边缘有锯齿状缺刻；雄性具单咽下内声囊；背面皮肤光滑有肤棱，两眼间有深褐色三角斑，上眼睑中部有 1 角状肤突；腹部光滑，从咽部至胸部为红棕色；股背面棕黑色，腹面米白色且具小疣粒，生活时背面红棕色；雄性无婚刺，背部无雄性线。

[生活习性] 该蟾生活在弱光或无光的喀斯特洞穴中，洞内有终年不干枯的地下河，雨水多时，河水能从洞口漫出，形成地表河流。3 月首次在洞穴内发现一只成体，6 月重复调查时发现超过 10 只成体，8 月在同一洞穴发现大量成体，但多次调查中均未在洞外发现过成体。因该物种为首次被发现，且多次调查中均未看到卵和幼体，故其繁殖和发育状况还有待进一步科考研究。

[地理分布] 目前仅见于贵州荔波，标本采集地为佳荣镇。

[模式标本] 正模标本 GNUG20160408008 保存于贵州师范大学生命科学学院动物标本室

[模式产地] 贵州荔波，25.4731°N，108.1054°E，634m

[保护级别] 未评估

[IUCN 濒危等级] 数据缺乏（DD）

[讨论] 该物种为 2017 年发表的两栖动物新种，目前仅在荔波世界自然遗产地范围内发现。经过多次重复调

查，结果表明，该物种种群数量有80～100只，仅分布于一个喀斯特溶洞中，在该调查区域内分布区极度狭窄，物种数量稀少，建议加强保护。

本书采纳Amphibia China（2017）和Chen等（2017）的相关记录与建议，接受将角蟾亚科划分为多个属的观点，并将此次发现的新种命名为荔波异角蟾 Xenophrys liboensis。

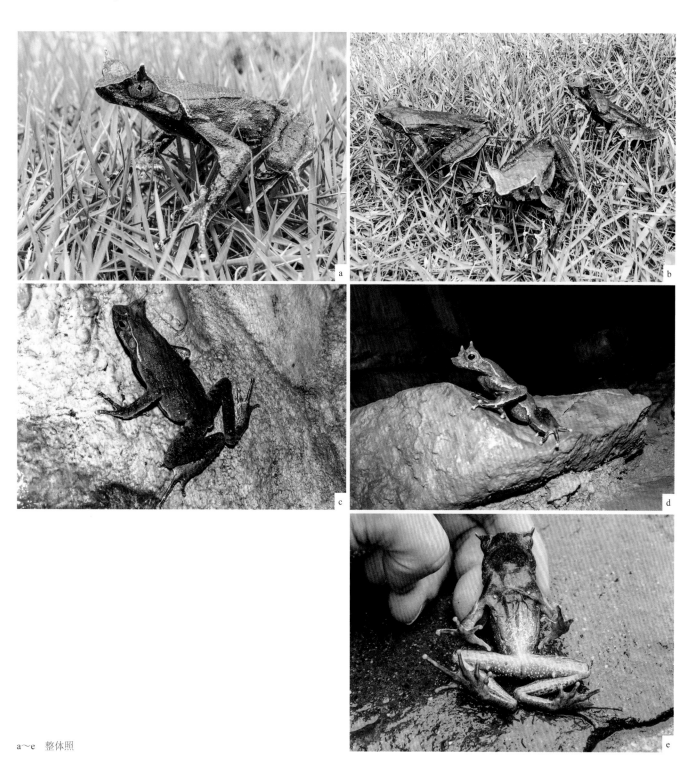

a～e 整体照

3. 蟾蜍科 Bufonidae Gray, 1825

Bufonina Gray, 1825, Ann. Philos., London, Ser. 2, 10: 214. Type genus: *Bufo* Laurenti, 1768.

Bufonidae: Bell, 1839, Hist. Brit. Rept.: 105.

Atelopoda Fitzinger, 1843, Syst. Rept.: 32. Type genus: *Atelopus* Duméril and Bibron, 1841. Synonymy by Duellman and Lynch, 1969, Herpetologica, 25: 239; McDiarmid, 1971, Sci. Bull. Nat. Hist. Mus. Los Angeles Co., 12: 52; Trueb, 1971, Contrib. Sci. Nat. Hist. Mus. Los Angeles Co., 216: 1-40.

Stephopaedini Dubois, 1987 "1986", Alytes, 5: 27. Type genus: *Stephopaedes* Channing, 1978.

肩带弧胸型（有的为拟固胸型，上喙骨相互愈合）。一般无肩胸骨，若有则为软骨质，中央很少钙化；正胸骨软骨质；肩胛骨长（小于锁骨的 2 倍），前端不与锁骨重叠。椎体前凹型，荐椎前椎骨 7～8 枚，有的属其寰椎与第一躯椎合并；无肋骨；荐椎横突宽大，通常有骨髁 2 个（尾杆骨与荐椎愈合例外）；尾杆骨近端无横突或偶尔有。大多数属头骨骨化程度很高，大部分种类皮肤与头骨相粘连；鼻骨大，两鼻骨相接触或分离；上颌无齿；有腭骨或无（小蟾属 *Parapelophryne* 和游蟾属 *Nectophryne*）；方轭骨、耳柱骨有或无。无副舌骨，舌骨前角无前突或有（如叶毒蛙属 *Atelopus*）；翼状突（即前侧突）小或宽大；无后侧突或有（如蟾蜍属 *Bufo*）；环状软骨完全。跟骨、距骨仅在两端并合；远列跗骨 2～3 枚；瞳孔横置；配对时抱握腋部。

本科动物大小相差很大，最小的对趾蟾属 *Oreophrynella* 体长仅 20mm，最大的蟾蜍属达 250mm。大多数属皮肤粗糙，少数属皮肤光滑（如斑蟾属 *Atelopus*）。舌呈长椭圆形，后端无缺刻。四肢较短；有的属指、趾节减少或变短（如小蟾属）；有外蹠突；指、趾末端正常；一般有毕德氏器。

成体多营陆栖或穴居，也有营树栖及水栖生活的。

本科大多数的属产出长形卵带，卵小，数量多，具色素。有的属卵呈单粒，卵大数少，无色素；为体外受精。有的卵在水外孵化，蝌蚪进入水中生活和完成变态；有的蝌蚪不进入水内，直接发育成幼蛙；有的蝌蚪在溪流中发育，具有大的口盘或腹吸盘。蝌蚪有唇齿和角质颌，出水孔单个位于体左侧。茂兰地区分布有 2 属 2 种（2 亚种）。

茂兰地区属检索表

眼角后方有鼓膜鼓环，头上无头棱···蟾蜍属 *Bufo*

眼角后方有鼓膜鼓环，头部具有头棱，棱上多角质化··头棱蟾属 *Duttaphrynus*

4）蟾蜍属 *Bufo* Garsault, 1764

Bufo Garsault, 1764, Fig. Plantes et Animaux: pl. 672. Type species: Not designated although animal in figure, assuming it is from France, is tentatively identifiable as *Rana bufo* Linnaeus, 1758 (DRF). Designated as *Rana bufo* Linnaeus, by Dubois and Bour, 2010, Zootaxa, 2447: 24. This disputed by Welter-Schultes and Klug, 2011, Zootaxa, 2814: 55, who suggested that the type species is *Bufo viridis* Laurenti, 1768, by subsequent designation of Fitzinger, 1843, Syst. Rept.: 32. Both sets of authors reject the apparent subsequent type species designation of *Bufo vulgaris* Laurenti, 1758 by Tschudi, 1838, Classif. Batr.: 88, as ambiguous. ICZN action is needed to finally resolve the issue, although DRF is inclined towards the position of Dubois and Bour. See comment under *Bufotes viridis*.

Bufo Laurenti, 1768, Synops. Rept.: 25. Type species: *Bufo viridis* Laurenti, 1768, by subsequent designation of Fitzinger, 1843, Syst. Rept.: 32.

Phryne Oken, 1816, Lehrb. Naturgesch., 3 (2): 210. Type species: *Bufo vulgaris* Laurenti, 1768, by original designation. Synonymy by Boulenger, 1882, Cat. Batr. Sal. Coll. Brit. Mus., Ed. 2: 281. Unavailable name by designation of Anonymous, 1956, Opin. Declar. Internatl. Comm. Zool. Nomencl., 14: 1-42.

Phryne Fitzinger, 1843, Syst. Rept.: 32. Type species: *Bufo vulgaris* Laurenti, 1768, by original designation.

Pegaeus Gistel, 1868, Die Lurche Europas: 161. Type species: *Rana bufo* Linnaeus, 1758, by subsequent designation of Dubois and Bour, 2010, Zootaxa, 2447: 24. Synonymy by Mertens, 1936, Senckenb. Biol., 18: 76.

Platosphus de l'Isle, 1877, J. Zool., Paris, 6: 473. Type species: *Platosophus gervais* d'Ilse, 1877, by monotypy. Synonymy by Sanchíz, 1998, Handb. Palaeoherpetol., 4: 122; Dubois and Bour, 2010, Zootaxa, 2447: 24.

Bufavus Portis, 1885, Atti Accad. Sci. Torino, Cl. Sci. Fis. Mat. Nat., 20: 1182. Type species: *Bufavus meneghinii* Portis, 1885, by monotypy. Fossil taxon. Synonymy by Sanchíz, 1998, Handb. Palaeoherpetol., 4: 125; Dubois and Bour, 2010, Zootaxa, 2447: 24.

Torrentophryne Yang, Liu and Rao, 1996, Zool. Res., Kunming, 17: 353. Type species: *Torrentophryne aspinia* Yang, Liu and Rao, 1996, by original designation (but see comment by Dubois and Bour, 2010, Zootaxa, 2447: 24, who regarded the type species to be by their subsequent designation). Synonymy by Liu et al., 2000, Mol. Phylogenet. Evol., 14: 423-435. Recognized subsequently without discussion by Fei, Ye and Jiang, 2003, Acta Zootaxon. Sinica, 28: 762-766. Considered without discussion, to be synonymous with *Phrynoidis* by Fei et al., 2005, in Fei et al. (eds.), Illust. Key Chinese Amph.: 92, 258-259. Synonymy with *Bufo* (*sensu stricto*) by Frost et al., 2006, Bull. Am. Mus. Nat. Hist., 297: 215; Dubois and Bour, 2010, Zootaxa, 2447: 24.

Schmibufo Fei and Ye, 2016, Amph. China, 1: 762. Type species: *Bufo stejnegeri* Schmidt, 1931. Coined as a subgenus of *Bufo*.

该属动物瞳孔水平；舌呈椭圆形或梨形，后端无缺刻；指间无蹼（除个别外），趾间蹼发达或无；指、趾末端正常或略膨大；外侧蹠间无蹼。皮肤粗糙具大小瘰粒，耳后腺大。

额顶骨正常，中央合并或略分开；蝶筛骨仅筛骨部分骨化；眶蝶骨未骨化；犁骨齿无；有腭骨和方轭骨；鼓环和耳柱骨有或无；咽鼓管孔有或无；舌骨前角无前突，有翼状突（前侧突）和后侧突。荐椎前椎骨8枚，寰椎与第一躯椎分离；荐椎与尾杆骨分离；荐椎横突适度扩大；关节髁2枚与尾杆骨相关节；肩带弧胸型，无肩胸骨，中胸骨为1片宽的软骨板。有毕德氏器。

卵群呈带状，卵粒的动物极黑色，植物极棕色，卵排列在带状胶质管内；行体外受精；卵在水中发育直至完成变态。蝌蚪仅两口角有唇乳突。

蟾蜍一般为陆栖或林栖，遍及广大地区，种群数量较大，在捕食害虫上具有重要的生态学意义，应该给予保护。

本属物种现有17种，据Amphibia China（2017）记载，中国现有分布11种，Fei和Ye（2016）记载有7种。茂兰地区分布有2个亚种。

（5）中华蟾蜍 *Bufo gargarizans* Cantor, 1842

Bufo gargarizans Cantor, 1842a, Ann. Mag. Nat. Hist., Ser. 1, 9: 483. Type(s): not stated, presumably originally in BMNH. Type locality: "Chusan ... Island", East China Sea, off northeastern coast of Zhejiang, China.

Bufo griseus Hallowell, 1861 "1860", Proc. Acad. Nat. Sci. Philadelphia, 12: 506. Holotype: Deposition not stated, presumably ANSP or USNM. Type locality: "Hong Kong, China ... in the marshes of Whampoa [= Huangpu]", actually Guangdong Province, China, according to Zhao and Adler, 1993, Herpetol. China: 128, who made the synonymy. Formerly placed in the synonymy of *Bufo bufo* (*sensu lato*) by Boulenger, 1881 "1880", Proc. Zool. Soc., London, 1880: 569, into *Bufo bufo gargarizans* by Stejneger, 1907, Bull. U.S. Natl. Mus., 58: 59; Liu and Hu, 1961, Tailless Amph. China: 119, and into *Bufo melanostictus* by Boettger, 1888, Ber. Offenbach. Ver. Naturkd., 1888: 99.

Bufo maculiventris Fitzinger, 1861 "1860", Sitzungsber. Akad. Wiss. Wien, Phys. Math. Naturwiss. Kl., 42: 415. Type(s):

Presumably NHMW, but not mentioned in recent type lists. Type locality: "Shanghai", China. Preoccupied by *Bufo maculiventris* Spix, 1824. Considered a junior synonym of *Bufo vulgaris* (*sensu lato*) by Steindachner, 1867, Reise Österreichischen Fregatte Novara, Zool., Amph.: 39, so (DRF) presumably associated with *Bufo gargarizans*.

Bufo sinicus Fitzinger, 1861 "1860", Sitzungsber. Akad. Wiss. Wien, Phys. Math. Naturwiss. Kl., 42: 415. Types species: Presumably NHMW, but not mentioned in recent type lists. Type locality: "Shanghai", China. Nomen nudum. Considered a junior synonym of *Bufo vulgaris* by Steindachner, 1867, Reise Österreichischen Fregatte Novara, Zool., Amph.: 39, so (DRF) presumably associated with *Bufo gargarizans*.

Bufo vulgaris var. *asiatica* Steindachner, 1867, Reise Österreichischen Fregatte Novara, Zool., Amph.: 39. Types: Presumably NHMW but not mentioned in recent type lists. Type locality: "Shanghai", China. Tentative synonymy by Schmidt, 1926, China J. Sci. Arts, Shanghai, 4: 76. Synonymy by Shannon, 1956, Herpetologica, 12: 30; Liu and Hu, 1961, Tailless Amph. China: 119; Zhao and Adler, 1993, Herpetol. China: 128.

Bufo vulgaris var. *sachalinensis* Nikolskii, 1905, Herpetol. Rossica: 389. Syntypes: ZISP 1934-1936 according to Kuzmin and Maslova, 2003, Adv. Amph. Res. Former Soviet Union, 8: 126 (given as 1935, 1936.1 and 1936.2 by Vedmederya, Zinenko and Barabanov, 2009, Russ. J. Herpetol., 16: 204), and MNKNU 26290 according to Vedmederya, Zinenko and Barabanov, 2009, Russ. J. Herpetol., 16: 204. Type locality: "Sachalin" [= Sakhalin], Russia. Synonymy with *Bufo asiaticus* by Gumilevskij, 1936, Trudy Zool. Inst. Akad. Nauk SSSR, Leningrad, 4: 167-171. Synonymy by implication of Shannon, 1956, Herpetologica, 12: 30, and specifically by Borkin and Roshchin, 1981, Zool. Zh., 60: 1802-1812; Matsui, 1986, Copeia, 1986: 561-579. Synonymy not noted by Zhao and Adler, 1993, Herpetol. China: 128.

[中文曾用名] 中华大蟾蜍

[同物异名] *Bufo griseus*；*Bufo maculiventris*；*Bufo sinicus*；*Bufo sachalinensis*；*Bufo andrewsi*（华西蟾蜍）；*Bufo minshanicus*（岷山蟾蜍）；*Bufo asiaticus*；*Bufo wright*；*Bufo tibetanus*（西藏蟾蜍）；*Bufo wolongensis*（卧龙蟾蜍）；*Bufo kabischi*（沙湾蟾蜍）

[俗名] 癞格包、癞蛤蟆

[鉴别特征] 本种与圆疣蟾蜍外形相似，但本种体腹面深色斑纹很明显，腹后部有一个深色大斑块。

本种依据费梁等（2009a，2010，2012）、Fei 和 Ye（2016）可分为3个亚种，即中华蟾蜍华西亚种 *Bufo gargarizans andrewsi* Schmidt, 1925、中华蟾蜍指名亚种 *Bufo gargarizans gargarizans* Cantor, 1842 和中华蟾蜍岷山亚种 *Bufo gargarizans minshanicus* Stejneger, 1926。茂兰地区分布有前两个亚种，检索表如下。

茂兰地区中华蟾蜍亚种检索表

一般无蹼褶，成体瘰粒多而密；腹面及体侧一般无土红色斑纹；蝌蚪唇齿式 I：1+1/ Ⅲ，体色黑，尾鳍色浅，尾末端较尖 ·· 中华蟾蜍指名亚种 *Bufo gargarizans gargarizans*

一般有蹼褶，成体瘰粒少而稀疏；腹面及体侧一般有土红色斑纹；蝌蚪唇齿式 Ⅱ / Ⅲ，尾鳍色黑，尾末端圆 ·· 中华蟾蜍华西亚种 *Bufo gargarizans andrewsi*

（5a）中华蟾蜍华西亚种 *Bufo gargarizans andrewsi* Schmidt, 1925

[鉴别特征] 一般有蹼褶，成体背面瘰粒少而稀疏；腹面及体侧一般有土红色斑纹。蝌蚪唇齿式 Ⅱ / Ⅲ，尾鳍色黑，尾末端圆。

[形态描述] 依据茂兰地区标本描述 雄蟾头体长 47.23～75.11mm，雌蟾头体长 99.72～117.75mm。头宽大于头长，吻端圆而高，吻棱明显；鼻间距与眼间距几乎相等；头部无骨质脊棱；瞳孔横椭圆形；鼓膜不显，呈椭圆形；耳后腺大，呈卵圆形，前宽后窄；上颌无齿，无犁骨齿。皮肤粗糙，头上分布有小疣粒，体及背后疣粒少且稀疏；前臂及手长约为体长之半；后肢粗短，无股后腺，前伸贴体时胫跗

关节达肩部，左右跟部不相遇；趾侧缘膜显著，第四趾具半蹼。体背面颜色变异颇大，一般雄性体背面棕色、橄榄绿色或褐绿色等，有不显著的黑斑点，体侧浅棕色，上面有黑色及棕红色斑点；雌性色较浅，背部有黑色或橘红色斑点或斑块，腹面浅黄色，有不规则的黑色斑块。

[第二性征]　雄性内侧3指有黑色婚刺，无声囊。

[生活习性]　该蟾生活于多种生态环境中。成蟾常栖于草丛间或石下及林间开阔地，常在路边或杂草间觅食昆虫及其他小动物，也捕食小型蛇类。雨后常常出现在路边。

据实际调查发现，茂兰地区的中华蟾蜍华西亚种在2月便开始繁殖产卵，6月中旬仍能发现幼体，因此可推断繁殖季节在1～6月，比费梁等（2012）的结果提前近一个月（3～6月）。繁殖季节时雄蟾常发出"咕、咕、咕"的连续鸣声；配对时雄性前肢抱握在雌性的腋下部位，卵产于山溪静水水塘内或岸边缓流处；卵群呈双行交错排列在圆管状胶质卵带内。蝌蚪成群生活于溪流或水塘内，以矽藻和腐物为食。同域分布的还有细痣疣螈和泽陆蛙。

[地理分布]　茂兰地区分布于翁昂、洞塘。国内分布于甘肃、陕西、四川、云南、贵州、湖北、广西、重庆等地。国外分布于俄罗斯、朝鲜等地。

[模式标本]　不详

[模式产地]　浙江舟山群岛

[保护级别]　三有保护动物

[IUCN濒危等级]　无危（LC）

[讨论]　该物种适应环境变化的能力极强，耐旱，成蟾可在离水环境很远的旱地中生活，这一特性使该物种能在多种环境中生存，降低了同类的生境竞争。该物种在茂兰地区分布广泛，数量众多，受干扰程度弱，主要干扰因素是该区域内四通八达的公路。来往车辆是所有野生动物的威胁，建议在物种丰富的区域公路上设置关卡，夜间禁止车辆通行。

成体量度　　　　　　　　　　　　　　　　　　　　　　　　　　（单位：mm）

编号	头体长	头长	头宽	前臂及手长	后肢全长	胫长	跗足长
GZNU20170510099 ♂	47.23	13.48	15.49	22.49	59.53	20.69	30.56
GZNU20170510007 ♀	106.28	41.47	40.54	52.74	133.67	43.59	66.60
GZNU20170510006 ♀	117.75	37.71	41.38	61.48	156.9	42.67	73.16
GZNU20170510003 ♀	99.72	33.15	31.89	49.55	118.83	41.25	58.45
GZNU20170510002 ♂	74.68	20.62	25.18	31.74	99.53	32.08	49.11
GZNU20170510001 ♂	75.11	29.87	31.79	47.73	106.99	37.78	52.63

a、b　整体照（罗涛摄）

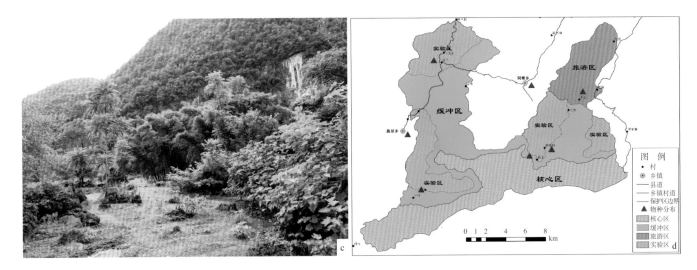

c 生境
d 在茂兰地区的分布

（5b）中华蟾蜍指名亚种 *Bufo gargarizans gargarizans* Cantor, 1842

[鉴别特征]　成体背面瘰粒多而密，一般无跗褶，腹面及体侧一般无土红色斑纹。蝌蚪唇齿式Ⅰ：1+1/Ⅲ，体色黑，尾鳍色浅，尾末端较尖。

[形态描述]　依据茂兰地区标本描述　雄蟾头体长73.9～79.82mm，雌蟾头体长87.72～103.1mm，头宽大于头长，吻圆而高，吻棱明显；鼻间距小于眼间距；上眼睑无显著的疣粒；头部无骨质脊棱；瞳孔横椭圆形，鼓膜显著，近圆形，耳后腺大，腺呈长圆形；上颌无齿，无犁骨齿。皮肤粗糙，背部布满大小不等的圆形瘰粒，仅头部平滑；腹部满布疣粒，胫部瘰粒大。后肢粗短，前伸贴体时胫跗关节达肩后，左右跟部不相遇，无股后腺，一般无跗褶，趾侧缘膜显著，第四趾具半蹼。体色变异颇大，随季节而异，一般雄性背面墨绿色、灰绿色或褐绿色，雌性背面多呈棕黄色；有的个体体侧有黑褐色纵列条纹，纹上方大疣乳白色；腹面乳黄色与棕色或黑色形成花斑，股基部有一团大棕色斑，体侧一般无土红色斑纹。

[第二性征]　雄蟾体略小，雄性内侧3指有黑色刺状婚垫，无声囊，无雄性线。雌蟾背、腹面瘰粒上有不同程度的黑色或棕色角质刺。

[生活习性]　成体生活于茂兰地区的多种生境中。除冬眠和繁殖期栖息于水中外，多在陆地草丛、地边、山坡石下或土穴等潮湿环境中栖息。黄昏后外出捕食，食性较广。于静水水域中繁殖产卵。

[地理分布]　茂兰地区分布于翁昂、洞塘、永康。国内广泛分布，贵州见于绥阳、江口、印江、金沙、荔波等地。

[模式标本]　不详

[模式产地]　浙江舟山群岛

[保护级别]　三有保护动物

成体量度　　　　　　　　　　　　　　　　　　　　　（单位：mm）

编号	头体长	头长	头宽	前臂及手长	后肢全长	胫长	跗足长
GZNU20170417018 ♀	87.72	30.23	30.91	39.13	100.10	28.32	48.24
GZNU20170418031 ♀	103.10	34.04	30.08	46.41	103.11	29.86	56.31
GZNU20170416007 ♀	90.02	26.97	27.64	44.86	108.83	35.45	56.47
GZNU20170418040 ♂	79.82	25.35	27.81	40.08	105.69	33.71	54.43
GZNU20170418042 ♂	73.90	22.01	23.02	39.72	91.23	30.44	48.41
GZNU20170418041 ♂	79.52	23.31	20.98	39.94	97.76	34.14	54.12

[IUCN 濒危等级] 无危（LC）

[讨论] 该物种在茂兰地区分布广泛，数量众多。与中华蟾蜍华西亚种同域分布，且生活习性极其相似，故两物种不易区别。

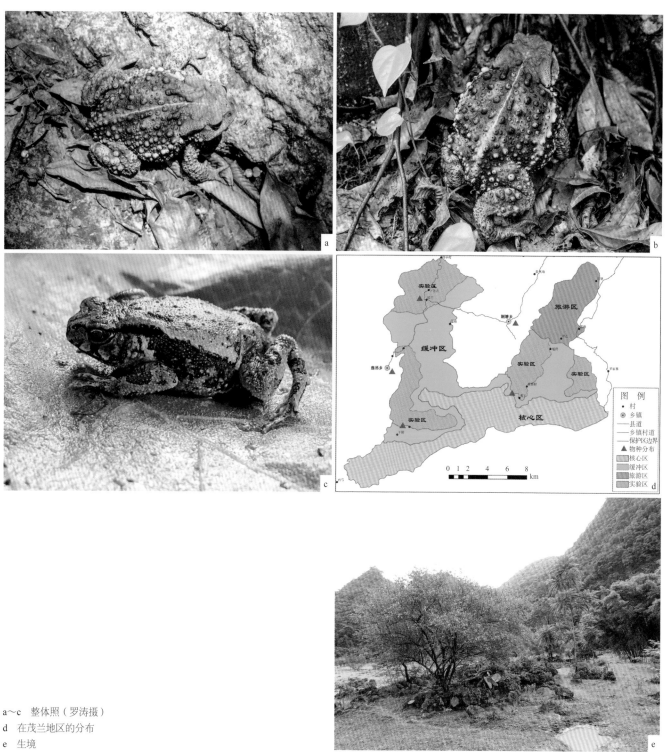

a～c 整体照（罗涛摄）
d 在茂兰地区的分布
e 生境

5）头棱蟾属 *Duttaphrynus* Frost et al., 2006

Duttaphrynus Frost et al., 2006, Bull. Am. Mus. Nat. Hist., 297: 219. Type species: *Bufo melanostictus* Schneider, 1799, by original designation.

瞳孔水平位；舌椭圆形或梨形，后端无缺刻；指间无蹼，趾间半蹼；指末端正常或略膨大；外侧蹠间无蹼。皮肤粗糙具大小瘰粒，头部两侧骨质棱强，耳后腺大。

本属现有 27 种，中国有 3 种。国内分布于西南和华南地区及台湾。国外分布于伊朗（东南部）、阿富汗（南部）、巴基斯坦（北部）和尼泊尔，以及印度（南部）和斯里兰卡等地。茂兰地区有黑眶蟾蜍 1 种。

（6）黑眶蟾蜍 *Duttaphrynus melanostictus* Schneider, 1799

Bufo melanostictus Schneider, 1799, Hist. Amph. Nat.: 216. Type(s): Museum Blochianum (= ZMB); ZMB 3461-63 are syntypes according to Peters, 1863, Monatsber. Preuss. Akad. Wiss. Berlin, 1863: 80; ZMB 3462 designated lectotype by Dubois and Ohler, 1999, J. South Asian Nat. Hist., 4: 139 (who redescribed this type). Type locality: India orientali.

Rana dubia Shaw, 1802, Gen. Zool., 3 (1): 157. Type locality: not designated. Neotype locality: "India orientali". Holotype: BMNH, now lost: Neotype: ZMB 3462, designated by Dubois and Ohler, 1999: 144.

Bufo scaber Daudin, 1802, Hist. Nat. Rain. Gren. Crap, Quarto: 94. Type locality: India orientali. Lectotype: ZMB 3462, designated by Dubois and Ohler, 1999: 145.

Bufo bengalensis Daudin, 1802, Hist. Nat. Rain. Gren Crap, Quarto: 96. Type locality: Bengale. Neotype: MNHNP 4967, designated by Dubois and Ohler, 1999: 145.

Bufo carinatus Gray, 1830, Illustrations Indian Zool: pl. 83. Type locality: Bengal. Neotype: MNHNP 4967, designated by Dubois and Ohler, 1999: 146.

Bufo isos Lesson, 1834, in Bélanger (ed.), Voy Indes-Orientales N. Eur. Caucase George Perse Zool.: 333. Type locality: Bengale. Holotype: MNHNP 4967, designated by Dubois and Ohler, 1999: 147.

Bufo gymnauchen Bleeker, 1858, Natuurkd. Tijdschr. Nederl. Indie, 16: 46. Type locality: Pulau Bintan (Birtang), Indonesia. Lectotype: BMNH 1947.2.21.71 (Holotype: Boulenger, 1882: 307). 2, SVL 82.4mm, in Dubois and Ohler, 1999: 148.

Bufo spinipes Steindachner, 1867, Reise osterreichischen Fregatte Novara, Zool.: 42. Type locality: Nikobare Nicobar Islands, India. Syntypes: NHMW 4 specimens. Lectotype: NHMW 5371.1, designated by Dubois and Ohler, 1999: 150.

Bufo longecristatus Werner, 1903, Zool. Anz., 26: 252. Type locality: Inneres von Borneo. Holotype: IRSNB 9422, designated by Dubois and Ohler, 1999: 155.

Bufo tienhoensis Bourret, 1937, Annexe Bull. Gen Instr. Publique, Hanoi, 1937: 6, 11. Type locality: Colde Tien-Ho (Route de Lang Son, Tonkin. alt. 300m), Vietnam. Syntypes: MNHNP 48: 123-124. Lectotype: MNHNP 48.0124, designated by Dubois and Ohler, 1999: 162.

Bufo camortensis Mansukhani and Sarkar, 1980, Bull. Zool. Surv. India, 3: 97. Type locality: Compound of Camorta Guest House. Camorta, Andaman and Nicobar Islanda, India. Holotype: ZSIC A6955, by original designation. Synonymy by Crombie, 1986. J. Bombay Nat. Hist. Soc., 83: 22.

Ansonia kamblei Ravichandran and Pillai, 1990, Rec. Zool. Surv. India, 86: 506. Type locality: Jeur, 29 km north of Tembhurni, Karnala, Maharashtra, India. Holotype: ZSIWRS V/198, by original designation. Synonymy by Dubois and Ohler, 1999: 167.

Bufo melanostictus hazarensis Kham, 2001, Pakistan J. Zool., 33: 297. Type localities: Ooghi, Manshera and Data. Hazara Division, eastern NWFP (Northwest Frontier Province), Pakistan. Syntypes: not designated by number or Museum. Synonymy by Frost, 2004.

Duttaphrynus melanostictus Frost et al., 2006, Bull. Am. Mus. Nat. Hist., 297: 365.

[同物异名] *Bufo chlorogaster*；*Rana dubia*；*Bufo scaber*；*Bufo bengalensis*；*Bufo flaviventris*；*Rana melanosticta*；*Bufo carinatus*；*Bufo dubius*；*Bufo isos*；*Bufo gymnauchen*；*Docidophryne isos*；*Docidophryne spinipes*；*Phrynoidis melanostictus*；*Bufo spinipes*；*Bufo longicristatus*；*Bufo tienhoensis*；*Bufo camortensis*；*Ansonia kamblei*；*Docidophryne melanostictus*

[俗名] 癞蛤蟆、蛤巴、癞疙疱、蟾蜍

[鉴别特征] 鼓膜大而显著；吻棱及上眼睑内侧棱为黑色骨质强棱；有鼓上棱，耳后腺不紧接眼后。

[形态描述] 依据茂兰地区标本描述 雄蟾头体长49.64～57.10mm，雌蟾头体长65.51～75.61mm，头宽大于头长；吻端钝圆形，吻棱明显（头部两侧有黑色骨质棱，该棱沿吻棱经上眼睑内侧直到鼓膜上方）；鼻间距明显小于眼间距，横椭圆形；鼓膜大，呈椭圆形；上、下颌无齿，舌呈椭圆形，无犁骨齿。成蟾背部为黄棕色，腹面乳黄色。皮肤粗糙，全身除头顶外，满布瘰粒或疣粒；背部瘰粒多，且疣粒基部黄棕色，顶部黑色；腹部密布小疣，四肢刺疣细小；部分疣粒顶部有黑色角质刺。

[第二性征] 雄性内侧3指有棕黑色婚刺，无雄性线。

[生活习性] 该蟾生活于茂兰地区的多种环境中，非繁殖期营陆栖生活，常活动在草丛、石堆、耕地、水塘边及住宅附近，行动缓慢，匍匐爬行。雨后或夜晚外出觅食。

[地理分布] 茂兰地区分布于翁昂、洞塘等地。国内分布于四川、云南、贵州、浙江、江西、湖南、福建、台湾、广东、香港、澳门、广西、海南。国外分布于东南亚各地。

[模式标本] 全模标本（ZMB 3461-63），选模标本（ZMB 3462）保存于德国柏林洪堡大学自然历史博物馆

[模式产地] 东印度群岛

[保护级别] 三有保护动物

[IUCN 濒危等级] 无危（LC）

[讨论] 该蟾在茂兰地区分布广泛，数量众多，适应性强，与中华蟾蜍同域分布。受放牧等一定的人为干扰，可在本区域内良好生活繁衍。

Wogan等（2016）通过对亚洲内陆和海岛的广泛采样，并基于1个线粒体基因（*ND3*）和2个核基因（*POMC*与*SOX9*），第一次对黑眶蟾蜍的系统发育地理学和种群结构进行了研究。结果显示，黑眶蟾蜍在遗传上至少存在3个明显的进化支系，分别对应于亚洲内陆支系（the Asian mainland）、缅甸沿海支系（the coastal Myanmar）和巽他群岛支系（the Sundaic islands），这3个支系的地理分布区域大部分没有重叠。在中国分布的黑眶蟾蜍属于亚洲内陆支系，具有很高的遗传多样性，并且分化出多个亚支。疣宁等（2011）基于线粒体DNA控制区，对福建、广东、海南、广西和台湾的黑眶蟾蜍的遗传变异和地理分化进行了研究，结果显示这些地区黑眶蟾蜍的遗传多样性较高，尤其是岛屿种群具有较高的遗传变异，但是，大陆种群和海岛种群之间并未发生明显的遗传分化。

成体量度 （单位：mm）

编号	头体长	头长	头宽	前臂及手长	后肢全长	胫长	跗足长
GZNU20170509014♀	65.51	19.23	21.02	29.92	74.11	17.81	36.15
GZNU20170509012♂	57.10	17.70	18.65	27.06	73.72	23.53	34.56
GZNU20170509013♂	52.08	15.27	18.17	23.42	62.97	22.01	31.70
GZNU20160826005♂	54.76	20.26	18.45	27.16	67.69	22.13	33.28
GZNU20160826004♂	49.64	15.16	15.56	23.83	63.02	20.39	30.99
GZNU20160816005♀	75.61	22.39	23.09	32.18	86.05	30.72	43.67

a~d 整体照
e 生境
f 在茂兰地区的分布

4. 雨蛙科 Hylidae Rafinesque, 1815

Hylae Laurenti, 1768, Spec. Med. Exhib. Synops. Rept.: 20. Unavailable plural of *Hyla* Laurenti, 1768, not apparently intended as a taxon name.

Hylarinia Rafinesque, 1815, Analyse Nat.: 78. Type genus: *Hylaria* Rafinesque, 1814 (an unjustified emendation of *Hyla* Laurenti, 1768).

Hylina Gray, 1825, Ann. Philos., London, Ser. 2, 10: 213. Type genus: *Hyla* Laurenti, 1768. Suggested as a subfamily.

Hylidae: Bonaparte, 1850, Conspect. Syst. Herpetol. Amph.: 1.

Phyllomedusidae Günther, 1858, Proc. Zool. Soc., London, 1858: 346. Type genus: *Phyllomedusa* Wagler, 1830.

Pelodryadidae Günther, 1858, Proc. Zool. Soc., London, 1858: 346. Type genus: *Pelodryas* Günther, 1858.

Hyloides: Bruch, 1862, Würzb. Naturwiss. Z., 3: 221.

肩带弧胸型，肩胸骨和正胸骨软骨质；肩胛骨长（小于锁骨的2倍），前端不与锁骨重叠。椎体前凹型，荐椎前椎骨8枚，第一与第二椎骨不合并；无肋骨；前后两个髓弓间的间隙大，多不呈覆瓦状（叶泡蛙亚科 Phyllomedusinae 呈覆瓦状）；大多数属荐椎横突宽大，骨髁2个，与尾杆骨相关节；尾杆骨近端无横突，背面无脊棱或有显著隆起。头骨骨化程度较强或弱，一般有腭骨；鼻骨或大或小，彼此相接触或分离，与额顶骨相接或不相接；上颌骨和前颌骨具齿；有方轭骨或缺；有耳柱骨。无副舌骨；舌骨前角无前突；无或有翼状突（雨蛙属 *Hyla*）；后侧突细长或呈短突状；环状软骨环完全。跟骨、距骨仅两端合并；远列跗骨2～3枚。指骨、趾骨末两节有介间软骨；指骨式3-3-4-4；趾骨式3-3-4-5-4；瞳孔横置（猴叶蛙 *Phyllomedusa sauvagii*）或纵置（雨夜蛙属 *Nyctimystes*）。

本科动物体型相差很大，最小的仅有17mm，最大的达140mm。指、趾端有吸盘，大多数雨蛙为树栖，也有水生（如蝗蛙属 *Acris*）和穴居（如圆蟾属 *Cyclorana*、跟掘蛙属 *Pternohyla*）的种类。多数属的卵和蝌蚪在水内发育；有唇齿［新热带区的某些雨蛙（如小头树蛙 *Dendropsophus microcephala*）无唇齿］和角质颌，偶尔角质颌缺如，出水孔单个，一般位于体左侧，属有唇齿左孔型。

抱对部位在腋部。卵生，体外受精。有的亲体将卵驮在背上（如扩角蛙属 *Hemiphractus*、蛙属 *Stefania* 和幽蛙属 *Cryptobatrachus*）；有的成蛙背上有囊状构造，卵植于囊内（如囊蛙属 *Gastrotheca*）；有的卵直接发育成幼蛙，或发育成摄食或不摄食的蝌蚪；有的卵产在陆地上或植物上，在水外发育，蝌蚪期在水中生活并完成变态。

雨蛙科目前包括50属970种（Amphibia Web，2017）。本科分为澳雨蛙亚科 Pelodryadinae、叶泡蛙亚科 Phyllomedusinae 和雨蛙亚科 Hylinae（Faivovich et al.，2005；Wiens et al.，2005，2010；Frost et al.，2006）。多数研究表明雨蛙属为一单系（Wiens et al.，2010；Pyron and Wiens，2011）。Hua 等（2009）基于线粒体基因片段（12S rRNA，ND1）和核基因片段（c-myc，POMC）的研究结果表明，雨蛙属可以划分为5个种组，并解决了各种组之间的分子系统关系。

我国仅有雨蛙亚科 Hylinae 中雨蛙属 *Hyla* 的8种。其中5种已有研究，不为单系。中国雨蛙、秦岭雨蛙和华西雨蛙聚为一支，属于雨蛙属基部的欧亚支系——*Hyla arborea* 种组，无斑雨蛙 *Hyla immaculata* 和日本雨蛙 *Hyla japonica* 聚为一支，属于东北亚支系——日本雨蛙 *Hyla japonicus* 种组（Hua et al.，2009；Wiens et al.，2010）。

6）雨蛙属 *Hyla* Laurenti, 1768

Rana viridis Linnaeus, 1761, Fauna Svec.: 102, 280. Type species: not stated. Type locality: Scania. Synonymy by Retzius, 1800, Fauna Svec., ed. 3, 1: 286; Merrem, 1820, Tent. Syst. Amph.: 170; Tschudi, 1838, Classif. Batr.: 74; Schinz, 1822, Thierr. Naturgesch., 2: 166; Schreiber, 1875, Herpetol. Eur.: 105.

Hyla Laurenti, 1768, Spec. Med. Exhib. Synops. Rept.: 32. Type species: *Hyla viridis* Laurenti, 1768 (= *Rana arborea* Linnaeus, 1758), by subsequent designation of Stejneger, 1907, Bull. U.S. Natl. Mus., 58: 75. Treatment as a subgenus of the genus *Hyla* by Fouquette and Dubois, 2014, Checklist N.A. Amph. Rept.: 331-332.

Hyla viridis Laurenti, 1768, Spec. Med. Exhib. Synops. Rept.: 33. Type species: by indication including frogs illustrated by Roesel von Rosenhof, 1758, Hist. Nat. Ran. Nost.: pls. 9-11 and frontispiece, as well as (var. b) specimen noted by Catesby, 1754, Nat. Hist. Carolina Florida Bahama Is.: 71, pl. 71 (Catesby's illustration looks to be of *Hyla cinerea* DRF). Lectotype is the calling male figured in middle left position on plate 9 of Roesel von Rosenhof, 1758, Hist. Nat. Ran. Nost., according to Dubois and Ohler, 1997 "1996", Bull. Mus. Natl. Hist. Nat., Paris, Sect. A, Zool., 18: 336. Type locality: "Europae arboribus"; restricted to vicinity of Nürnburg by Dubois and Ohler, 1997 "1996", Bull. Mus. Natl. Hist. Nat., Paris, Sect. A, Zool., 18: 336. Named (see page 138 of original publication) as a junior synonym of *Rana arborea* Linnaeus. Synonymy by Daudin, 1800, Hist. Nat. Quad. Ovip., Livr., 1: 13; Daudin, 1802 "An. XI", Hist. Nat. Rain. Gren. Crap., Quarto: 14.

Hyla vulgaris Lacépède, 1788, Hist. Nat. Quadrup. Ovip. Serpens, Quarto ed., 2: 459; Lacépède, 1788, Hist. Nat. Quadrup. Ovip. Serpens, Quarto ed., 1: table following page 618 and referencing account starting on page 550. Substitute name for *Hyla viridis* Laurenti, 1768 and *Rana arborea* Linnaeus. Rejected as published in a nonbinominal work by Opinion 2104, Anonymous, 2005, Bull. Zool. Nomencl., 62: 55.

Calamita arboreus Schneider, 1799. Hist. Amph. Nat.: 153; Merrem, 1820, Tent. Syst. Amph.: 170.

Calamita viridis Schneider, 1799, Hist. Amph. Nat.: 153.

Hylaria viridis Rafinesque, 1814, Specchio Sci., 2: 103.

Hyla arborea Cuvier, 1817, Regne Animal., 2: 94; Schinz, 1822, Thierr. Naturgesch., 2: 166; Schinz, 1833, Naturgesch. Abbild Rept.: 223; Lesson, 1841, Act. Soc. Linn. Bordeaux, Ser. 3, 12: 60; Boulenger, 1882, Cat. Batr. Sal. Coll. Brit. Mus., Ed. 2: 379.

Hyas arborea Wagler, 1830, Nat. Syst. Amph.: 201.

Dendrohyas arborea Tschudi, 1838, Classif. Batr.: 74.

Dendrohyas arboreus Fitzinger, 1843, Syst. Rept.: 30.

Dendrohyas arborea var. *daudinii* Gistel, 1868, Die Lurche Europas: 161. Types: not stated or known to exist. Type locality: southern France. Synonymy by Mertens and Wermuth, 1960, Amph. Rept. Europas: 49.

Hyla arborea arborea Boulenger, 1882, Cat. Batr. Sal. Coll. Brit. Mus., Ed. 2: 381, by implication; Nikolskii, 1918, Fauna Rossii, Zemnovodnye: 132.

Hyla arborea kretensis Ahl, 1931, Ann. Naturhist. Mus. Wien, 45: 161. Syntypes: ZMB 31569 (according to Duellman, 1977, Das Tierreich, 95: 31-33), NHMW 18413.1-5 (according to Häupl and Tiedemann, 1978, Kat. Wiss. Samml. Naturhist. Mus. Wien, 2: 18, and Häupl, Tiedemann and Grillitsch, 1994, Kat. Wiss. Samml. Naturhist. Mus. Wien, 9: 23). Type localities: Canea, Psychro auf der Lasithi-Hochebene, Chania, westlich von Canea, Neapolis, nord-westl. v. St. Nikolo, Crete, Greece. Data associated with NHMW syntypes are "Canea, Kreta" and "Psychro auf der Lasithi-Hochebene". Distinctiveness from *Hyla arborea arborea* rejected by Schneider, 1974, Occologia, Berlin, 14: 109; Schneider, 2004, Z. Feldherpetol., Suppl., 5: 9. But, Valakos, Pafilis, Sotiropoulos, Lymberakis, Maragou and Foufopoulos, 2008, Amph. Rept. Greece: 95, provided diagnostic characters. Stöck et al., 2008, Mol. Phylogenet. Evol., 49: 1019-1024, rejected the distinctiveness of this taxon.

Hyla arborea cretensis Stugren and Lydataki, 1986, Zool. Abh. Staatl. Mus. Tierkd. Dresden, 42: 57.

Hyla (Hyla) arborea Fouquette and Dubois, 2014, Checklist N.A. Amph. Rept.: 331.

瞳孔横置；舌卵圆且大，后端游离具有缺刻，鼓膜显著或不显；背面皮肤多光滑无疣粒；指、趾末端多膨大成吸盘状，有边缘沟（马蹄形沟）；指间无蹼或有蹼迹，外侧趾间有蹼或有蹼迹。上颌有齿并且有犁骨齿。

成体多为树栖；卵群多产于静水水域，卵粒一般有色素；蝌蚪的上尾鳍高而薄，一般起始于背中部；上唇缘无乳突，下唇及口角部有乳突；肛孔多位于下尾鳍基部右侧。

本属现有15种，分布甚广，几乎遍及全球，我国有13种7亚种，贵州有4种。茂兰地区有2种，其中1种为保护区内新纪录物种——三港雨蛙 *Hyla sanchiangensis*。茂兰地区种检索表如下。

茂兰地区种检索表

颞褶细，其上无疣粒；体侧及股前后有大小不等的黑色斑点；颞部、鼓膜上、下方的黑色细纹在肩部不会合，几乎成两条平行线或不显··三港雨蛙 *Hyla sanchiangensis*

颞褶粗厚，其上一般有疣粒；体侧和股前后黑色斑点多；颞部、鼓膜的上、下方没有到达肩部的黑色细纹························华西雨蛙 *Hyla gongshanensis*

（7）华西雨蛙 *Hyla gongshanensis* Li and Yang, 1985

Hyla annectans annectans: Yang, Su and Li, 1983, Acta Herpetol. Sinica, Chengdu, N.S., 2 (3): 40; Yang, 1991, Amph. Fauna of Yunnan: 107.

Hyla annectans gongshanensis Yang, Su and Li, 1983, Acta Herpetol. Sinica, Chengdu, N.S., 2 (3): 40; Types: KIZ. Type locality: Outskirts of Gongshan County, Yunnan, China; 1500m, Nomen nudum.

Hyla annectans gongshanensis Li and Yang, 1985, Zool. Res., Kunming, 6: 23, 27. Holotype: KIZ 730059, by original designation. Type locality: "outskirts of Gongshan Xian, Yunnan, China; 1500m".

[同物异名]　*Hyla annectans*

[鉴别特征]　成体上眼睑外侧和颞部疣粒多；吻端一般有"Y"形斑，不镶细线纹；体侧和股前后有黑斑；蝌蚪的肛孔位于下尾鳍基部中央。

依据费梁等（2009b）可分为5个亚种，依次是华西雨蛙川西亚种 *Hyla gongshanensis chuanxiensis* Ye and Fei, 2000、华西雨蛙贡山亚种 *Hyla gongshanensis gongshanensis* Li and Yang, 1985、华西雨蛙景东亚种 *Hyla gongshanensis jingdongensis* Fei and Ye, 2000、华西雨蛙腾冲亚种 *Hyla gongshanensis tengchongensis* Fei and Li, 2000 和华西雨蛙武陵亚种 *Hyla gongshanensis wulingensis* Shen, 1997。贵州有华西雨蛙景东亚种、华西雨蛙武陵亚种，茂兰地区仅有华西雨蛙武陵亚种。

（7a）华西雨蛙武陵亚种 *Hyla gongshanensis wulingensis* Shen, 1997

Hyla annectans gongshanensis Li and Yang, 1985, Zool. Res., Kunming, 6: 23, 27. Holotype: KIZ 730059, by original designation. Type locality: outskirts of Gongshan County, Yunnan, China; 1500m.

Hyla annectans wulingensis Shen, 1997, Zool. Res., Kunming, 18: 177, 182. Holotype: HNNU 86.646, by original designation. Type locality: "Tianping Mountain (29°49′N, 110°9′E), Sangzhi, Hunan Province, China; 1350m".

Hyla gongshanensis wulingensis: Fei, Hu, Ye and Huang, 2009, Fauna Sinica, Amph., 2: 634.

[鉴别特征]　体侧、股前后及胫后黑色斑点较多；吻端有"Y"形棕色斑，不镶细黑线，下达口缘，且宽于鼻间距；吻棱下方疣粒明显；背面皮肤光滑，呈翠绿色。

[形态描述]　依据茂兰地区标本描述　雄性头体长30.26～34.80mm，头宽小于头长，吻宽而圆，瞳孔横椭圆形，鼓膜圆而明显；有上颌齿，有犁骨齿且呈两小团。背面皮肤光滑，呈翠绿色；体和股腹面满布颗粒疣粒。指、趾端均有吸盘和边缘沟，蹼不发达；后肢前伸贴体时胫跗关节达眼部，左右跟部相遇；吻端沿吻棱经上眼睑、颞褶、鼓膜达体侧，为棕黄色，与背部绿色交界处有一条带状疣粒，吻部棕色斑呈"Y"形；前臂背面绿色达肘部；体腹面密布肉红色疣粒。

[第二性征]　雄性小于雌性，雄性第一指有棕色婚垫，咽喉部棕褐色，具单咽下外声囊。

[生活习性]　该蛙生活于茂兰地区的山区稻田或山间凹地静水水塘及其附近草丛中，攀附能力强，常栖于灌丛、

雄性成体量度 （单位：mm）

编号	头体长	头长	头宽	前臂及手长	后肢全长	胫长	跗足长
GZNU20170418612 ♂	30.81	12.20	9.95	15.09	43.51	14.16	21.10
GZNU20170418052 ♂	30.26	10.17	8.75	15.08	42.57	16.36	19.42
GZNU20170422009 ♂	34.80	10.89	10.45	17.68	52.80	17.82	22.96
GZNU20170404049 ♂	32.48	9.82	9.51	15.64	47.61	13.38	21.74
GZNU20170418048 ♂	32.42	10.64	10.91	17.02	49.88	12.28	23.28
GZNU20170418060 ♂	34.05	10.51	10.41	17.18	48.76	13.68	22.46

a～d 整体照

农田水稻丛中及山地玉米叶片之上。

[地理分布] 茂兰地区分布于翁昂、洞塘等地。国内分布于重庆、四川、云南、贵州、湖南、湖北、广西等地。国外无记录。

[模式标本] 正模 HNU 86-646（♂）保存于湖南师范大学生命科学学院

[模式产地] 湖南桑植天平山（29°49′N，110°9′E），海拔 1350m

[保护级别] 无

[IUCN 濒危等级] 无危（LC）

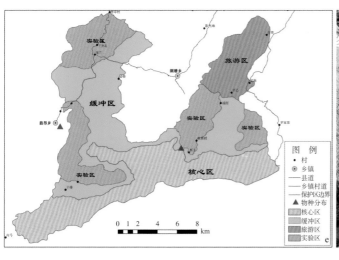

e 在茂兰地区的分布
f 生境

（8）三港雨蛙　*Hyla sanchiangensis* Pope, 1929

Hyla sanchiangensis Pope, 1929, Am. Mus. Novit., 352: 2. Holotype: AMNH 30198, by original designation. Type locality: San Chiang [= Sangang], Chungan Hsien [now Wuyishan], northwestern Fukien [= Fujian] Province, China, 3000-3500 feet altitude.

Hyla sanchiangensis Boring, 1930, Peking Nat. Hist. Bull., 5: 43; Boring, 1938 "1938-1939", Peking Nat. Hist. Bull., 13: 94; Liu, 1950, Fieldiana, Zool. Mem., 2: 225.

Hyla chinensis sanchiangensis Bourret, 1942, Batr. Indochine: 224.

Hyla (*Hyla*) *sanchiangensis* Fouquette and Dubois, 2014, Checklist N.A. Amph. Rept.: 331, by implication.

［同物异名］　*Hyla chinensis sanchiangensis*

［鉴别特征］　颞褶较细，上面一般无疣；眼下方至口角有一块明显的灰白斑，眼后鼓膜上、下方两条深棕色线纹在肩部不会合；体侧后段、股前后、胫腹面有黑色斑点。

［形态描述］　依据茂兰地区内2只标本描述（1♂1♀）　雄蛙体长31.25mm，雌蛙体长35.17mm，头宽略大于头长；吻短圆而高，吻棱明显，吻端和颊部平直向下；鼻孔位于鼻眼之间而近于吻端上方；鼓膜圆而显著；舌圆且厚，后端缺刻稍显；上颌有齿，犁骨齿两小团。

指、趾端均有吸盘和马蹄形边缘沟，第三指吸盘与鼓径几乎相等；第二、第四指几乎等长，第一指短小；指基部具微蹼，外侧2指间蹼较发达，指式为3＞2≈4＞1；趾端与指端同，但趾端吸盘较小；趾间几乎为全蹼，趾式为4＞5≈3＞2＞1。后肢长，后肢贴体时胫跗关节达眼；左右跟部相重叠。

背面黄绿色或绿色，皮肤表面光滑而无疣粒；颞褶细，其上无着生疣粒；四肢背面绿色，前肢自上臂达前臂约1/2处，后肢自股部达胫跗关节处；内跗褶起棱。胸、腹及股腹面密布颗粒疣，咽喉部较少。眼前下方自鼓膜下方至口角有一明显的灰白色斑，略呈"△"形，眼后鼓膜上、下方两条深棕色线纹在肩部不会合而成两平行线；体侧前段棕色，后段和股前后及体腹面浅黄色；体侧后段及四肢有不同数量的黑圆斑，体侧前段无黑斑点；手和跗足棕色。

［第二性征］　雄性第一指有深棕色婚垫；咽喉部色深，皮肤松弛，具单咽下外声囊；有雄性线。

［生活习性］　该蛙生活在海拔565～863m的山区稻田及其附近灌丛。傍晚外出鸣叫，尤以晴朗的夜晚鸣声特多。鸣叫时前肢直立，发出"格阿、格阿、格阿"连续的声音，声音低而缓慢，受惊扰后，鸣声即停。

［地理分布］　茂兰地区分布于洞塘，为茂兰保护区首次记录。国内分布于贵州、安徽、浙江、江西、湖北、湖南、福建、广州、广西。国外无分布。

［模式标本］　正模标本（AMNH 30198）保存于美国自然历史博物馆（American Museum of Natural History,

New York，USA，AMNH）

[模式产地] 福建武夷山市三港，海拔914.4～1524m

[保护级别] 三有保护动物

[IUCN濒危等级] 无危（LC）

[讨论] 三港雨蛙生活环境为山下农田周围清澈溪水中，溪水两岸植被茂密，多为常绿阔叶林。据相关文献记载，三港雨蛙仅分布于贵州东部及东南部地区。三港雨蛙在茂兰地区的发现，拓宽了人们对三港雨蛙地理分布的认识，新分布点位于贵州南部，表明三港雨蛙在贵州的分布点已经向西向南延伸。这一发现对研究该物种的地理分布格局具有重要意义。

成体量度 （单位：mm）

编号	体长	头长	头宽	前臂及手长	后肢全长	胫长	跗足长
GZNU2017050802 ♂	31.25	8.58	9.73	13.00	52.22	14.56	21.26
GZNU2017050800 ♀	35.17	10.45	10.92	16.60	51.95	18.81	22.09

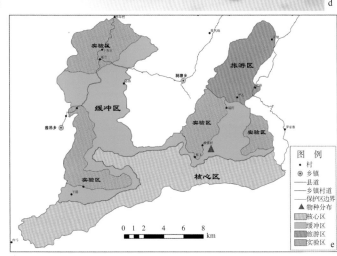

a～d 整体照
e 在茂兰地区的分布

5. 姬蛙科 Microhylidae Günther, 1858

Hylaedactyli Fitzinger, 1843, Syst. Rept.: 33. Type genus: *Hylaedactylus* Duméril and Bibron, 1841.

Gastrophrynae Fitzinger, 1843, Syst. Rept.: 33. Type genus: *Gastrophryne* Fitzinger, 1843. Synonymy by Parker, 1934, Monogr. Frogs Fam. Microhylidae: 71.

Microhylidae Günther, 1858, Proc. Zool. Soc., London, 1858: 346. Type genus: *Micrhyla* Duméril and Bibron, 1841, an incorrect subsequent spelling of *Microhyla* Tschudi, 1838.

Microhylidae: Parker, 1934, Monogr. Frogs Fam. Microhylidae: 128.

肩带固胸型，肩胸骨很小或缺如，正胸骨软骨质；前喙骨及锁骨不同程度地退化以致缺失，若有锁骨，则不与肩胛骨重叠。椎体参差型，少数为前凹型，荐椎前椎骨 8 枚，前 7 枚为前凹，多数亚科第 8 椎骨为双凹；荐椎为双凸，横突宽大；无肋骨；大多数髓弓不呈覆瓦状排列，少数属呈覆瓦状排列；关节髁 2 枚，与尾杆骨相关节；尾杆骨无横突。大多数类群腭骨缩小或无；通常有耳柱骨；上颌骨一般无齿；多数无犁骨齿。无副舌骨；环状软骨环完全。跟骨、距骨仅两端并合；远列小跗骨 2 枚。大多数类群指骨节、趾骨节正常，个别类群骨节减少或增加。瞳孔圆形或横椭圆形（暴蛙属 *Dyscophus* 纵椭圆形）。配对时抱握于腋部。

本科动物一般体型小，小的体长 20mm 左右，大者可达 100mm。体型各异，头小，体短胖，有的呈球状或蟾状；树栖类群的指、趾末端膨大。外蹠突有或无；在上腭部有 2～3 个腭褶。

许多类群在陆地上产卵，直接发育，孵化出非摄食性的蝌蚪；有些类群有水生性的幼体。蝌蚪的口较为特殊，一般在吻顶端，无唇齿和角质颌（除鼓蛙属 *Otophryne* 外）。

本科动物成体多为陆生和树栖类群，目前本科共包括 687 种，广泛分布于亚洲、非洲、大洋洲和美洲。我国有 5 属 19 种。我国的物种不是单系，其中，狭口蛙属、小姬蛙属、小狭口蛙属和姬蛙属隶属于姬蛙亚科 Microhylinae，细狭口蛙属隶属于细狭口蛙亚科 Kalophryninae（Peloso et al., 2015）。茂兰地区仅有姬蛙亚科物种 4 种。

7）姬蛙属 *Microhyla* Tschudi, 1838

Microhyla Tschudi, 1838, Classif. Batr.: 71. Type species: "*Hylaplesia achatina* Boie, 1827" (Nomen nudum) (= *Microhyla achatina* Tschudi, 1838), by monotypy.

Micrhyla Duméril and Bibron, 1841, Erp. Gen., 8: 28, 613. Unjustified emendation of *Microhyla* Tschudi, 1838.

Siphneus Fitzinger, 1843, Syst. Rept.: 19. Type species: *Engystoma ornatum* Duméril and Bibron, 1841, by original designation. Preoccupied by Siphneus Brants, 1827. Synonymy by Stejneger, 1907, Bull. U.S. Natl. Mus., 58: 87.

Dendromanes Gistel, 1848, Naturgesch. Thierr.: xi. Substitute name for *Microhyla* Tschudi, 1838.

Diplopelma Günther, 1859 "1858", Cat. Batr. Sal. Coll. Brit. Mus.: 50. Replacement name for *Siphneus* Fitzinger, 1843. Synonymy by Boulenger, 1882, Cat. Batr. Sal. Coll. Brit. Mus., ed. 2: 164. Synonymy by Boulenger, 1882, Cat. Batr. Sal. Coll. Brit. Mus., ed. 2: 164.

Scaptophryne Fitzinger, 1861 "1860", Sitzungsber. Akad. Wiss. Wien, Phys. Math. Naturwiss. Kl., 42: 416. Type species: *Scaptophryne labyrinthica* Fitzinger, 1861 "1860", Nomen nudum. Synonymy by Boulenger, 1882, Cat. Batr. Sal. Coll. Brit. Mus., ed., 2: 164.

Copea Steindachner, 1864, Verh. Zool. Bot. Ges. Wien, 14: 286. Type species: *Copea fulva* Steindachner, 1864, by monotypy. Synonymy by Parker, 1932, Ann. Mag. Nat. Hist., Ser. 10, 10: 341.

Ranina David, 1872 "1871", Nouv. Arch. Mus. Natl. Hist. Nat., Paris, 7: 76. Type species: *Ranina symetrica* David, 1871, by monotypy. Junior homonym of Ranina Lamarck, 1801. Synonymy by Boulenger, 1882, Cat. Batr. Sal. Coll. Brit. Mus., ed. 2: 164.

上颌无齿，无犁骨齿，犁骨分为前、后两个部分，在内鼻孔后缘的部分消失。无锁骨、前喙骨或肩胸骨，只有喙骨及软骨质的胸骨。喉前缩肌前端附着在舌骨的腹面，左、右不重叠。舌圆形，后端无缺刻。腭部有1排横肤棱。指、趾末端膨大或不膨大。掌突2个或3个，较扁平；指、趾关节下瘤较低平，其间无肤棱或肤突，无指（趾）基下瘤。

成体以陆栖为主，常在小水塘或稻田附近活动，可远离水域栖息在潮湿的草丛中。卵群产在稻田或有水草的小水塘内，分散成小群附于叶茎上。蝌蚪在静水水塘中，常在水的中层活动。

本属已知49种，主要分布于亚洲南部和东南部。我国现有9种，多分布于秦岭以南各省区。贵州除缅甸姬蛙 *Microhyla berdmorei*、穆氏姬蛙 *Microhyla mukhlesuri* 和北仑姬蛙 *Microhyla beilunensis* 外均有分布。茂兰地区有4种，分种检索表如下。

茂兰地区种检索表

1 { 有趾吸盘，吸盘背面有纵沟 ·· 2
   无趾吸盘，如有吸盘，则背面无纵沟 ··· 3
2 { 背面小疣显著，呈纵列排列；背部有镶浅黄边的深褐色粗大花斑 ··· 粗皮姬蛙 *Microhyla butleri*
   背面无显著小疣；背正中有深色小弧形斑1个或2个 ··· 小弧斑姬蛙 *Microhyla heymonsi*
3 { 体背面有若干个相套叠的"Λ"形斑，整个背面花斑色彩醒目、美丽 ······································· 花姬蛙 *Microhyla pulchra*
   体背面"Λ"形斑少，其中第一"Λ"形斑始自两眼间，斜向体后两侧，色彩不醒目 ···················· 饰纹姬蛙 *Microhyla fissipes*

（9）粗皮姬蛙 *Microhyla butleri* Boulenger, 1900

Microhyla butleri Boulenger, 1900, Ann. Mag. Nat. Hist., Ser. 7, 6: 188. Holotype: Selangor Mus. (now MNM), according to original publication; BMNH 1901.3.20.5, according to Parker, 1934, Monogr. Frogs Fam. Microhylidae: 132. Type locality: Larut Hills at 4000 feet, Perak, Malaya (Malaysia). Given in error by Gee and Boring, 1929, Peking Nat. Hist. Bull., 4: 26, as Tonkin, Vietnam.

Microhyla boulengeri Vogt, 1913, Sitzungsber. Ges. Naturforsch. Freunde Berlin, 1913: 222. Syntypes: ZMB 23334 (2 specimens, according to Parker, 1934, Monogr. Frogs Fam. Microhylidae: 132; Bauer, Günther and Robeck, 1996, Mitt. Zool. Mus. Berlin, 72: 265, although the original publication only mentions one specimen). Type locality: "Hainan", China. Tentative synonymy by Parker, 1928, Ann. Mag. Nat. Hist., Ser. 10, 2: 484.

Microhyla latastii Boulenger, 1920, Ann. Mag. Nat. Hist., Ser. 9, 6: 107. Syntypes: BMNH 1920.1.20.4054a (2 specimens), according to Parker, 1934, Monogr. Frogs Fam. Microhylidae: 132. Type locality: Saigon, Cochin China [= Vietnam]. Synonymy by Smith, 1922, J. Nat. Hist. Soc. Siam, 4: 214; Parker, 1928, Ann. Mag. Nat. Hist., Ser. 10, 2: 484.

Microhyla grahami Stejneger, 1924, Occas. Pap. Boston Soc. Nat. Hist., 5: 119. Holotype: USNM 65936, by original designation. Type locality: Suifu [= Yibin], Province of Szechwan [= Sichuan], China. Synonymy by Parker, 1928, Ann. Mag. Nat. Hist., Ser. 10, 2: 484.

Microhyla sowerbyi Stejneger, 1924, Occas. Pap. Boston Soc. Nat. Hist., 5: 119. Holotype: USNM 65309, by original designation. Type locality: Near Yen-ping-fu [= Nanping], Province of Fukien [= Fujian], China. Tentative synonymy by Parker, 1928, Ann. Mag. Nat. Hist., Ser. 10, 2: 484; this synonymy confirmed by Pope, 1931, Bull. Am. Mus. Nat. Hist., 61: 593.

Microhyla cantonensis Chen, 1929, China J. Sci. Arts, Shanghai, 10: 338. Holotype: Zool. Laboratory of Sun Yat-sen Univ., 1201, by original designation; status of this specimen currently not known (DRF). Type locality: in a pond in the northern part about one mile from [north of] Cantoncity [= Guangzhou], Guangdong, China. Provisional synonymy by Bourret, 1942, Batr. Indochine: 514; Zhao and Adler, 1993, Herpetol. China: 162.

[同物异名] *Microhyla boulengeri*；*Microhyla latastii*；*Microhyla grahami*；*Microhyla sowerbyi*；*Microhyla cantonensis*

[鉴别特征] 体型甚小，趾间具微蹼；指、趾末端均具吸盘，背面均有纵沟；背部棕黑色且皮肤粗糙；背部、股、胫背面均有疣粒。

[形态描述] 依据茂兰地区标本描述 体较小，雄蛙体长20mm左右，雌蛙体长23mm左右，体略呈三角形且头宽大于头长；吻端钝尖，突出于下唇；鼻孔近吻端，鼻间距小于眼间距；鼓膜不显；无犁骨齿；舌后端圆。前肢细弱，指端具小吸盘，背面有小纵沟；后肢较粗壮，前伸贴体时胫跗关节达眼部，左、右跟部重叠；趾端具吸盘，背面有明显的纵沟。体背面皮肤粗糙，满布疣粒；四肢背面均有疣粒；枕部有肤沟，向两侧斜达口角后绕至腹面，在咽喉部相连形成咽褶；腹面皮肤光滑。背上许多疣粒上有红色小点。背部中央有镶黄边的黑褐色大花斑，此斑自上眼睑内侧至躯干中央汇成宽窄相间的主干，在背后端，主干向两侧分叉，形成"Λ"形的深色花斑，斜向胯部。

[第二性征] 雄蛙具单咽下外声囊，咽喉部色深；雄性线明显。

[生活习性] 该蛙生活于海拔100～1300m的山区。成蛙常栖息于山坡水田、水坑边土隙或草丛中。在繁殖季节（5～6月），雄蛙发出"歪！歪！歪！"的鸣叫声。

[地理分布] 茂兰地区分布于翁昂、洞塘等地。国内从四川东部及云南到浙江都有分布，海南和台湾也有分布。国外分布于马来半岛、越南、印度（东北部）。

[模式标本] 正模标本（BMNH 1901.3.20.5）保存于英国伦敦自然历史博物馆（The Natural History Museum, London, UK, BMNH）

[模式产地] 马来西亚霹雳州，1219m

[保护级别] 三有保护动物

[IUCN 濒危等级] 无危（LC）

a 背面照
b 腹面照
c 在茂兰地区的分布

（10）饰纹姬蛙 *Microhyla fissipes* (Duméril and Bibron, 1841)

Engystoma ornatum Duméril and Bibron, 1841, Erpet. Gen., 8: 745. Type locality: India.

Microhyla fissipes Boulenger, 1884, Ann. Mag. Nat. Hist., Ser. 5, 13: 397. Holotype: BMNH 84.3.11.6, according to Parker, 1934, Monogr. Frogs Fam. Microhylidae: 141, now BM 1947.2.11.85, according to museum records. Type locality: Taiwan foo [= Tainan], S. Formosa* [= Taiwan].

Microhyla eremita Barbour, 1920, Occas. Pap. Mus. Zool. Univ. Michigan, 76: 3. Holotype: MCZ 5114, by original designation. Type locality: Nanking [= Nanjing], Jiangsu, China. Synonymy with *Microhyla ornata* by Parker, 1928, Ann. Mag. Nat. Hist., Ser. 10, 2: 494. Synonymy with *Microhyla fissipes implied* by Matsui, Ito, Shimada, Ota, Saidapur, Khonsue, Tanaka-Ueno, and Wu, 2005, Zool. Sci., Tokyo, 22: 489-495.

[同物异名] *Engystoma ornatum*；*Microhyla eremita*；*Microhyla ornata*

[俗名] 犁头拐、土地公蛙

[鉴别特征] 趾间具蹼迹；指、趾末端圆而无吸盘及纵沟；背部有两个前后连续的深褐色"Λ"形斑。

[形态描述] 依据茂兰地区标本描述 体型小，略呈三角形，雄蛙体长17～20mm，雌蛙体长19～21mm，头小，体宽，头长与头宽几乎相等；吻端尖圆，突出于下唇，吻棱不显；鼻孔近吻端，鼻间距小于眼间距而大于上眼睑之宽；鼓膜显；无犁骨齿；舌呈长椭圆形，后端无缺刻。前肢细弱，指末端圆，无吸盘也无纵沟；后肢较粗短，前伸贴体时胫跗关节达肩前方，左、右跟部有重叠；趾间具蹼迹。皮肤粗糙，背部有许多小疣粒；由眼后至胯部有一斜行大长疣；枕部有一横肤沟，并在两侧延伸至肩部；肛周围小圆疣较多；腹面皮肤光滑，呈白色。背部有两个前后连续的深褐色"Λ"形斑。

[第二性征] 雄蛙咽喉部色深；具单咽下外声囊；有雄性线。

[生活习性] 该蛙生活在茂兰地区丘陵和山地的泥窝或土穴内，或在水域附近的草丛中，也会在雨后的玉米地中捕食蚂蚁等昆虫。

[地理分布] 茂兰地区分布于翁昂、洞塘等地。国内在黄河以南各省区均有分布。国外分布于克什米尔地区、巴基斯坦、印度、斯里兰卡、尼泊尔、马来半岛、柬埔寨、越南、日本。

[模式标本] 正模标本（BMNH 84.3.11.6）保存于英

a 整体照
b 在茂兰地区的分布

---

* 为保持史料原貌，编者（出版者）对史料中因时代背景、政治立场不同而形成的称谓等，在引用时未做处理，相信读者通过研读，对编者（出版者）立场自有明鉴。

国伦敦自然历史博物馆

[模式产地] 台湾

[保护级别] 三有保护动物

[IUCN 濒危等级] 无危（LC）

生境

（11）小弧斑姬蛙 *Microhyla heymonsi* Vogt, 1911

Microhyla heymonsi Vogt, 1911, Sitzungsber. Ges. Naturforsch. Freunde Berlin, 1911: 181. Syntypes: ZMB 21944 (9 specimens), according to Bauer, Günther and Robeck, 1996, Mitt. Zool. Mus. Berlin, 72: 266. Type locality: Formosa [= Taiwan], China; restricted to Kosempo, Formosa [= Taiwan], by Parker, 1934, Monogr. Frogs Fam. Microhylidae: 135.

Microhyla (*Microhyla*) *heymonsi* Dubois, 1987, Alytes, 6: 3.

Microhyletta heymonsi Lue, Tu and Hsiang, 1999, Atlas Taiwan Amph. Rept.: 58.

[同物异名] 无

[鉴别特征] 趾间具蹼迹；指、趾末端均具吸盘，背面均有纵沟；体背中线上有1个或2个棕褐色的小弧形斑。

[形态描述] 依据茂兰地区标本描述 体型略小而呈三角形；雄蛙头体长为19.57～21.14mm，雌蛙头体长21.07～25.40mm。头小，头长小于头宽；吻端钝尖，突出于下唇；吻棱明显，与颊部几近垂直；鼻孔近吻端，鼓膜不显；无犁骨齿；舌窄长，后端无缺刻。前肢细弱，指末端有小吸盘，背面有纵沟；后肢较粗壮，向前伸贴体时胫跗关节达眼部，左、右跟部有重叠；趾吸盘大于指吸盘，背面有明显的纵沟；趾间具蹼迹；关节下瘤明显；内蹠突大，长椭圆形；外蹠突略小，圆球形；蹠外侧有肤棱。体背面皮肤较光滑，呈灰白色，散有细痣粒；由眼后至胯部有明显的斜行肤棱将背面和侧面颜色区分开。股基部腹面光滑，有较大的痣粒。从吻端至肛部常有1条米黄色的细脊线，线上有一棕褐色斑呈弧形；从头侧沿体侧至胯部各有1条宽的黑棕色斜线纹；四肢有黑棕色横纹，股前方有黑棕色纵纹，恰与体侧的黑棕色线纹略衔接。体腹部白色。

[第二性征] 雄蛙具单咽下外声囊，声囊孔长裂状；雄性线明显。

[生活习性] 该蛙常栖息于茂兰地区的山区稻田、水坑边、沼泽泥窝、土穴或草丛中。繁殖季节雄蛙发出低而慢的"嘎、嘎"鸣叫声，低沉而慢。繁殖旺季卵产于静水水域中，卵群成片。蝌蚪集群浮游于水体表层，受惊时潜入水下。

[地理分布] 茂兰地区全区分布。国内从云南南部到浙江都有分布，海南和台湾也有分布。国外分布于印度、老挝、越南、柬埔寨、马来半岛、印度尼西亚。

[模式标本] 全模标本 [ZMB 21944（9 specimens）] 保存于德国柏林洪堡大学自然历史博物馆

[模式产地] 台湾甲仙

[保护级别] 三有保护动物

[IUCN 濒危等级] 无危（LC）

成体量度 （单位：mm）

编号	头体长	头长	头宽	前臂及手长	后肢全长	胫长	跗足长
GZNU201705026007 ♀	25.40	4.07	6.38	10.31	45.97	15.46	19.46
GZNU20170509033 ♂	21.14	5.04	6.39	7.79	36.89	12.44	17.08
GZNU20170509032 ♂	19.98	4.96	6.05	8.18	36.35	13.16	17.51
GZNU20170508029 ♂	19.57	4.04	5.24	7.51	30.43	10.28	14.31
GZNU20170509023 ♀	21.07	4.86	5.54	7.57	34.73	12.17	16.85
GZNU20170509028 ♀	23.88	6.38	6.69	9.01	41.26	15.11	20.25

a～d 整体照
e 在茂兰地区的分布

（12）花姬蛙 *Microhyla pulchra* (Hallowell, 1860)

Engystoma pulchrum Hallowell, 1861 "1860", Proc. Acad. Nat. Sci. Philadelphia, 12: 506. Holotype: Deposition not stated, presumably USNM or ANSP. Type locality: Hong Kong, China. Common in the brackish water marshes between Hong Kong and Whampoa [= Huangpu], Guangdong, China.

Scaptophryne labyrinthica Fitzinger, 1861 "1860", Sitzungsber. Akad. Wiss. Wien, Phys. Math. Naturwiss. Kl., 42: 416. Types: not stated, though likely NHMW. Type locality: Hong Kong, China. Nomen nudum. Synonymy by Steindachner, 1867, Reise Österreichischen Fregatte Novara, Zool., Amph.: 36; Boulenger, 1882, Cat. Batr. Sal. Coll. Brit. Mus., ed. 2: 166; Bourret, 1942, Batr. Indochine: 522.

Diplopelma pulchrum Günther, 1864, Rept. Brit. India: 417.

Ranina symetrica David, 1872 "1871", Nouv. Arch. Mus. Natl. Hist. Nat., Paris, 7: 76. Types: not stated, presumably MNHNP but not reported in type lists. Type locality: les collines voisines du Yantzékiang, et devient abondante vers le Setchuan (= Hills near the Yangtzékiang [= Changjiang River]), and toward Sichuan Province, China. Synonymy by Boulenger, 1882, Cat. Batr. Sal. Coll. Brit. Mus., ed. 2: 166; Bourret, 1942, Batr. Indochine: 522.

Microhyla pulchra Boulenger, 1882, Cat. Batr. Sal. Coll. Brit. Mus., ed. 2: 165.

Microhyla hainanensis Barbour, 1908, Bull. Mus. Comp. Zool., 51: 322. Syntypes: MCZ 2435 (4 specimens), by original designation. Type locality: Mt. Wuchi [= Mt. Wuzhi], Central Hainan [Island], Hainan Province, China. Synonymy by Smith, 1923, J. Nat. Hist. Soc. Siam, 6: 211; Parker, 1928, Ann. Mag. Nat. Hist., Ser. 10, 2: 484.

Microhyla melli Vogt, 1914, Sitzungsber. Ges. Naturforsch. Freunde Berlin, 1914: 101. Holotype: ZMB 24097 (according to Parker, 1934, Monogr. Frogs Fam. Microhylidae: 138; Bauer, Günther and Robeck, 1996, Mitt. Zool. Mus. Berlin, 72: 266). Type locality: Umbegung der Stadt Canton (= Environs of the city of Canton [= Guangzhou]), Guangdong Province, China. Synonymy by Parker, 1934, Monogr. Frogs Fam. Microhylidae: 137; Bourret, 1942, Batr. Indochine: 522.

Microhyla (*Diplopelma*) *pulchrum* Bourret, 1927, Fauna Indochine, Vert., 3: 263.

Microhyla major Ahl, 1930, Sitzungsber. Ges. Naturforsch. Freunde Berlin, 1930: 317. Holotype: ZMB, unnumber (according to the original publication and Parker, 1934, Monogr. Frogs Fam. Microhylidae: 138). Type locality: not stated; presumably Yaoshan (= Mt. Dayao, Kwangsi [= Guangxi]). Synonymy by Parker, 1934, Monogr. Frogs Fam. Microhylidae: 137; Bourret, 1942, Batr. Indochine: 522.

Microhyla (*Diplopelma*) *pulchra*: Dubois, 1987, Alytes, 6: 4.

[同物异名] *Engystoma pulchrum*; *Scaptophryne labyrinthica*; *Diplopelma pulchrum*; *Ranina symetrica*; *Microhyla hainanensis*; *Microhyla melli*; *Microhyla major*

[俗名] 犁头蛙、三角蛙

[鉴别特征] 趾间半蹼；指、趾末端无吸盘及纵沟；体背面有若干相套叠的"Λ"形斑；胯部及股后多为柠檬黄色。

[形态描述] 依据茂兰地区标本描述 体略呈三角形；雄蛙头体长27.73～28.92mm，雌蛙头体长29.14～30.19mm。头小，头宽略大于头长；吻端钝尖，突出于下唇；吻棱不显；鼻孔近吻端；鼓膜不显；无犁骨齿；舌后端圆。指端圆，无吸盘，背面无纵沟；后肢粗壮，向前伸贴体时胫跗关节抵达眼，左、右跟部重叠；趾端圆，无吸盘也无纵沟；趾间半蹼，趾侧缘膜达趾端。体背面皮肤较光滑，散有少量小疣粒，有若干相套叠的"Λ"形斑；胯部及股后多为柠檬黄色。两眼后方有一横沟，向两外侧斜伸至肩部并绕至腹面横贯咽喉部而形成咽褶。后腹部、股下方及肛孔附近小疣颇多；其余腹面光滑。

[第二性征] 雄蛙体略小，具单咽下外声囊；咽喉部色深；雄性线明显。

[生活习性] 该蛙生活于茂兰地区的丘陵和山区，常栖息于水田、水坑附近的泥窝、洞穴或草丛中。5月可见其抱对繁殖。在繁殖季节，雄蛙发出"嘎！嘎嘎嘎嘎！"清脆悦耳的鸣叫声，卵产于水田或静水坑内，卵群粘连成片状，

单层漂浮于水面略呈圆形。蝌蚪生活于稻田或静水坑内，常集群浮游于水表层。

[地理分布]　茂兰地区仅分布于洞塘。国内分布于云南、贵州、江西、浙江、湖南、福建、广东、香港、澳门、海南、广西。国外分布于印度（东北部）、泰国。

[模式标本]　不详
[模式产地]　香港和广东之间
[保护级别]　三有保护动物
[IUCN 濒危等级]　无危（LC）

成体量度　（单位：mm）

编号	头体长	头长	头宽	前臂及手长	后肢全长	胫长	跗足长
GZNU20170509015 ♂	28.39	7.97	8.63	10.84	49.47	17.79	22.76
GZNU20170513012 ♂	28.92	8.40	8.14	11.09	45.13	17.63	17.72
GZNU20170513017 ♂	27.99	10.51	6.00	11.65	49.43	18.47	21.67
GZNU20170509018 ♂	27.73	8.40	8.50	11.82	47.39	17.11	21.67
GZNU20170509022 ♀	30.19	9.13	9.95	11.09	48.85	18.77	23.23
GZNU20170509019 ♀	29.14	9.35	9.73	10.59	50.90	18.71	23.75

a　背面观
b　腹面观
c　生境
d　在茂兰地区的分布

6. 叉舌蛙科 Dicroglossidae Anderson, 1871

Dicroglossini Anderson, 1871, Jour. Asiat. Soc. Bengal, 40: 38. Type genus: *Euphlyctis* Fitzinger, 1843.
Dicroglossini Dubois, 1987 "1986", Alytes, 5: 57; Dubois, 1992, Bull. Mens. Soc. Linn. Lyon, 61: 314-315; Dubois, 2005, Alytes, 23: 16.
Occydozyginae Fei, Ye and Huang, 1990, Key to Chinese Amph.: 123. Type genus: *Occidozyga* Kuhl and van Hasselt, 1822.
Dicroglossinae Dubois, 1992, Bull. Mens. Soc. Linn. Lyon, 61: 313; Dubois, 2005, Alytes, 23: 16.
Paini Dubois, 1992, Bull. Mens. Soc. Linn. Lyon, 61: 317. Type genus: *Paa* Dubois, 1975. Synonymy by Roelants, Jiang and Bossuyt, 2004, Mol. Phylogenet. Evol., 31: 734-735.
Limnonectini Dubois, 1992, Bull. Mens. Soc. Linn. Lyon, 61: 315. Type genus: *Limnonectes* Fitzinger, 1843.
Occydozyginae Dubois, Ohler and Biju, 2001, Alytes, 19: 55.
Occydozygini Dubois, 2005, Alytes, 23: 16.
Annandini Fei, Ye and Jiang, 2010, Herpetol. Sinica, 12: 19. Type genus: *Annandia* Dubois, 1992.
Hoplobatrachini Fei, Ye and Jiang, 2010, Herpetol. Sinica, 12: 19. Type genus: *Hoplobatrachus* Peters, 1863.
Ingeranini Fei, Ye and Jiang, 2010, Herpetol. Sinica, 12: 19. Type genus: *Ingerana* Dubois, 1986.
Nannophryini Fei, Ye and Jiang, 2010, Herpetol. Sinica, 12: 19. Type genus: *Nannophrys* Günther, 1869.
Quasipaini Fei, Ye and Jiang, 2010, Herpetol. Sinica, 12: 22. Type genus: *Quasipaa* Dubois, 1992.

犁骨齿发达或无，一般鼻骨大；左右在中线相接触，与蝶筛骨或额顶骨相接触或略分开；蝶筛骨在背面多隐蔽；肩胸骨基部深度分叉或不分叉；舌骨的前角有前突，呈环状或长形，向内弯；指、趾骨末端尖或略大，呈圆形；指、趾末端尖或钝尖或膨大成球状，不形成吸盘状；腹侧无沟。

蝌蚪口部有唇齿和唇乳突；体腹面无大的腹吸盘，背面及腹面无腺体。

本科已知14属214种，分布于亚洲和非洲。我国现有7属41种，主要分布于秦岭以南各省区。茂兰地区有4属6种，分属检索表如下。

茂兰地区属检索表

1 {雄性胸部无刺团；指、趾端一般较尖，不呈球状，肩胸骨分叉 ·· 2
  {雄性胸部或胸腹部有刺团；体型甚肥硕，肩胸骨不分叉 ··· 3
2 {体大；下颌前端齿状骨突甚显；肩带弧固型（上喙骨部分重叠，部分愈合）；舌角前突膨大成环状 ············ 虎纹蛙属 *Hoplobatrachus*
  {体较小；下颌无明显齿突；肩带固胸型；舌角前突短小 ·· 陆蛙属 *Fejervarya*
3 {雄蛙胸部的刺分成左右两团 ··· 双团棘胸蛙属 *Gynandropaa*
  {雄蛙胸部或胸腹部刺大而稀疏，呈锥状 ·· 棘胸蛙属 *Quasipaa*

8）陆蛙属 *Fejervarya* Bolkay, 1915

Fejervarya Bolkay, 1915, Anat. Anz., 48: 181. Type species: *Rana limnocharis* Gravenhorst, 1829, by subsequent designation of Dubois, 1981, Monit. Zool. Ital., N.S., Suppl., 15: 238. Coined as a subgenus under *Rana*.
Minervarya Dubois, Ohler and Biju, 2001, Alytes, 19: 58. Type species: *Minervarya sahyadris* Dubois, Ohler and Biju, by original designation. Synonymy by Dinesh et al., 2015, Zootaxa, 3999: 79.
Zakerana Howlader, 2011, Bangladesh Wildl. Bull., 5: 1. Type species: *Rana limnocharis syhadrensis* Annandale, 1919, by original designation. Synonymy by Dinesh, et al., 2015, Zootaxa, 3999: 79.

鼻骨大，两内缘相接，与额顶骨相接触或略分离；蝶筛骨完全隐蔽或少部分显露；额顶骨窄长；鳞骨颧枝长；前耳骨大。舌角前突短小。肩胸骨基部分叉，呈"人"字形；上胸软骨略小于剑胸软骨；中胸骨杆状或哑铃状；剑胸软骨伞状，后端钝尖，无缺刻；锁骨纤细。指、趾骨末端钝尖。

指、趾末端尖，无沟；无指基下瘤；趾间全蹼或半蹼，外侧蹠间蹼几乎达蹠基部或达蹠部的1/2。无背侧褶，背部和体侧有疣粒或肤棱，腹面皮肤光滑。雄蛙第一指基部有婚垫。

成体以陆栖生活为主，产卵于稻田、池塘及临时水坑内。蝌蚪在静水水域内生活。

本属世界现有13种，我国有4种。国内分布于南方各省区。据 Amphibia China（2017）记载贵州有泽陆蛙 *Fejervarya multistriata* 和川村陆蛙 *Fejervarya kawamurai* 2 种，茂兰地区有1种——泽陆蛙。

（13）泽陆蛙 *Fejervarya multistriata* (Hallowell, 1860)

Rana multistriata Hallowell, 1861 "1860", Proc. Acad. Nat. Sci. Philadelphia, 12: 505. Syntypes: 2 specimens, presumably in ANSP or USNM, considered lost by Dubois and Ohler, 2000, Alytes, 18: 43, who designated ZMB 3255 (the holotype of *Rana gracilis* Wiegmann) as neotype. Type locality: Hong Kong, China. Neotype from China (bei Cap Syng-more). Considered a nomen nudum by Boulenger, 1882, Cat. Batr. Sal. Coll. Brit. Mus., ed. 2: 7; this corrected by the neotype designation.

Fejervarya multistriata Ohler, 2004, The IUCN red list of threatened species. IUCN, 2004: e.T58277A11748427.

Fejervarya multistriata Frost, 2014, Amphibian Species of the World: an Online Reference. Version 6.0. American Museum of Natural History.

［中文曾用名］ 泽蛙

［同物异名］ *Rana gracili*；*Rana multistriata*

［俗名］ 梆声蛙、乌蟆、泥噶度、噶度、狗污田鸡

［鉴别特征］ 体长小于60mm；第五趾无缘膜或极不明显；有外蹠突；雄蛙有单咽下外声囊。

［形态描述］ 依据茂兰地区标本描述 体型较小，雄蛙头体长36.23～39.01mm，雌蛙头体长42.51～45.42mm。头长略大于头宽；吻部尖，末端钝圆，突出于下唇，吻棱不显，鼻孔位于吻眼之间；眼间距很窄，小于鼻间距，鼓膜圆而明显；犁骨齿两团，小而突出；舌宽厚，卵圆形，后端有缺刻。后肢较粗短，前伸贴体时胫跗关节达肩或仅达鼓膜，左、右跟部不相遇；趾端钝尖。背面皮肤粗糙，有数行长短不一的纵肤棱，在肤棱之间散布许多小疣粒，无背侧褶；体侧及体后端疣粒圆而明显；股、胫背面有零星的小疣粒，肛周及股腹面密布扁平疣粒，腹面除体腹后端有扁平疣外，其余各部光滑。背部青灰色、橄榄色和深灰色杂斑均有分布，从吻端至肛部有浅色脊纹，脊纹淡黄色；上、下唇缘有6条黑纵纹，两眼间有深色倒"Λ"形斑。

［第二性征］ 雄蛙体型略小；第一指内侧有浅色婚垫；咽喉部两侧深灰色呈皱褶状，有单咽下外声囊；有雄性线。

［生活习性］ 该蛙生活于茂兰地区的多种生境。稻田、沼泽、水塘、水沟等静水水域或其附近的旱地草丛均有该蛙

成体量度 （单位：mm）

编号	头体长	头长	头宽	前臂及手长	后肢全长	胫长	跗足长
GZNU20170508015♀	42.51	16.73	11.86	35.23	71.33	21.69	31.35
GZNU20170509010♀	45.42	16.16	9.09	17.04	67.41	21.15	27.71
GZNU20170508021♂	36.54	12.86	10.37	15.66	67.32	18.97	31.29
GZNU20170508020♂	39.01	12.36	10.57	15.92	60.18	17.70	27.82
GZNU20170508024♂	38.23	12.37	10.46	16.39	65.02	19.01	28.81
GZNU20170508022♂	36.23	14.69	10.44	12.21	52.77	15.55	24.29

出没。昼夜活动，主要在夜间觅食。

[地理分布]　广布种，茂兰地区全区分布。国内分布于河北、天津、山东、河南、陕西、甘肃、湖北、安徽、江苏、浙江、上海、江西、湖南、福建、台湾、四川、重庆、贵州、云南、西藏。国外分布于印度（北部）、泰国、缅甸、老挝、越南、日本（南部岛屿）等地。

[模式标本]　全模标本（两号）可能保存于美国费城自然博物馆（Academy of Natural Sciences, Philadelphia, Pennsylvania, USA, ANSP）或美国国家自然历史博物馆（National Museum of Natural History, Washington D.C., USA, USNM）；新模标本（ZMB 3255）保存于德国柏林洪堡大学自然历史博物馆

[模式产地]　新模标本产地为香港大屿山

[保护级别]　三有保护动物

[IUCN 濒危等级]　无危（LC）

a~d　整体照
e　在茂兰地区的分布

9）虎纹蛙属 *Hoplobatrachus* Peters, 1863

Hoplobatrachus Peters, 1863, Monatsber. Preuss. Akad. Wiss. Berlin, 1863: 449. Type species: *Hoplobatrachus ceylanicus* Peters, 1863 (= *Rana tigerina* Daudin, 1802), by monotypy.

Hydrostentor Steindachner, 1867, Reise Österreichischen Fregatte Novara, Zool., Amph.: 18. Type species: *Hydrostentor tigrina* var. *pantherina* Steindachner, 1867, by monotypy. Synonymy by Dubois, 1981, Monit. Zool. Ital., N.S., Suppl., 15: 240.

Ranosoma Ahl, 1924, Arch. Naturgesch., Abt. A, 90: 250. Type species: *Ranosoma schereri* Ahl, 1924, by original designation. Synonymy with *Euphlyctis* by Dubois, 1981, Monit. Zool. Ital., N.S., Suppl., 15: 240.

Tigrina Fei, Ye and Huang, 1990, Key to Chinese Amph.: 144. Type species: *Rana tigerina* Daudin, 1803, by original designation. Preoccupied by Tigrina Grevé, 1894.

鼻骨大，似三角形，两内缘相接，与额顶骨相接触；蝶筛骨完全被覆盖或仅小部分显露；额顶骨窄长，前窄后宽，个体大者，后部愈合成峪。鳞骨颧枝细长，耳枝较大，前耳骨较小；下颌前部齿状骨突发达。舌角前突呈环状膨大。肩带弧固型（上喙骨部分重叠，部分愈合）；肩胸骨基部分叉，呈"人"字形，上胸软骨较小；中胸骨粗短，剑胸骨宽大，后端无缺刻。指、趾骨节游离端钝尖。

趾末端钝尖，无沟；无掌突，无指基下瘤；趾间全蹼，外侧趾间蹼达蹠基部；有内跖褶。体大，无背侧褶，背面有纵肤棱，腹部皮肤光滑。雄蛙婚垫位于第一指基部。

成体和蝌蚪生活于稻田、池塘，成体很少离开水域。本属现有5种，我国有1种，即虎纹蛙 *Hoplobatrachus chinensis*。

（14）虎纹蛙 *Hoplobatrachus chinensis* (Osbeck, 1765)

Rana chinensis Osbeck, 1765, Reise Ostindien China: 244. Type: now lost. Neotype: CIB 980505, ♂, SVL 92.3mm, by present desjgnation of Fei and Ye. Type locality: near Guangzhou City, Guangdong, China. Kosuch, Vences, Dubois, Ohler and Bdhme, 2001, Mol. Phylogenetics Evol., 21 (3): 398-407.

Rana rugulosa Wiegmann, 1835, in Meyen (ed.), Reise in die Erde K. Preuss. Seehandl., 3 (Zool.): 508. Subsequently published by Wiegmann, 1835, Nova Acta Phys. Med. Acad. Caesar Leopold Carol., Halle, 17: 258. Holotype: ZMB 3721, according to Peters, 1863, Monatsber. Preuss. Akad. Wiss. Berlin, 1863: 78. Type locality: Cap Syng-more (= Kap Shui Mun, Lantau I., Hong Kong, China).

Rana tigrina var. *pantherina* Steindachner, 1867, Reise Österreichischen Fregatte Novara, Zool., Amph.: pl. 1, figs. 14-17. Types: not stated, presumably NHMW. Type locality: Hong Kong, China. Synonymy by Stejneger, 1925, Proc. U.S. Natl. Mus., 66: 29.

Rana burkilli Annandale, 1910, Rec. Indian Mus., 5: 79. Syntypes: ZSIC 16569-70, according to XXX. Chanda, Das and Dubois, 2001 "2000", Hamadryad, 25: 108, suggested additional specimens (from "Mandalay" as syntypes, although Annandale clearly regarded only the two specimens as primary types. Type locality: Tavoy upper Myanmar. Synonymy with *Rana rugulosa* by Annandale, 1917, Mem. Asiat. Soc. Bengal, 6: 126. Synonymy with *Hoplobatrachus tigerina* by Boulenger, 1918, Rec. Indian Mus., 15: 58.

Rana tigerina var. *burkilli* Boulenger, 1918, Rec. Indian Mus., 15: 58.

Rana rugulosa Annandale, 1918, Rec. Indian Mus., 15: 60; Okada, 1927, Copeia, 158: 165.

Rana (Rana) tigerina var. *pantherina* Boulenger, 1920, Rec. Indian Mus., 20: 6-17.

Rana tigrina rugulosa Smith, 1930, Bull. Raffles Mus., 3: 93; Fang and Chang, 1931, Contrib. Biol. Lab. Sci. Soc., China, Zool. Ser., 7: 107.

Rana tigerina rugulosa Fang and Chang, 1931, Contrib. Biol. Lab. Sci. Soc., China, Zool. Ser., 7: 65-114.
Rana tigerina pantherina Taylor and Elbel, 1958, Univ. Kansas Sci. Bull., 38: 1050.
Rana (*Euphlyctis*) *rugulosa* Dubois, 1981, Monit. Zool. Ital., N.S., Suppl., 15: 240.
Euphlyctis tigerina rugulosa Poynton and Broadley, 1985, Ann. Natal Mus., 27: 124.
Limnonectes (*Hoplobatrachus*) *rugulosus* Dubois, 1987 "1986", Alytes, 5: 60.
Tigrina rugulosa Fei, Ye and Huang, 1990, Key to Chinese Amph.: 145.
Hoplobatrachus rugulosus Dubois, 1992, Bull. Mens. Soc. Linn. Lyon, 61: 315.
Hoplobatrachus chinensis Ohler, Swan and Daltry, 2002, Raffles Bull. Zool., 50: 467; Fei, Ye and Jiang, 2010, Herpetol. Sinica, 12: 28.

［同物异名］ *Rana chinensis*；*Rana rugulosa*；*Hydrostentor pantherinus*；*Rana burkilli*；*Hoplobatrachus rugulosus*；*Tigrina rugulosa*

［俗名］ 水鸡、青鸡、虾蟆、田鸡

［鉴别特征］ 体长可达100mm以上；下颌前侧方有两个骨质齿状突；鼓膜明显；雄蛙声囊内壁黑色。

［形态描述］ 依据茂兰地区标本描述 雄蛙头体长34.62～68.50mm，头长略大于头宽；吻端钝尖，吻棱钝；鼻孔略近吻端或于吻眼之间，颊部向外倾斜；鼻间距大于眼间距，而小于上眼睑，瞳孔横椭圆形；鼓膜明显，鼓膜约为眼径的3/4；上颌齿锐利，下颌前缘有两个齿状骨突；恰与上颌的两个凹陷相吻合；舌后端缺刻深；犁骨齿极强。

背侧褶无；前肢短，指短，指式为1≈3＞4＞2，第二、第三指指侧具缘膜，关节下瘤明显；指间无蹼，内掌突略显，无外掌突。后肢较短，后肢前伸贴体时胫跗关节达眼至鼓膜，左、右跟部相遇或略重叠，胫长小于体长之半；有内跗褶，趾末端钝尖，关节下瘤小；趾间全蹼，第一、第五趾游离侧缘膜发达，外侧蹠间蹼达蹠基部，蹠突窄长，具游离刃，无外蹠突；体背面粗糙，背部有长短不一、多断续排列成纵列的肤棱，其间散有小疣粒，胫部纵列肤棱明显；头侧、手、足背面和体腹面光滑。背面多为黄绿色或灰棕色，散有不规则的深绿褐色斑纹；四肢横纹明显；体和四肢腹面肉色，咽、胸部有棕色斑，胸后和腹部略带浅蓝色，有斑或无斑。

［第二性征］ 雄蛙前肢粗壮；第一指内侧有肥厚的灰色婚垫，垫上密布细小的颗粒；具1对咽侧下外声囊，声囊内壁黑色；有雄性线。

［生活习性］ 该蛙生活于茂兰保护区山区地带的稻田、鱼塘、水坑和沟渠内。白天隐匿于水域岸边的洞穴内，夜间外出活动。跳跃能力很强，稍有响动即迅速跳入深水中。成蛙捕食各种昆虫，也捕食蝌蚪、小蛙及小鱼等。卵单粒至数粒粘连成片，漂浮于水面。蝌蚪栖息于水塘底部。种群数量较大。

［地理分布］ 茂兰地区分布于翁昂、洞塘等地。国内分布于河南、陕西、安徽、江苏、上海、浙江、江西、湖南、福建、台湾、四川、云南、贵州、湖北、广东、香港、澳门、海南、广西。国外分布于缅甸、泰国、越南、柬埔寨、老挝。

［模式标本］ 正模标本（ZMB 3721）保存于德国柏林洪堡大学自然历史博物馆

［模式产地］ 广东广州

［保护级别］ 国家Ⅱ级重点保护野生动物；贵州省重点保护野生动物

［IUCN濒危等级］ 濒危（EN）

成体量度 （单位：mm）

编号	头体长	头长	头宽	前臂及手长	后肢全长	胫长	蹠足长
GZNU20160726052 ♂	64.96	25.05	18.03	24.20	97.62	25.15	48.74
GZNU20160726058 ♂	57.12	23.11	15.23	24.78	82.17	23.79	47.10
GZNU20160726057 ♂	53.04	21.13	15.07	22.73	77.13	22.83	50.64
GZNU20160726055 ♂	53.42	20.51	14.24	27.86	73.05	18.80	43.37
GZNU20160804071 ♂	34.62	14.82	9.70	13.11	50.37	13.55	31.72
GZNU20160726056 ♂	68.50	25.69	22.01	30.48	98.32	34.21	50.28

[讨论] 虎纹蛙属 Hoplobatrachus 在中国只记录有 1 种，但是其学名存在争议。费梁等（2012）收录为 Hoplobatrachus chinensis，模式产地为中国广州。而 Frost（2017）及 Amphibia Web（2017）收录为 Hoplobatrachus rugulosus，模式产地为中国香港。

Werner（1903）与 Boulenger（1918）等曾认为 Hoplobatrachus chinensis 是黑斑侧褶蛙 Pelophylax nigromaculata 的同物异名。但是，费梁等（2009b）根据中国学者的调查，指出广州附近未发现过黑斑侧褶蛙，将两者作为同物异名是对 Hoplobatrachus chinensis 的错误鉴定，Hoplobatrachus chinensis 是虎纹蛙的有效学名。Kosuch 等（2001）认为 Hoplobatrachus chinensis 发表早于 Hoplobatrachus rugulosus，后者为前者的同物异名。

基于以上研究，本书收录 Hoplobatrachus chinensis 为虎纹蛙的学名。

a～d 整体照
e 在茂兰地区的分布

10）双团棘胸蛙属 *Gynandropaa* Dubois, 1992

Paa Dubois, 1992, Bull. Mens. Soc. Linn. Lyon, 61: 319. Elevation from subgenus of *Rana* (at which rank it had been originally proposed) to generic rank.

Gynandropaa Dubois, 1992, Bull. Mens. Soc. Linn. Lyon, 61: 319. Type species: *Rana yunnanensis* Anderson, 1879, by original designation. Proposed as a subgenus of *Paa*. Synonymy with *Nanorana* by implication of Frost et al., 2006, Bull. Am. Mus. Nat. Hist., 297: 318. See comment.

Gynandropaa Fei, Ye and Jiang, 2010. Asian Herpetological Research, 12: 1-43.

鼻骨大，两内缘相接，额顶骨前、后几乎等宽；前耳骨大；鳞骨颧枝呈刀状；舌骨体宽几乎为长的 2 倍；舌角前突粗短，并向外弯曲几乎成环状。肩胸骨基部不分叉，上胸软骨略小于剑胸软骨；中胸骨粗短，剑胸软骨宽大，呈圆盘状，后端略有浅缺刻。雄蛙前肢粗壮，肱骨具脊棱，其长约为最宽处的 4 倍；内侧第一、第二指粗大。

体型肥硕，无背侧褶；体背面和体侧皮肤粗糙，有长肤棱和疣粒，指、趾末端呈球状且无沟；无指基下瘤；内掌突大而突出；趾间全蹼或满蹼，外侧趾间蹼较弱。鼓膜隐蔽或不显（个别种类较为明显）。雄蛙前臂甚为粗壮，内侧 2 指或 3 指及内掌突上着生粗大婚刺；胸部有两团椎状黑色刺，有声囊。雄蛙肱骨远端外侧 2 条脊棱极为隆起；第一掌、指骨及前拇指骨粗壮。

蝌蚪下唇乳突 2 排，间距宽且排列规则、整齐，内排较为稀疏。出水孔位于身体左侧，肛孔位于尾基部右侧，均无游离管。唇齿式多为 I : 4+4/1+1 : II。成体和幼体多生活于山涧溪流。一次产卵的卵群含 680～2550 粒卵，成串悬于溪流内的石块下。卵径 2.0～2.5mm，卵表面具有色素，个别种类为乳黄色。

本属世界有 28 种，我国已知有 19 种。我国主要分布在秦岭以南各地。茂兰地区有 1 种，即双团棘胸蛙 *Gynandropaa phrynoides*。

（15）双团棘胸蛙 *Gynandropaa phrynoides* Boulenger, 1917

Rana yunnanensis Anderson, 1879 "1878", Anat. Zool. Res.: 839. Syntypes: BMNH 2 specimens (now long lost-not mentioned by Boulenger, 1882, Cat. Batr. Sal. Coll. Brit. Mus., ed. 2: 21); BMNH 1947.2.3.76 (lectotype of *Rana phrynoides* Boulenger, 1917) designated neotype by Dubois, 1987 "1986", Alytes, 5: 45. Type locality: Hotha [= Husa], 5000 feet above the sea, Yunnan, China; neotype from Tongchuan Fu [= Dongchuan], Yunnan, China.

Rana phrynoides Boulenger, 1917, Ann. Mag. Nat. Hist., Ser. 8, 20: 413. Syntypes: BMNH 1947.2.3.76-82; BMNH 1947.2.3.76 designated lectotype by Dubois, 1987 "1986", Alytes, 5: 45. Type locality: Yunnan at Tongchuan Fu, Yunnan, China. Synonymy by Liu and Hu, 1961, Tailless Amph. China: 162; Dubois, 1987 "1986", Alytes, 5: 45.

Rana (Rana) phrynoides Boulenger, 1920, Rec. Indian Mus., 20: 8.

Rana (Paa) phrynoides Dubois, 1975, Bull. Mus. Natl. Hist. Nat., Paris, Ser. 3, Zool., 324: 1097; Dubois, 1976, Cah. Nepal., Doc., 6: 24.

Rana (Paa) yunnanensis Dubois, 1975, Bull. Mus. Natl. Hist. Nat., Paris, Ser. 3, Zool., 324: 1098; Dubois, 1976, Cah. Nepal., Doc., 6: 24; Dubois, 1987 "1986", Alytes, 5: 43.

Rana muta Su and Li, 1986, Acta Herpetol. Sinica, Chengdu, N.S., 5 (2): 152-154. Holotype: KIZ 79006, by original designation. Type locality: Ninglang County, Yunnan, alt. 2650m, China. Preoccupied by *Rana muta* Laurenti, 1768 (= *Rana temporaria* Linnaeus, 1758). Synonymy by Che, Hu, Zhou, Murphy, Papenfuss, Chen, Rao, Li and Zhang, 2009, Mol. Phylogenet. Evol., 50: 71; Chen and Hu, 2009, Sichuan J. Zool., 28: 696-699.

Rana (Paa) liui Dubois, 1987 "1986", Alytes, 5: 150. Replacement name for *Rana muta* Su and Li, 1986.

Paa (*Paa*) *muta* Fei, Ye and Huang, 1990, Key to Chinese Amph.: 158.

Paa (*Paa*) *yunnanensis* Fei, Ye and Huang, 1990, Key to Chinese Amph.: 158; Ye, Fei and Hu, 1993, Rare and Economic Amph. China: 279; Fei and Ye, 2001, Color Handbook Amph. Sichuan: 187.

Paa (*Gynandropaa*) *yunnanensis* Dubois, 1992, Bull. Mens. Soc. Linn. Lyon, 61: 319.

Paa (*Gynandropaa*) *liui* Dubois, 1992, Bull. Mens. Soc. Linn. Lyon, 61: 319.

Rana liui Zhao and Adler, 1993, Herpetol. China: 144.

Rana yunnanensis Zhao and Adler, 1993, Herpetol. China: 151.

Paa (*Paa*) *liui* Fei, 1999, Atlas Amph. China: 216.

Nanorana liui Chen, Murphy, Lathrop, Ngo, Orlov, Ho and Somorjai, 2005, Herpetol. J., 15: 239, by implication; Frost et al., 2006, Bull. Am. Mus. Nat. Hist., 297: 367.

Nanorana yunnanensis Chen et al., 2005, Herpetol. J., 15: 239, by implication; Frost et al., 2006, Bull. Am. Mus. Nat. Hist., 297: 367.

Gynandropaa (*Gynandropaa*) *liui* Ohler and Dubois, 2006, Zoosystema, 28: 781.

Gynandropaa (*Gynandropaa*) *yunnanensis* Ohler and Dubois, 2006, Zoosystema, 28: 781.

Nanorana (*Chaparana*) *yunnanensis* Che et al., 2010, Proc. Natl. Acad. Sci. USA, Suppl. Inform., doi: 10.1073/pnas.1008415107 /-/DC Supplemental: 2.

Gynandropaa liui Fei, Ye and Jiang, 2010, Herpetol. Sinica, 12: 25.

Gynandropaa yunnanensis Fei, Ye and Jiang, 2010, Herpetol. Sinica, 12: 25.

Gynandropaa phrynoices Huang et al., 2016, Integrative Zool., 11: 144.

[同物异名] *Rana phrynoides*; *Rana yunnanensis*; *Rana muta*; *Rana*（*Paa*）*liui*; *Paa*（*Paa*）*muta*; *Paa*（*Gynandropaa*）*yunnanensis*; *Paa*（*Gynandropaa*）*liui*; *Nanorana liui*; *Nanorana yunnanensis*

[鉴别特征] 全蹼或满蹼；两眼间无大疣粒；体背面疣粒较小而稀疏；胸部有2刺团，略呈"Λ"形；背部长条形疣粒排列成行，上有少许小黑刺；腹面淡黄色，咽喉部有点状紫色斑纹。

[形态描述] 依据茂兰地区标本描述 体型健硕，雄蛙头体长65.65～84.33mm，雌蛙头体长87.33～100.37mm，吻端圆，吻棱不显；鼻孔近眼，鼻间距大于眼间距，略大于上眼睑宽；瞳孔菱形，鼓膜可见，但不甚清晰，约为眼径3/5，鼓膜光滑，上无疣粒，颞褶显著，自眼后经鼓膜上方至肩部；背侧褶无。前肢短，雄性成体前臂肌肉甚发达，前臂及手长小于体长之半，指关节下瘤大，较突出；内掌突大，外掌突窄长。雄性前2指有稀疏小黑刺；雄性成体胸部具1对刺团，两刺团中间边缘清晰，外侧边缘不规则，向外扩散。后肢肥壮，较短；左、右跟部不相遇；指端钝圆，手指端部膨大；指式为3＞4＞2＞1；趾式为4＞3＞5＞2＞1；趾间全蹼，趾关节下瘤明显，内蹠突长椭圆形，无外蹠突；头顶和头侧、前肢皮肤较为光滑，无大疣；体侧、后肢背面及蹠部有黑刺疣。背部部分疣粒较长，前后排列成与脊椎平行的长线；其余疣粒多为圆点状。背面多为棕褐色或黄棕色，前肢横纹不显，后肢横纹较为清晰。腹面淡黄色，咽喉部、四肢腹面有紫色斑纹。

[第二性征] 雄性前2指或3指有稀疏小黑刺；雄性成体胸部具1对刺团，略呈"Λ"形，两刺团中间边缘清晰，外侧边缘不规则，常向外扩散。无雄性线。

[生活习性] 该蛙生活于茂兰地区山区林间石块较多的溪流内。在5月和6月调查时发现该蛙抱对产卵。白天躲藏在水下石洞或泥洞中，能够待在水下很长时间，夜晚出来水边捕食。

[地理分布] 茂兰地区分布于洞塘等地。国内分布于云南、贵州。

[模式标本] 全模标本（BMNH 1947.2.3.76～82）保存于英国伦敦自然历史博物馆

[模式产地] 云南东川

[保护级别] 三有保护动物

[IUCN 濒危等级] 易危（VU）

[讨论] 双团棘胸蛙 *Gynandropaa phrynoides* 的分类地位一直在变动。1871年法国学者David根据江西九江的标本，将棘胸蛙 *Paa spinosa* 定名为 *Rana latrans*，开始了对棘蛙群的研究工作。后因 *latrans* 早为人所用，故于1975年改为棘胸蛙 *Paa spinosa*。1975年法国学者Dubois把棘

蛙群物种从广义蛙属 Rana 中分离出来，独立为一属——棘胸蛙属 Paa。而双团棘胸蛙是 Anderson 根据云南陇川县户撒的标本发表的新种，定名为 Paa yunnanensis (Anderson, 1879)，但是现在模式标本已经遗失。Boulenger 于 1917 年根据云南洱源县邓川的标本定名为 P. phrynoiddes，是 Paa yunnanensis 的同物异名。由于模式标本丢失，Dubois 指定新模将模式标本产地定为云南邓川。

国内早期的蛙科分类研究，见于 1961 年《中国无尾两栖类》一书。该书共记录中国蛙科动物 7 属 50 种，其中仅蛙属就有约 34 种，其中包括了双团棘胸蛙 Rana phrynoides。

20 世纪 80 年代以后，蛙科内部的系统分类问题受到越来越多的关注。费梁等（1990）基于对蛙类外部形态、骨骼构造及生活史的系统研究，确认棘蛙属有效。

江建平和周开亚（2001）基于线粒体 12S rRNA 片段对中国代表性蛙科物种的系统发育关系进行了分析，其结果支持费梁等（1990）的观点，认为把棘蛙属 Paa、大头蛙属 Limnonectes、虎纹蛙属 Hoplobatrachus、陆蛙属 Fejervarya、臭蛙属 Odorrana、粗皮蛙属 Rugosa 及侧褶蛙属 Pelophylax 等从原来的广义蛙属 Rana 中分离出来成为独立的属更为科学合理。

Roelants 等（2004）采用线粒体和核基因的联合数据对 55 种蛙科动物进行的分子系统学分析表明，蛙科（包括后来独立出去的树蛙科）可以分成 11 亚科。根据该文的分析，分布于我国的现生蛙科物种分属 3 亚科：①浮蛙亚科 Occidozyginae，包括浮蛙属 Occidozyga；②叉舌蛙亚科 Dicroglossinae，包括棘蛙属 Paa、大头蛙属 Limnonectes、虎纹蛙属 Hoplobatrachus、陆蛙属 Fejervarya 和倭蛙属 Nanorana；③蛙亚科 Raninae，包括湍蛙属 Amolops 和广义蛙属 Rana。其中，广义蛙属又被进一步分成侧褶蛙亚属 Pelophylax、水蛙亚属 Hylarana、粗皮蛙亚属 Rugosa 和蛙亚属 Rana 等，这与费梁等（1990）、江建平和周开亚（2001）对广义蛙属的分类观点一致。Roelants 等（2004）对中国蛙科动物的分类与江建平和周开亚（2001）等学者的研究结果非常接近。

Jiang 和 Zhou（2005）基于线粒体 12S rRNA 和 16S rRNA 分析了中国 29 种蛙科动物的系统进化关系，结果赞同把中国蛙科动物划分为两个亚科：蛙亚科 Raninae 和叉舌蛙亚科 Dicroglossinae。其中，蛙亚科包括林蛙属 Rana、湍蛙属 Amolops、侧褶蛙属 Pelophylax、臭蛙属 Odorrana、腺蛙属 Glandirana、水蛙属 Hylarana 和粗皮蛙属 Rugosa 等；叉舌蛙亚科包括棘蛙属 Paa、陆蛙属 Fejervarya、虎纹蛙属 Hoplobatrachus、大头蛙属 Limnonectes、倭蛙属 Nanorana 和高山蛙属 Altirana 等。该研究结果支持将棘蛙属 Paa 从广义蛙属 Rana 分离成为独立一属。

费梁等（2009b）根据近年来蛙科系统分类学研究成果，将中国蛙科动物进一步分成 4 亚科：蛙亚科 Raninae、叉舌蛙亚科 Dicroglossinae、湍蛙亚科 Amolopinae 和浮蛙亚科 Occidozyginae。其中，蛙亚科包括林蛙属 Rana、侧褶蛙属 Pelophylax、水蛙属 Hylarana、腺蛙属 Glandirana、趾沟蛙属 Pseudorana、粗皮蛙属 Rugosa 和臭蛙属 Odorrana 7 个属；叉舌蛙亚科包括陆蛙属 Fejervarya、棘蛙属 Paa、倭蛙属 Nanorana、虎纹蛙属 Hoplobatrachus、大头蛙属 Limnonectes、隆肛蛙属 Feirana、肛刺蛙属 Yerana、棘肛蛙属 Unculuana 和舌突蛙属 Liurana 9 个属；湍蛙亚科包括拟湍蛙属 Pseudoamolops 与湍蛙属 Amolops；浮蛙亚科包括浮蛙属 Occidozyga 和蟾舌蛙属 Phrynoglossus。

至此，双团棘胸蛙经历了由 Rana phrynoides 到 Paa phrynoiddes 再到 Paa yunnanensis 的地位变更。其后，费梁等（2012）将 Paa yunnanensis 纳入双团棘蛙属 Gynandropaa 并改名为 Gynandropaa yunnanensis。但是 Amphibia Web（2017）、Frost（2017）及 Amphibia China（2017）将双

成体量度 （单位：mm）

编号	头体长	头长	头宽	前臂及手长	后肢全长	胫长	跗足长
GZNU20160730101 ♂	65.65	24.34	25.16	34.39	121.83	35.88	68.87
GZNU20170808033 ♂	84.33	30.22	30.64	40.95	154.57	43.44	85.94
GZNU20170809001 ♂	82.27	33.69	34.21	37.60	141.88	40.34	77.01
GZNU20170808031 ♀	87.33	35.87	37.61	41.23	159.76	45.29	88.12
GZNU20170808032 ♀	100.37	33.92	34.12	48.29	190.02	54.91	81.94
GZNU20160726056 ♂	68.50	25.69	22.01	30.48	98.32	34.21	50.28

团棘胸蛙属物种收录在倭蛙属 Nanorana，即双团棘胸蛙 Nanorana phrynoides，原来的 Gynandropaa yunnanensis 和 Rana yunnanensis 为云南棘蛙 Nanorana yunnanensis 的同物异名。最近，Hu 等（2016）通过采集的贵州西部、云南和四川共计 37 个地区的样品，并整合形态、分子系统学等多种研究方法，系统解析了该属的系统发育关系、物种鉴别特征及地理分布，恢复了双团棘蛙属 Gynandropaa 的有效性，同时证明了云南棘蛙 Gynandropaa yunnanensis、双团棘胸蛙 Gynandropaa phrynoides 和四川棘蛙 Gynandropaa sichuanensis 的有效性，主要分类变动包括：①Gynandropaa yunnanensis 为有效种，中文名为"云南棘蛙"，而 Gynandropaa bourreti（布氏棘蛙）是 Gynandropaa yunnanensis 的亚种；②恢复双团棘胸蛙 Gynandropaa phrynoides 为有效种；③Gynandropaa liui（无声囊棘胸蛙）是 Gynandropaa sichuanensis 的次级同物异名。此外，由于该属模式种 Gynandropaa yunnanensis 的模式标本丢失，该研究重新指定了新模（CIBYN09060612）。而在其研究中仅仅采用了贵州威宁和水城两地区的标本，并且在 Hu 等（2016）的文中显示，双团棘胸蛙 Gynandropaa phrynoides 的分布地包含贵州西部和云南中东部地区，四川棘蛙的分布地包含四川西南部和云南中部地区，云南棘蛙的分布地包含云南西部和南部地区。

鉴于贵州分布的双团棘胸蛙的拉丁学名曾经被记为 Rana phrynoides、Paa yunnanensis 和 Gynandropaa yunnanensis，而 Hu 等（2016）并未收集贵州足够多的样品进行分析，所以暂将本次茂兰地区发现的双团棘胸蛙记为 Gynandropaa phrynoides。但贵州其他地区的双团棘胸蛙是否为双团棘胸蛙 Gynandropaa phrynoides，还有待进一步考证。

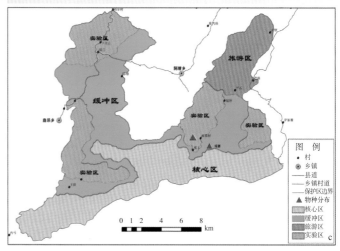

a 腹面照
b 背面照
c 在茂兰地区的分布

11) 棘胸蛙属 *Quasipaa* Dubois, 1992

Quasipaa Dubois, 1992, Bull. Mens. Soc. Linn. Lyon, 61: 319. Type species: *Rana boulengeri* Günther, 1889, by original designation. Proposed as a subgenus of *Paa*.

Eripaa Dubois, 1992, Bull. Mens. Soc. Linn. Lyon, 61: 319. Type species: *Rana fasciculispina* Inger, 1970, by original designation. Proposed as a subgenus of *Paa*. Synonymy by Ohler and Dubois, 2006, Zoosystema, 28: 781.

Annandia Dubois, 1992, Bull. Mens. Soc. Linn. Lyon, 61: 317-318. Type species: *Rana delacouri* Angel, 1928, by original designation. Proposed originally as a subgenus of *Chaparana*. Synonymy with *Nanorana* by Che et al., 2009, Mol. Phylogenet. Evol., 50: 69. Synonymy with *Quasipaa* by Che et al., 2009, Mol. Phylogenet. Evol., 50: 69.

Annandia Dubois, 2005, Alytes, 23: 16. Consideration as a genus.

Yerana Jiang, Chen and Wang, 2006, J. Anhui Normal Univ., Nat. Sci., 29: 468. Type species: *Paa* (*Feirana*) *yei* Chen, Qu and Jiang, 2002, by original designation. Synonymy by implication of placement of *Nanorana yei* in tree of Che et al., 2009, Mol. Phylogenet. Evol., 50: 66.

鼻骨大，两内缘相接，并与额顶骨相接触（多数种）；蝶筛骨在背面少部分显露或不显露；额顶骨前、后几乎等宽；前耳骨大；鳞骨颧枝刀状；舌骨体宽几乎为长的 2 倍；舌角前突粗短，向外弯曲，几乎呈环状。肩胸骨基部不分叉，上胸软骨略小于剑胸软骨；中胸骨粗短，剑胸软骨宽大，呈圆盘状，后端有浅缺刻。雄蛙前肢肱粗壮，肱骨具脊棱，其长约为最宽处的 4 倍；内侧第一、第二指粗大。

体型肥硕，一般无背侧褶，仅个别种有；体背面和体侧皮肤粗糙，有长肤棱和疣粒。指、趾末端呈球状，无沟；无指基下瘤；内掌突大而突出；趾间全蹼或满蹼，外侧蹠间蹼较弱；有或无跗褶。鼓膜隐蔽或不显。雄蛙前臂甚粗壮，内侧 2 指或 3 指及内掌突上有粗大婚刺，胸部或胸腹部或前臂内侧有锥状黑刺，多数种有声囊。雄蛙肱骨远端外侧 2 脊棱极为隆起；第一掌、指骨及前拇指骨粗壮。

成体和蝌蚪生活于林间山溪；卵群成串悬于溪内石块下或树根间（如棘腹蛙等）。

本属世界现有 11 种，广泛分布于中国（南部）、越南（中部和西南部）、泰国（东南部）、柬埔寨（西南部）等地。中国有 7 种，贵州有 3 种，茂兰地区有 3 种，即棘腹蛙 *Quasipaa boulengeri*（Günther，1889）、棘侧蛙 *Quasipaa shini*（Ahl，1930）、棘胸蛙 *Quasipaa spinosa*（David，1875），分种检索表如下。

茂兰地区种检索表

1 { 雄蛙胸部、腹部均有刺 ······ 棘腹蛙 *Quasipaa boulengeri*
   { 雄蛙仅胸部有刺 ······ 2
2 { 体侧刺疣甚多，体背面皮肤粗糙；雄蛙胸部刺疣大而稀疏并延伸至腹前消失，大者胸疣上一般有刺 3～8 枚 ······ 棘侧蛙 *Quasipaa shini*
   { 体侧无刺疣，体背面皮肤不太粗糙；胸疣上只有 1 枚小黑刺 ······ 棘胸蛙 *Quasipaa spinosa*

（16）棘腹蛙 *Quasipaa boulengeri* (Günther, 1889)

Rana Boulengeri Günther, 1889, Ann. Mag. Nat. Hist., Ser. 6, 4: 222. Syntypes: BM (2 specimens); BMNH 1947.2.3.86 designated lectotype by Dubois, 1987 "1986", Alytes, 5: 44. Type locality: Ichang [= Yichang], Hubei, China.

Rana tibetana Boulenger, 1917, Ann. Mag. Nat. Hist., Ser. 8, 20: 414. Holotype: BMNH 1947.2.3.63. Type locality: Yin tsin wau [= Yintsinwan], Wassu State [= Yin Xiu Wan], Tibet, now in part of Sichuan, China. Synonymy by Liu and Hu, 1961, Tailless Amph. China: 153; Dubois, 1987 "1986", Alytes, 5: 44.

Rana (*Rana*) *tibetana* Boulenger, 1920, Rec. Indian Mus., 20: 8.

Rana (*Paa*) *boulengeri* Dubois, 1975, Bull. Mus. Natl. Hist. Nat., Paris, Ser. 3, Zool., 324: 1098; Dubois, 1987 "1986", Alytes, 5: 43.

Rana boulengeri Inger et al., 1990, Fieldiana, Zool., N.S. 58: 9.

Paa (*Paa*) *boulengeri* Fei, Ye and Huang, 1990, Key to Chinese Amph.: 156; Ye, Fei and Hu, 1993, Rare and Economic Amph. China: 273; Fei and Ye, 2001, Color Handbook Amph. Sichuan: 182.

Paa (*Quasipaa*) *boulengeri* Dubois, 1992, Bull. Mens. Soc. Linn. Lyon, 61: 320.

Rana robertingeri Wu and Zhao, 1995, in Zhao (ed.), Amph. Zoogeograph. Div. China: 52, 54. Holotype: CIB 6885, by original designation. Type locality: Tiantangba Village, Hechuan Co., Sichuan Province, China, 900 meters. Synonymy by Che, Hu, Zhou, Murphy, Papenfuss, Chen, Rao, Li and Zhang, 2009, Mol. Phylogenet. Evol., 50: 71; Zhang, Zhang, Yu, Storey and Zheng, 2018, BMC Evol. Biol., 18 (26): 12.

Paa (*Paa*) *robertingeri* Fei and Ye, 2001, Color Handbook Amph. Sichuan: 184.

Nanorana robertingeri Chen et al., 2005, Herpetol. J., 15: 239, by implication; Frost et al., 2006, Bull. Am. Mus. Nat. Hist., 297: 367.

Nanorana boulengeri Chen et al., 2005, Herpetol. J., 15: 239, by implication.

Quasipaa robertingeri Jiang et al., 2005, Zool. Sci., Tokyo, 22: 358; Ohler and Dubois, 2006, Zoosystema, 28: 781.

Quasipaa boulengeri Jiang et al., 2005, Zool. Sci., Tokyo, 22: 358.

Quasipaa tibetana Ohler and Dubois, 2006, Zoosystema, 28: 781.

Quasipaa (*Quasipaa*) *boulengeri* Che et al., 2010, Proc. Natl. Acad. Sci. USA, Suppl. Inform., 107 (31): 13 765-13 770.

[同物异名] *Rana boulengeri*；*Rana tibetana*；*Rana robertingeri*（合江棘蛙）；*Nanorana robertinger*；*Nanorana boulengeri*；*Quasipaa robertingeri*；*Quasipaa tibetana*

[俗名] 石棒、石坎

[鉴别特征] 体肥壮；雄蛙胸、腹部满布大小黑刺疣。

[形态描述] 依据茂兰地区标本描述 体型肥硕，雄蛙头体长66.76～100.61mm，雌蛙头体长82.07～106.01mm。头宽大于头长；吻端圆，略突出于下唇，吻棱略显；鼻孔位于吻眼之间，眼间距与鼻间距几乎等宽；鼓膜略显；犁骨齿短，呈"V"形，自内鼻孔内侧向中线倾斜，齿列后端间距窄；舌椭圆形，后端缺刻深。

前肢短，前臂及手长不到体长之半；雄蛙前臂极粗壮；指略扁，指端圆球状；指式为3＞1≈4＞2；第二指两侧及第三指内侧具缘膜；原拇指发达，关节下瘤甚明显；内掌突大，卵圆形，外掌突窄长。后肢肥壮，前伸贴体时胫跗关节达眼部，左、右跟部仅相遇；胫长超过体长之半；趾端圆球状；第一、第五趾游离侧缘膜发达，达蹠基部；趾间几乎全蹼，第四、第五蹠基部超过蹠长之半；关节下瘤明显；内蹠突窄长，无外蹠突；跗褶清晰，超过跗足长之半。

体背面皮肤粗糙，背部有纵列的长形小刺疣；眼后有1横肤沟；四肢背面疣少，长形肤棱明显；背面为深褐色，两眼间有1黑横纹，背部具不规则的黑斑；四肢背面有黑横纹；体和四肢腹面肉色，咽喉部有棕色斑，下颌缘更明显。

[第二性征] 雄蛙胸、腹部满布大小刺疣；前臂极粗壮，内侧3指有黑色锥状刺；有单咽下内声囊，声囊孔大，长裂状；背面有两条紫色雄性线。

[生活习性] 该蛙生活于山区的溪流或其附近的水塘中。白天隐匿于溪底的石块下、溪边大石缝或瀑布下的石洞内；晚间外出，蹲于石块上或伏于水边捕食周边昆虫。卵多产于溪流瀑布下水坑内，卵群成串，似葡萄状，一般黏附在水中石块下、倒木或枯枝上。

[地理分布] 茂兰地区分布于翁昂、洞塘等地。国内分布于陕西、山西、甘肃、四川、重庆、云南、贵州、湖北、江西、湖南、广西等地。国外无记录。

[模式标本] 全模标本（BM 2号），选模标本（BMNH 1947.2.3.86）保存于英国伦敦自然历史博物馆

[模式产地] 湖北宜昌

[保护级别] 三有保护动物

[IUCN 濒危等级] 濒危（EN）

[讨论] 棘腹蛙的分类地位目前已明确，但有争议的是合江棘蛙是否为棘腹蛙的同物异名。费梁等（2009b，2010，2012）基于形态数据、染色体核型认为合江棘蛙为有效种。而 Che 等（2009）基于线粒体基因（12S rRNA、16S rRNA）和核基因片段（Rhodopsin、Tyrosinase）的分子系统学研究显示，"合江棘蛙"位于棘腹蛙支系的基部位置，很可能不是一个有效种。Yan 等（2013）对棘腹蛙进行

了种内系统地理学研究,结果显示,"合江棘蛙"与其他地区的棘腹蛙群体间共享核基因等位基因(GCC),具有广泛的基因流。基于以上研究,合江棘蛙不是一个有效种,建议作为棘腹蛙的同物异名处理。本书采用Frost(2017)和Amphibia Web(2017)的结果将合江棘蛙作为棘腹蛙的同物异名。

成体量度 (单位:mm)

编号	头体长	头长	头宽	前臂及手长	后肢全长	胫长	跗足长
GZNU20170512002 ♂	70.13	26.82	27.26	36.42	123.68	38.48	75.62
GZNU20170512001 ♂	79.42	29.48	28.53	38.72	141.47	42.26	81.65
GZNU20170417020 ♂	66.76	24.43	24.34	33.26	119.43	35.38	68.75
GZNU20170512004 ♂	100.61	36.21	36.95	54.10	185.96	51.81	104.01
GZNU20170417021 ♀	82.07	29.90	30.63	41.87	148.44	43.37	85.64
GZNU20170512003 ♀	83.95	29.23	28.66	39.42	144.90	41.01	81.49
GZNU20150404002 ♀	105.01	34.79	36.84	51.18	190.18	54.28	111.45
GZNU20150404002 ♀	106.01	35.79	37.84	51.19	187.89	55.28	111.45
GZNU20140404001 ♀	100.72	22.08	27.45	51.68	159.49	60.85	75.22

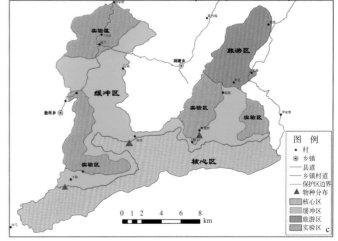

a 背面照
b 腹面照
c 在茂兰地区的分布

(17) 棘侧蛙 *Quasipaa shini* (Ahl, 1930)

Rana shini Ahl, 1930, Sitzungsber. Ges. Naturforsch. Freunde Berlin, 1930: 315. Syntypes: ZMB (originally 4 specimens), unnumbered according to the original publication; MCZ 17651 (on exchange from ZMB, is a syntype according to Barbour and Loveridge, 1946, Bull. Mus. Comp. Zool., 96: 184). Type locality: Yao-schan [= Mt. Dayao], Nordteil der Provinz Kwangsi [= Guangxi], China, 1500m.

Rana (*Paa*) *shini* Dubois, 1987 "1986", Alytes, 5: 43.

Paa (*Paa*) *shini* Fei, Ye and Huang, 1990, Key to Chinese Amph.: 156; Ye, Fei and Hu, 1993, Rare and Economic Amph. China: 281.

Paa (*Quasipaa*) *shini* Dubois, 1992, Bull. Mens. Soc. Linn. Lyon, 61: 320.

Rana shini Zhao and Adler, 1993, Herpetol. China: 149.

Nanorana shini Chen et al., 2005, Herpetol. J., 15: 239.

Quasipaa shini Frost, 2006, Amph. Spec. World, Vers. 4.0: 358; Frost et al., 2006, Bull. Am. Mus. Nat. Hist.: 297.

Quasipaa (*Quasipaa*) *shini* Che et al., 2010, Proc. Natl. Acad. Sci. USA, Suppl. Inform., 107 (31): 13 765-13 770.

[同物异名] *Rana shini*; *Nanorana shini*; *Paa shini*

[鉴别特征] 该种与棘胸蛙 *Quasipaa spinosa* 很相似，但棘侧蛙背面皮肤极粗糙，背部满布长形疣，体侧疣刺很多。

[形态描述] 依据茂兰地区标本描述 雄性头体长 60.65～78.97mm；吻端钝圆，稍突出于下颌，吻棱不显；颊部略向外侧倾斜；鼻孔位于吻眼之间，距眼较近；鼓膜略显；犁骨齿自内鼻孔前缘向中线斜行，呈"V"形，后端齿列间距窄，不到齿列的1/2；舌椭圆形，后端缺刻深；前肢短，前臂及手长不到体长之半，雄蛙前肢极粗壮，指端球状；第二指短于第一指，指式为 3 > 4 ≈ 1 > 2，趾式为 4 > 3 > 5 > 2 > 1，原拇指发达；第二、第三指微具缘膜；指近端关节下瘤大而圆，远端者小；内掌突卵圆形，外掌突窄长。后肢长，前伸贴体时胫跗关节达眼前角，左、右跟部相遇；胫长超过体长之半；趾端圆球形，趾间全蹼；外侧蹠间蹼达蹠基部；第一、第五趾外侧缘膜宽，达蹠基部；第五趾略短于第三趾，达第四趾的第三关节下瘤下方；关节下瘤明显；背部及体侧皮肤很粗糙。背部长短疣排列成纵列，其间散有圆形小刺；自口角经体侧至胯部有密集的大小不等的圆疣，每个疣上有一颗小黑刺；头背面、后枕部及四肢背面皮肤较光滑；上颌缘及股部小刺疣疏少；颞褶粗厚，枕后横肤沟明显。雄蛙腹面胸部有大小不等的肉质疣，大疣分布由胸中线向两外侧至臂基部略呈两个三角区，大疣前后肉质疣渐次变小，肉质疣前端不超过臂基的水平线，向后延至腹前部并变小至逐渐消失；小疣上一般有小黑刺1枚，大疣上刺可多达6枚。雌蛙腹面皮肤光滑。

生活时背面深棕黑色，两眼间有黑色宽的横纹；上、下唇有浅色纵纹；体侧自眼后至胯部有浅黄色宽纵纹；四肢背面隐约可见浅色横纹。腹面灰白色，咽喉部及股、胫部浅棕色，下颌缘深棕色；雌蛙胸部色深，对应雄蛙刺疣生长部位有分散的小白点。

[第二性征] 雄蛙胸部及前腹部有大小肉质刺疣，胸疣上一般着生 3～8 枚黑刺；前臂极粗壮；内侧 3 指及原拇指均有黑色锥状黑刺；有 1 对咽下内声囊；有雄性线。雌蛙前臂相对较细。

[生活习性] 该蛙生活于山间溪流内，所在环境潮湿，植被繁茂，溪水清澈。成蛙白天隐藏在溪边石下，或在岸上大石上，受惊扰后跳入深水处隐伏于水底石下。夜晚出没于溪边石上，在手电筒光照射下蹲在原地不动。

[地理分布] 茂兰地区分布于翁昂。国内见于贵州、湖南、广西。

[模式标本] 全模标本[4号未编号（MCZ 17651 来自 ZMB 的交换）]保存于德国柏林洪堡大学自然历史博物馆

[模式产地] 广西金秀大瑶山，1500m

[保护级别] 三有保护动物

[IUCN 濒危等级] 易危（VU）

成体量度 （单位：mm）

编号	头体长	头长	头宽	前臂及手长	后肢全长	胫长	跗足长
GZNU20160920003 ♂	67.35	30.73	34.64	41.86	143.30	41.30	61.01
GZNU20160918001 ♂	78.97	25.58	22.80	29.25	112.70	36.78	46.27
GZNU20160920002 ♂	60.65	22.65	21.21	26.65	112.20	35.29	47.29

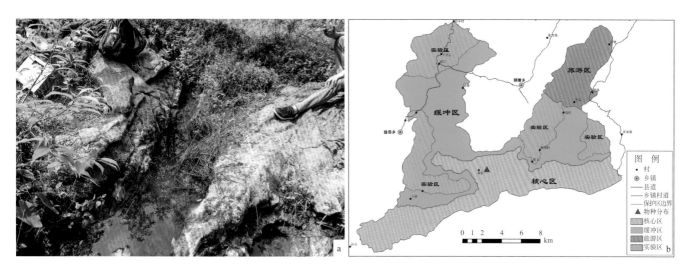

a 生境
b 在茂兰地区的分布

（18）棘胸蛙 *Quasipaa spinosa* (David, 1875)

Rana latrans David, 1872 "1871", Nouv. Arch. Mus. Natl. Hist. Nat., Paris, 7: 76. Types: not stated; presumably in MNHNP but not reported in type lists. Type locality: torrent des montagnes, Jiangxi, China. Junior homonym of *Rana latrans* Steffen, 1815. Synonymy by David, 1875, J. Trois. Voy. Explor. Emp. Chinoise, 2: 253. Considered by Thurston, 1888, Cat. Batr. Sal. Apoda S. India: 23, to be a synonym of *Rana tigerina*.

Rana spinosa David, 1875, J. Trois. Voy. Explor. Emp. Chinoise, 1: 253. Type(s): not stated, presumably deposited originally in MNHNP. CIB 641280 designated neotype by Fei and Ye in Fei, Hu, Ye and Huang, 2009, Fauna Sinica, Amph., 3: 1375. Type locality: Ouang-mao-tsae [= Wangmaozhai], a mountain village in Jiangxi near the Fujian boundary, China. Neotype from Guadun, Wuyishan, Fujian., China; 1100m.

Nyctibatrachus sinensis Peters, 1882, Sitzungsber. Ges. Naturforsch. Freunde Berlin, 1882: 146. Holotype: ZMB 10373 (according to Bauer, Günther and Klipfel, 1995, in Bauer et al. (eds.), Herpetol. Contr. W.C.H. Peters: 49). Type locality: Lofau-Gebirge [= Mt. Luofu], Provinz Canton [actually Guangzhou], China. Synonymy by Liu and Hu, 1961, Tailless Amph. China: 156. See also synonymy of *Limnonectes fujianensis*.

Rana (*Rana*) *spinosa* Boulenger, 1920, Rec. Indian Mus., 20: 8; Guibé, 1950 "1948", Cat. Types Amph. Mus. Natl. Hist. Nat.: 35.

Rana duboisreymondi Vogt, 1921, Sitzungsber. Ges. Naturforsch. Freunde Berlin, 1921: 75. Types: Presumably ZMB (2 specimens). Type locality: chinesischen Bade Kuling auf dem rechten Yangtzeufer bei der Stadt Kin Kinag = Kuling [= Guling], right bank of Yangtze Rvier, Kin Kiang [= Jin Jiang], Jiangxi], China. Synonymy by Pope, 1931, Bull. Am. Mus. Nat. Hist., 61: 502; Bourret,

1942, Batr. Indochine: 287; Liu and Hu, 1961, Tailless Amph. China: 156.

Rana spinosa spinosa Bourret, 1937, Annexe Bull. Gen. Instr. Publique, Hanoi, 1937: 26.

Hylorana dubois-reymondi Deckert, 1938, Sitzungsber. Ges. Naturforsch. Freunde Berlin, 1938: 144.

Rana chekiensis Angel and Guibé, in Angel, Bertin and Guibé, 1947 "1946", Bull. Mus. Natl. Hist. Nat., Paris, Ser. 2, 18: 473. Syntypes: MNHNP 1923.16 and 1923.22, by original designation. Type locality: not given; given as Changaï [= Shanghai], China, by Guibé, 1950 "1948", Cat. Types Amph. Mus. Natl. Hist. Nat.: 35. Synonymy by Zhao and Adler, 1993, Herpetol. China: 149.

Rana (Rana) chekiensis Guibé, 1950 "1948", Cat. Types Amph. Mus. Natl. Hist. Nat.: 35.

Rana (Paa) spinosa Dubois, 1975, Bull. Mus. Natl. Hist. Nat., Paris, Ser. 3, Zool., 324: 1098; Dubois, 1976, Cah. Nepal., Doc., 6: 24; Dubois, 1987 "1986", Alytes, 5: 43.

Paa (Paa) spinosa Fei, Ye and Huang, 1990, Key to Chinese Amph.: 157; Ye, Fei and Hu, 1993, Rare and Economic Amph. China: 284; Fei, 1999, Atlas Amph. China: 208.

Paa (Quasipaa) spinosa Dubois, 1992, Bull. Mens. Soc. Linn. Lyon, 61: 320.

Nanorana spinosa Chen et al., 2005, Herpetol. J., 15: 239.

Quasipaa spinosa Frost, 2006, Amph. Spec. World, Vers. 4.0: 358.

[同物异名] *Rana latrans*; *Rana spinose*; *Nyctibatrachus sinensis*; *Rana duboisreymondi*; *Hylorana dubois-reymondi*; *Rana chekiensis*; *Nanorana spinosa*

[俗名] 石鸡、棘蛙、石鳞、石蛙、石蛤

[鉴别特征] 外形与棘侧蛙 *Quasipaa shini* 相似，但本种胸部每个肉质疣上仅 1 枚小黑刺；体侧无刺疣，皮肤不太粗糙。

[形态描述] 体肥大，头长小于头宽，吻端钝圆，吻棱、鼓膜均不明显；背部皮肤不太粗糙，长短不一疣粒排列成行，背面、前臂及胫股背部密布小细疣，疣上有黑刺，颞褶明显，无背侧褶；雄蛙胸部布满小肉疣，疣上有小黑刺。雌蛙腹面光滑。前臂及手长约为体长之半，指、趾末端球状；后肢贴体前伸时胫跗关节达眼部，趾间全蹼；外侧蹠间蹼达蹠长中部，第五趾外侧缘膜达蹠基部。体背棕黑色，两眼间有深色横纹，上、下唇缘有浅纵纹，四肢有褐色横纹，体腹面浅黄色无斑。

[生活习性] 该蛙生活于植被繁茂，溪水清澈的山间溪流内。成蛙白天隐藏在溪边石下或岸上草丛中，受惊扰后跳入深水处隐伏于水底石下。夜间出没于溪边石上捕食昆虫。

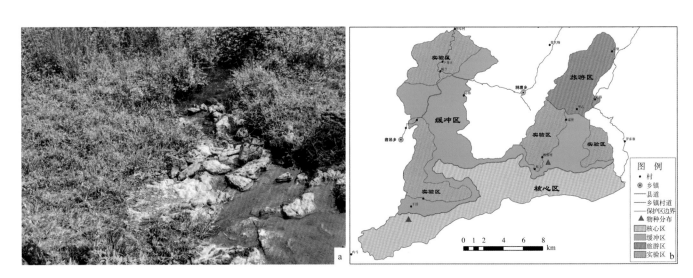

a 生境
b 在茂兰地区的分布

[地理分布]　茂兰地区分布于翁昂、洞塘等地。据《贵州两栖类志》记载的棘胸蛙在贵州分布，本次在茂兰地区的发现属于该种在贵州的新分布区。该种在国内分布于安徽、湖北、福建、广东、广西、江苏、浙江、云南、贵州、江西、湖南、香港。

[模式标本]　模式标本（CIB 641280）保存于中国科学院成都生物研究所

[模式产地]　江西，接近福建边界王茂斋村

[保护级别]　三有保护动物

[IUCN 濒危等级]　易危（VU）

7. 蛙科 Ranidae Batsch, 1796

Ranae Laurenti, 1768, Spec. Med. Exhib. Synops. Rept.: 20. Unavailable plural of *Rana* Linnaeus, 1758, not apparently intended as a taxon name.

Ranina Batsch, 1796, Umriss der gesammten Naturgeschichte: 179. Type genus: *Rana* Linnaeus, 1758. See discussion by Dubois and Bour, 2011, Alytes, 27: 154-160.

Ranaridia Rafinesque, 1814, Specchio Sci., 2: 102. Type genus: *Ranaridia* Rafinesque, 1814 (= unjustified emendation of *Rana* Linnaeus, 1758).

Ranarinia Rafinesque, 1815, Analyse Nat.: 78.

Ranae Goldfuss, 1820, Handb. Zool., 2: 131.

Ranadae Gray, 1825, Ann. Philos., London, Ser. 2, 10: 213.

Ranoidea Fitzinger, 1826, Neue Class. Rept.: 37, explicit family; Laurent, 1967, Acta Zool. Lilloana, 22: 208; Lynch, 1973, in Vial (ed.), Evol. Biol. Anurans: 162; Duellman, 1975, Occas. Pap. Mus. Nat. Hist. Univ. Kansas, 42: 5.

Ranidae Boie, 1828, Isis von Oken, 21: 363.

Raniformes Duméril and Bibron, 1841, Erp. Gen., 8: plate opposite page 53 and page 491. Explicit non-Latinized family-group name.

Limnodytae Fitzinger, 1843, Syst. Rept.: 31. Type genus: *Limnodytes* Duméril and Bibron, 1841.

Ranoides Bruch, 1862, Würzb. Naturwiss. Z., 3: 221.

Ranida Haeckel, 1866, Gen. Morphol. Organ., 2: 132.

Ranidi Acloque, 1900, Fauna de France, 1: 489.

Ranoidae Dubois, 1992, Bull. Mens. Soc. Linn. Lyon, 61: 309.

肩带固胸型，个别属为弧固型。肩胸骨和正胸骨发达，成为骨质柱；肩胛骨长（长小于锁骨的2倍），前端不与锁骨重叠。椎体前凹型或参差型（第八椎体双凹，而荐椎双凸），荐椎前椎骨8枚；无肋骨；后面的椎骨横突延长；大部分属髓弓不呈覆瓦状排列，少数属呈覆瓦状排列；荐椎横突圆柱状，关节髁2枚，与尾杆骨相关节；尾杆骨无横突。少有头骨骨片缺失的情况，有腭骨；大多数类群上颌有齿，有耳柱骨者（倭蛙属 *Nanorana* 物种）有不同程度的退化；通常有方轭骨。无副舌骨；环状软骨环完全。跟骨、距骨仅两端并合；远列小跗骨2或3枚；指、趾末两节间无介间软骨；指、趾端部尖或圆，贵州分布的除侧褶蛙属外均呈吸盘状；瞳孔多横置，配对时抱握于腋部。成体有多种多样的栖居习性和繁殖习性。

大多数种类在静水中产卵，卵小，具色素；部分种类在溪流内产卵。蝌蚪肥硕，尾肌弱。茂兰地区有4属，分属检索表如下。

茂兰地区属检索表

1. 趾末端不呈吸盘状，钝尖或尖，腹侧无沟 ·· 侧褶蛙属 *Pelophylax*
   趾末端呈吸盘状，膨大或不明显膨大，腹侧有沟 ··· 2
2. 背侧褶弱或无，若有，亦极细或者不明显；肩胸骨不呈叉状 ··· 臭蛙属 *Odorrana*
   背侧褶甚宽或较宽，一般有内外2趾褶；肩胸骨浅度分叉 ·· 3
3. 雄蛙肱部前方一般有发达的腺体 ··· 水蛙属 *Hylarana*
   雄蛙肩上方有大而扁平的腺体 ··· 琴蛙属 *Nidirana*

12）琴蛙属 *Nidirana* Dubois, 1992

Babina Thompson, 1912, Herpetol. Notices, 1: 1. Type species: *Rana holsti* Boulenger, 1892, by original designation. Considered synonymous with *Hylarana* by Dubois, 1981, Monit. Zool. Ital., N.S., Suppl., 15: 225-284, but considered a distinct genus by Okada, 1966, Fauna Japon., Anura: 138-143; considered a subgenus of *Rana* by Nakamura and Ueno, 1963, Japan. Rept. Amph. Color: 54. Equivalent to the *Rana holstii* group of Boulenger, 1920, Rec. Indian Mus., 20: 1-226. Resurrected as a subgenus by Dubois, 1992, Bull. Mens. Soc. Linn. Lyon, 61: 523. Equivalent to the *Rana* (*Hylorana*) *holsti* group of Boulenger, 1920, Rec. Indian Mus., 20: 129-130. Barbour, 1917, Occas. Pap. Mus. Zool. Univ. Michigan, 44: 6 (footnote) says that he has no evidence to doubt the printed dates of publication but he did note that he received all three of Thomson's papers at the same time and implied on the basis of no clear evidentiary basis that they may be erroneous (DRF).

Babina van Denburgh, 1912, Adv. Diagn. New Rept. Amph. Loo Choo Is. Formosa: 3. Type species: *Rana holsti* Boulenger, 1892, by original designation. Objective synonym and preoccupied by *Babina* van Denburgh, 1912, according to Barbour, 1917, Occas. Pap. Mus. Zool. Univ. Michigan, 44: 1-9. Synonymy with *Hylarana* by Boulenger, 1917, C.R. Hebd. Séances Acad. Sci., Paris, 165: 989; Boulenger, 1918, Ann. Mag. Nat. Hist., Ser. 9, 1: 238.

Nidirana Dubois, 1992, Bull. Mens. Soc. Linn. Lyon, 61: 324. Type species: *Rana psaltes* Kuramoto, 1985, by original designation. Originally proposed as a subgenus of *Rana*. Recognition as a genus by Chen et al., 2005, Herpetol. J., 15: 237. Synonymy with *Babina* by Frost et al., 2006, Bull. Am. Mus. Nat. Hist., 297: 248.

Dianrana Fei, Ye and Jiang, 2010, Herpetol. Sinica, 12: 21. Type species: *Rana pleuraden* Boulenger, 1904.

鼻骨大，内缘长，左、右鼻骨相接触，与额顶骨相接触或略分离；肩胸骨基部浅度分叉；舌骨前突长，向外弯曲。指端有腹侧沟或不显或无；趾间近全蹼或半蹼。体型不窄长；吻钝圆；四肢适中；背侧褶明显；有内跖褶。雄蛙肩上方有大而扁平的腺体，无肱腺，有声囊，第一指基部有婚垫，个别种无声囊和无婚垫（如竖琴蛙 *Nidirana psaltes*）。

蝌蚪体背侧或腹面无腺体；口部位于吻部腹面，下唇乳突2排，两排间距窄，外排长，呈须状；上唇齿2排，下唇齿3排；出水孔位于体左侧，无游离管；肛孔斜开口于尾基部右侧。

每个卵囊内含卵1～30粒，卵直径为1.4～2.0mm，动物极具色素。

本属世界已知8种，分布于越南、琉球群岛和中国的热带和亚热带地区。中国已知7种，主要分布于长江以南各省区。茂兰地区分布有1种，即滇蛙 *Nidirana pleuraden*。

（19）滇蛙 *Nidirana pleuraden* (Boulenger, 1904)

Rana pleuraden Boulenger, 1904, Ann. Mag. Nat. Hist., Ser. 7, 13: 131. Syntypes: BMNH. Type locality: Yunnan Fu (altitude about 6000 feet), Yunnan, China.

Rana (*Rana*) *pleuraden* Boulenger, 1920, Rec. Indian Mus., 20: 9.

Pelophylax pleuraden Fei, Ye and Huang, 1990, Key to Chinese Amph.: 133-134; Ye, Fei and Hu, 1993, Rare and Economic Amph. China: 228.

Rana (*Nidirana*) *pleuraden* Dubois, 1992, Bull. Mens. Soc. Linn. Lyon, 61: 324.

Nidirana pleuraden Chen et al., 2005, Herpetol. J., 15: 237; Lyu et al., 2017, Amphibia Reptilia, 38: 494.

Babina pleuraden Frost et al., 2006, Bull. Am. Mus. Nat. Hist., 297: 368.

Dianrana pleuraden Fei, Ye and Jiang, 2010, Herpetol. Sinica, 12: 35. Not mentioned by the most recent revisors of the group, Lyu et al., 2017, Amphibia Reptilia, 38: 483-502.

[同物异名] *Rana pleuraden*；*Pelophylax pleuraden*；*Dianrana pleuraden*

[鉴别特征] 外形与黑斑侧褶蛙 *Pelophylax nigromaculatus* 相近，但本种背侧褶间疣粒明显，无长形肤棱，外蹠突小，内蹠突大而无游离刃；雄蛙体背肩上方有扁平肩腺，有1对咽侧下外声囊。

[形态描述] 头体长55～66mm，头略平扁，头长、宽几乎相等；吻端钝尖；吻棱不显，颊部稍向外侧倾斜；鼻孔位于吻眼之间或略近吻端，眼间距窄，几乎与上眼睑等宽；鼓膜明显，约与眼间距等宽；犁骨齿两小团在鼻孔内侧。

前肢较为粗短；指端钝圆，指细长，指式为3≈1＞4≈2；关节下瘤明显；掌突3个。后肢较长，前伸时胫跗关节略超过眼，左、右跟部相重叠，胫长略大于体长之半，足长于胫；趾端钝尖，趾细长，趾式为4＞5≈3＞2＞1；蹼明显，但不达趾端。头部皮肤较光滑；背部及体侧有较明显的疣粒；背侧褶较窄而清晰，自眼后角起直达胯部；背部及体侧有较明显的疣粒；腹面皮肤一般光滑，生活时背面为橄榄绿色略带黄色，背面的斑纹变异大，在疣粒上有分散的小黑斑点，背正中有略宽的浅色脊纹，脊纹两侧的斑点连成明显的黑纹。上唇缘、颞褶、背侧褶有浅黄色线纹；头侧及浅色背侧褶下方棕褐色，体侧深色斑点较多，后肢横纹清晰，腹后缘斑纹不规则。

[第二性征] 雄蛙前肢较粗壮，第一指有灰色婚垫；肩上方具扁平大腺体；有1对咽侧下外声囊，声囊孔长裂形；仅体背侧有雄性线。

[生活习性] 该蛙生活于山区低洼地的水塘、水沟、稻田内。成蛙以多种昆虫及其他小动物为食。捕食不分昼夜，多隐于稻田中或田埂的杂草丛中，受惊扰后跳到水田内躲避。

[地理分布] 茂兰地区分布于翁昂、永康等地。据《贵州两栖类志》记载的滇蛙在贵州的分布，茂兰地区为贵州新纪录。国内分布于四川、云南、贵州。

[模式标本] 模式标本保存于英国伦敦自然历史博物馆

[模式产地] 云南昆明

[保护级别] 无

[IUCN 濒危等级] 无危（LC）

[讨论] 滇蛙的属级分类存在争议。费梁等（2009b）将其归为侧褶蛙属 *Pelophylax*，并命名为滇侧褶蛙 *Pelophylax pleuraden*；之后费梁等（2012）将滇蛙在内的其他5个水蛙属 *Hylarana* 物种分别划分为滇蛙属 *Dianrana* Fei, Ye and Jiang, 2010（滇蛙）和琴蛙属 *Nidirana* Dubois, 1992［弹琴蛙、仙琴蛙、海南琴蛙、林琴蛙、竖琴蛙（现为琉球琴蛙的次级同物异名）］。Frost（2017）将滇蛙归为琴蛙属 *Nidirana*。该属目前有8个物种，其中中国分布7种（弹琴蛙 *Nidirana adenopleura*、仙琴蛙 *Nidirana daunchina*、海南琴蛙 *Nidirana hainanensis*、林琴蛙 *Nidirana lini*、琉球琴蛙 *Nidirana okinavana*、滇蛙 *Nidirana pleuraden*、南昆山琴蛙 *Nidirana nankunensis*）。Lyu等（2017）依据分子系统学、形态及求偶鸣叫的声学证据，从拇棘蛙属 *Babina* 中恢复了琴蛙属 *Nidirana* 的有效性，同时将腹斑蛙 *Nidirana caldwelli* 作为弹琴蛙 *Nidirana adenopleura* 的同物

异名。依据其研究结果，琴蛙属目前包括8种，即琉球琴蛙 *Nidirana okinawana*、弹琴蛙 *Nidirana adenopleura*、海南琴蛙 *Nidirana hainanensis*、沙巴琴蛙 *Nidirana chapaensis*、仙琴蛙 *Nidirana daunchina*、林琴蛙 *Nidirana lini*、南昆山琴蛙 *Nidirana nankunensis* 及滇蛙 *Nidirana pleuraden*。

本书采用Frost（2017）及Lyu等（2017）的分类建议，记录为滇蛙 *Nidirana pleuraden*。

a～d 整体照（李家堂摄）
e 在茂兰地区的分布

13）水蛙属 *Hylarana* Tschudi, 1838

Hylarana Tschudi, 1838, Classif. Batr.: 37. Type species: *Hyla erythraea* Schlegel, 1827, by monotypy.
Limnodytes Duméril and Bibron, 1841, Erp. Gen., 8: 510. Substitute name for *Hylarana* Tschudi, 1838.
Zoodioctes Gistel, 1848, Naturgesch. Thierr.: xi. Substitute name for *Hylarana* Tschudi, 1838.
Hylorana Günther, 1864, Rept. Brit. India: 425. Incorrect subsequent spelling of *Hylarana* Tschudi, 1838.
Tenuirana Fei, Ye and Huang, 1990, Key to Chinese Amph.: 139. Type species: *Rana taipehensis* van Denburgh, 1909, by original designation. Coined as a subgenus of *Hylarana*. Considered a subjective synonymy of *Hylarana* by Ohler and Mallick, 2003 "2002", Hamadryad, 27: 62.

鼻骨小，两内缘间距或宽或窄，与蝶筛骨和额顶骨分离或连接；前耳骨大或较大；额顶骨前、后几乎等宽；鳞骨颧枝或长或短。舌角前突长，向外弯或不弯。肩胸骨分叉，上胸软骨极小；中胸骨细长，基部粗；剑胸软骨远大于上胸软骨，后端有缺刻。

指端吸盘状或略膨大，有腹侧沟，不显或无；指基下瘤明显；趾端膨大成小吸盘，横径略大于趾节宽，有腹侧沟；趾间近全蹼，一般第四趾蹼的凹陷处达第二关节下瘤；外侧蹠间蹼达蹠基部或略逊；有内、外跗褶或仅有内蹠褶；背侧褶多明显或甚宽；鼓膜清晰。

雄蛙一般有肱前腺或肩上腺（无肱前腺者头体窄长，四肢细长，背侧褶细）；第一指基部有婚垫。

蝌蚪口部位于吻部腹面，下唇乳突2排完整，外排长，呈须状；唇齿行少，上唇齿1～2排，下唇齿1～3排。茂兰地区有2种，即沼水蛙 *Hylarana guentheri* 和阔褶水蛙 *Hylarana latouchii*，分种检索表如下。

茂兰地区种检索表

背侧褶正常，不宽厚 ··· 沼水蛙 *Hylarana guentheri*
背侧褶甚明显，甚宽厚 ··· 阔褶水蛙 *Hylarana latouchii*

（20）沼水蛙 *Hylarana guentheri* (Boulenger, 1882)

Rana guentheri Boulenger, 1882, Cat. Batr. Sal. Coll. Brit. Mus., ed. 2: 48. Syntypes: BMNH (3 specimens, including animal figured in pl. 4, fig. 2 of the original publication). Type locality: Amoy (2 specimens) and China (1 specimen). Restricted to "Amoy" by Gee and Boring, 1929, Peking Nat. Hist. Bull., 4: 29.
Rana elegans Boulenger, 1882, Cat. Batr. Sal. Coll. Brit. Mus., ed. 2: 59. Syntypes: BMNH (3 specimens, including one figured in pl. 5, fig. 1 of the original), by original designation. Type locality: W. Africa (2 specimens) and unknown (1 specimen). Synonymy by Boulenger, 1907, Proc. Zool. Soc., London, 1907: 481. Synonymy questioned by Bourret, 1942, Batr. Indochine: 309.
Limnodytes elegans Rochebrune, 1884, Fauna Senegambie, Amph.: 23.
Rana (Hylorana) guentheri Boulenger, 1920, Rec. Indian Mus., 20: 123.
Hylorana güntheri Deckert, 1938, Sitzungsber. Ges. Naturforsch. Freunde Berlin, 1938: 145.
Hylarana guentheri Bourret, 1939, Annexe Bull. Gen. Instr. Publique, Hanoi, 1939: 46; Fei and Ye, 2001, Color Handbook Amph. Sichuan: 193; Song et al., 2002, Herpetol. Sinica, 9: 71.
Rana (Hylarana) guentheri Dubois, 1987 "1986", Alytes, 5: 42.
Hylarana (Hylarana) guentheri Fei, Ye and Huang, 1990, Key to Chinese Amph.: 140; Ye, Fei and Hu, 1993, Rare and Economic

Amph. China: 241; Fei et al., 2005, in Fei et al. (eds.), Illust. Key Chinese Amph.: 115.

Rana (*Sylvirana*) *guentheri* Dubois, 1992, Bull. Mens. Soc. Linn. Lyon, 61: 326.

Hylarana guentheri Chen et al., 2005, Herpetol. J., 15: 237, by implication; Frost et al., 2006, Bull. Am. Mus. Nat. Hist., 297: 370; Che et al., 2007, Mol. Phylogenet. Evol., 43: 3.

Rana guentheri Yang, 2008, in Yang and Rao (eds.), Amph. Rept. Yunnan: 68.

Hylarana (*Sylvirana*) *guentheri* Fei et al., 2009, Fauna Sinica, Amph. 3: 1128.

Boulengerana guentheri Fei, Ye and Jiang, 2010, Herpetol. Sinica, 12: 35. See comment under Ranidae record.

Sylvirana guentheri Oliver et al., 2015, Mol. Phylogenet. Evol., 90: 191.

［同物异名］ *Boulengerana guentheri*（沼蛙）；*Limnodytes elegans*；*Rana elegans*；*Rana guentheri*；*Sylvirana guentheri*

［鉴别特征］ 本种指端没有腹侧沟；雄蛙前肢基部有肱腺；有1对咽侧下外声囊。

［形态描述］ 依据茂兰地区标本描述 雄蛙头体长65.58～71.59mm，体型大而狭长；头部较扁平，头长大于头宽；吻长而略尖，末端钝圆，突出于下唇，吻棱明显；颊部略向外倾斜，有深凹陷，鼻孔近吻端，鼻间距大于眼间距；眼大，上眼睑宽几乎与眼间距、鼓膜相等；鼓膜圆而明显；犁骨齿2斜列；舌大，后端缺刻深。前臂及手长不到体长的一半；指长，末端钝圆，不膨大，腹侧无沟，指式为3＞1＞4＞2；关节下瘤发达，指基下瘤略小；后肢全长大于体长的1.5倍，前伸贴体时胫跗关节达眼部，左、右跟部相重叠；胫长约为体长的一半；趾长，趾端钝圆，腹侧有沟；背部皮肤光滑，背侧褶平直而明显，自眼后直达胯部；体背后部有分散的小痣粒；口角后至肩部有2个明显的颌腺；颞褶不显。雄蛙前肢基部的前方有发达的肱腺。体侧皮肤有小痣粒；肛后和股内侧痣粒密集；胫部背面有细肤棱；体腹面除雄蛙的咽侧下外声囊处有褶皱外，其余各部光滑。

生活时背面为淡棕色；沿背侧褶下缘有黑纵纹，体侧有不规则的黑斑；鼓膜后沿颌腺上方有一斜行的细黑纹；鼓膜周围有一淡黄色小圈；颌腺淡黄色；后肢背面有3～4条深色宽横纹，股后有黑白相间的云斑；外声囊灰色；体腹面淡黄色，两侧黄色稍深。

［第二性征］ 雄蛙有肱腺，第一指内侧婚垫不明显；有1对咽侧下外声囊；体背侧雄性线明显。

［生活习性］ 生活在茂兰地区的丘陵和山区，成蛙多栖息于稻田、池塘或水坑内，喜出没于田埂和稻田中。白天藏于杂草丛中，夜间出来捕食，常隐藏在水生植物丛间、土洞或杂草丛中，不易被发现，受惊吓后急忙跳入水中。

［地理分布］ 茂兰地区全区分布。国内分布于河南、四川、重庆、云南、贵州、湖北、安徽、湖南、江西、江苏、上海、浙江、福建、台湾、广东、香港、澳门、广西、海南等地。国外分布于越南，后被引入关岛。

［模式标本］ 模式标本保存于英国伦敦自然历史博物馆

［模式产地］ 福建厦门

［保护级别］ 三有保护动物

［IUCN 濒危等级］ 无危（LC）

成体量度 （单位：mm）

编号	头体长	头长	头宽	前臂及手长	后肢全长	胫长	蹠足长
GZNU20170508001 ♂	69.83	24.76	18.62	34.79	104.63	38.27	41.94
GZNU20170513019 ♂	71.59	25.72	18.91	31.02	128.22	36.58	57.50
GZNU20170508003 ♂	70.36	24.99	19.54	32.97	128.83	37.63	59.27
GZNU20170513020 ♂	69.13	23.87	18.34	30.37	123.98	37.05	55.05
GZNU20170508004 ♂	65.58	24.42	18.21	33.41	127.89	35.55	58.15
GZNU20170508002 ♂	69.36	25.11	18.71	33.30	131.90	39.16	73.28

a~d 整体照
e 在茂兰地区的分布

（21）阔褶水蛙 *Hylarana latouchii* (Boulenger, 1899)

Rana latouchii Boulenger, 1899, Proc. Zool. Soc., London, 1899: 167. Syntypes: BMNH (2 specimens), by original designation; Boulenger, 1920, Rec. Indian Mus., 20: 138, mentions 3 types. Type locality: Kuatun, a village about 270 miles from Foochow [= Fuzhou, Chong'an County], in the mountains at the Northwest of the Province of Fokien [= Fujian], at an altitude of 3000 to 4000 feet or more, China.

Rana (*Hylorana*) *latouchii* Boulenger, 1920, Rec. Indian Mus., 20: 127-130.

Hylorana latouchii Deckert, 1938, Sitzungsber. Ges. Naturforsch. Freunde Berlin, 1938: 144.

Rana latouchii Liu and Hu, 1961, Tailless Amph. China: 188; Tian et al., 1986, Handb. Chinese Amph. Rept.: 59.

Rana (*Hylarana*) *latouchii* Dubois, 1987 "1986", Alytes, 5: 42.

Hylarana (*Hylarana*) *latouchii* Fei, Ye and Huang, 1990, Key to Chinese Amph.: 140; Ye, Fei and Hu, 1993, Rare and Economic Amph. China: 243.

Rana (*Sylvirana*) *latouchii* Dubois, 1992, Bull. Mens. Soc. Linn. Lyon, 61: 326.

Hylarana latouchii Chen et al., 2005, Herpetol. J., 15: 237; Che et al., 2007, Mol. Phylogenet. Evol., 43: 3.

Sylvirana latouchii Frost et al., 2006, Bull. Am. Mus. Nat. Hist., 297: 370; Fei, Ye and Jiang, 2010, Herpetol. Sinica, 12: 34.

Hylarana (*Sylvirana*) *latouchii* Fei et al., 2009, Fauna Sinica, Amph., 3: 1134.

"*Hylarana*" *latouchii* Provisional treatment here (version 6.0) due to not being assigned to any of the hylaranine genera by Oliver et al., 2015, Mol. Phylogenet. Evol., 90: 176-192.

[同物异名] *Rana latouchii*；*Sylvirana latouchii*

[鉴别特征] 本种背侧褶宽厚，褶宽度约等于上眼睑宽，褶间距窄；颌腺甚明显。

[形态描述] 依据茂兰地区标本描述　雄蛙头体长37.74～50.15mm，头长大于头宽；吻较短而钝，末端略圆，吻棱明显；颊部凹陷；鼻孔近吻端，位于吻侧，鼻间距较宽，略大于眼间距；鼓膜明显，与上眼睑等宽；犁骨齿两小团，舌长，呈卵圆形，后端缺刻深。前臂及手长小于体长之半；指纤细而长，末端钝圆略扁，无腹侧沟；指式为3 > 1 > 4 > 2；关节下瘤小而清晰，有指基下瘤；掌突3个；后肢全长约为体长的1.5倍，胫长约为体长之半；前伸贴体时胫跗关节达眼部，左、右跟部重叠；趾末端略膨大成吸盘，趾腹侧有沟；趾间半蹼，均不达趾端；关节下瘤小而明显；有不明显的跗褶。皮肤粗糙；背面有稠密的小刺粒；吻端、头侧、前肢及腹面的皮肤光滑；股部近肛周疣粒扁平；两眼前角之间有凸出的小白点；生活时体背面金黄色夹杂少量的灰色斑，背侧褶上的金黄色更加明显；从吻端开始通过鼻孔沿背侧褶下方有黑带；吻缘淡黄色有灰色斑，颌腺黄色；体侧有形状和大小不等的黑斑，疣粒黄色；四肢背面有黑横纹，股后方有黑斑点及云斑。体腹部淡黄色，两侧黄色稍淡而无斑。

[第二性征] 雄蛙体较小；吻端明显，比较尖；有1对咽侧内声囊，声囊孔小，长裂基部有光滑小臂腺；第一指内侧有浅色婚垫；体背侧有雄性线。

[生活习性] 该蛙生活于茂兰地区的丘陵和山区，生活环境植被茂密，碎枝落叶较多，水环境（如山间细小溪流、林间雨后小水塘等）短期内相对稳定。白天隐于林下碎枝乱叶中，夜间出来捕食、求偶等活动。

[地理分布] 茂兰地区分布于森林植被保存较好的区域，几乎全区分布。国内分布于贵州、河南、安徽、江苏、浙江、江西、湖南、湖北、福建、台湾、广东、香港、广西、云南等地。国外无分布记录。

[模式标本] 模式标本保存于英国伦敦自然历史博物馆

[模式产地] 福建南坪挂墩

[保护级别] 三有保护动物

[IUCN 濒危等级] 无危（LC）

[讨论] 关于该属物种的分类，目前稍有一些混乱，据 Amphibia China（2017）记载，该属包括沼水蛙和

成体量度 （单位：mm）

编号	头体长	头长	头宽	前臂及手长	后肢全长	胫长	蹠足长
GZNU20170508026♂	37.74	11.82	11.27	18.15	70.65	21.04	31.17
GZNU20170509008♂	41.61	14.11	12.46	18.84	71.07	22.10	30.97
GZNU20170509009♂	39.96	13.70	11.97	18.33	71.98	21.39	31.61
GZNU20170509007♂	50.15	13.33	12.24	24.01	85.80	26.98	37.70
GZNU20170508025♂	40.20	13.75	11.82	19.88	70.78	21.32	31.69
GZNU20170508099♂	44.53	15.63	12.64	21.88	83.53	26.76	33.97

阔褶水蛙在内共计12种，而Frost（2017）记录4种，即 *Hylarana erythraea*（Schlegel，1837）、长趾纤蛙 *Hylarana macrodactyla*（Günther，1858）、台北纤蛙 *Hylarana taipehensis*（van Denburgh，1909）、*Hylarana tytleri*（Theobald，1868）。Amphibia Web（2017）记录12种，包括 Amphibia China（2017）中记录的阔褶水蛙、长趾纤蛙、台北纤蛙。有关该属物种的分类变动较多，刘承钊等（1961）将当时的多数蛙科物种均归为广义蛙属 *Rana*，包括现在的一些水蛙属物种；费梁等（2009b）在水蛙属下设立水蛙亚属 *Hydarana*（*Sylvirana*）8种、纤蛙亚属 *Hydarana*（*Hylarana*）

2种和琴蛙亚属 *Hylaruna*（*Nidirana*）5种，并将当时的15个水蛙属物种列入3亚属当中；而后费梁等（2010）沿用了费梁等（2009b）的分类系统，将16个水蛙属物种列入3亚属当中；费梁等（2012）将原来的3亚属全部提升为属（水蛙亚属改为肱腺蛙属 *Sylvirana*），并将沼水蛙单独提升为属级，即沼蛙属 *Boulengerana*。

关于该属分子生物学研究多是单个物种，鉴于此，Oliver等（2015）基于2个线粒体基因片段和4个核基因位点，重建了 *Hylarana* 物种的系统关系。尽管结果显示广义的 *Hylarana* 呈单系，但内部分化支持率很低，有待进一步研究。

a～d 整体照
e 在茂兰地区的分布

Oliver 等（2015）将 *Hylarana* 划分为 10 属，其中命名了 2 属，但该划分可能存在一些问题，有些属如 *Indosylvirana* 的支持率过低；有些物种未使用模式产地样品等，其研究中涉及的很多中国物种的划分有待进一步商榷。阔褶水蛙与黑斜线水蛙分类地位不确定，Oliver 等（2015）认为它们属于 *Hydrophylax* 或 *Indosylvirana*。米尔水蛙位于 *Indosylvirana* 基部，但支持率极低，其中最大似然法构建的系统发育树支持率为 36，贝叶斯系统发育树后验概率仅为 0.83，小于常规认为的阈值（0.95），其演化地位值得进一步的研究。河口水蛙与勐腊水蛙尽管未包含在研究中，但 Oliver 等（2015）认为二者应属于 *Sylvirana*。

综上所述，考虑到水蛙类相关物种目前的研究还存在一些问题，本书参考 Amphibia China（2017）的意见，将沼水蛙和阔褶水蛙保持 *Hylarana* 的划分，将其作为一个属（指端膨大，有腹侧沟，若无腹侧沟，则雄性肩部或肱前具腺体；趾端呈吸盘状或较明显膨大，或至少具腹侧沟；具明显背侧褶；肩胸骨基部浅分叉或不分叉）。

14）臭蛙属 *Odorrana* Fei, Ye and Huang, 1990

Odorrana Fei, Ye and Huang, 1990, Key to Chinese Amph.: 147. Type species: *Rana margaretae* Liu, 1950, by original designation. Considered a subgenus of *Rana* by Dubois, 1992, Bull. Mens. Soc. Linn. Lyon, 61: 329; Matsui, 1994, Zool. J. Linn. Soc., 111: 385-415.

Eburana Dubois, 1992, Bull. Mens. Soc. Linn. Lyon, 61: 328. Type species: *Rana narina* Stejneger, 1901, by original designation. Proposed as a subgenus of *Rana*. Synonymy with *Odorrana* by Matsui, 1994, Zool. J. Linn. Soc., 111: 385-415; Bain, Lathrop, Murphy, Orlov and Ho, 2003, Am. Mus. Novit., 3417: 6.

Bamburana Fei et al., 2005, in Fei et al. (eds.), Illust. Key Chinese Amph.: 124. Type species: *Rana versabilis* Liu and Hu, 1962, by original designation. Named as a subgenus of *Odorrana*.

Wurana Li, Lu and Lü, 2006, Sichuan J. Zool., 25: 206, 209. Type species: *Rana tormotus* Wu, 1977, by original designation. Synonymy by Cai, Che, Pang, Zhao and Zhang, 2007, Zootaxa, 1531: 49.

Bamburana Fei, Ye and Jiang, 2010, Herpetol. Sinica, 12: 21. Treatment as a genus.

Matsuirana Fei, Ye and Jiang, 2010, Herpetol. Sinica, 12: 21. Type species: *Rana ishikawae* Stejneger, 1901.

鼻骨小，两内缘间距宽，与额顶骨不相接；蝶筛骨显露甚多，与鼻骨连接或不连接，个别种前达鼻骨之间；前耳骨大；鳞骨耳枝小。上胸软骨小，肩胸骨基部不分叉；中胸骨细长，基部较粗，剑胸软骨较大或甚大，大于上胸软骨，剑胸软骨后端缺刻深或无缺刻。舌角前突细长。雄蛙前拇指粗大；指、趾骨节末端略膨大或略呈"T"形，膨大的宽度小于该指、趾节基部的宽度。

指、趾端吸盘一般纵径大于横径，腹侧具沟，背面有横凹陷；第三、第四指有指基下瘤；趾间全蹼（第四趾蹼达远端关节下瘤或趾端；外侧蹠间蹼达基部），无跗褶。体扁平，皮肤光滑，无背侧褶或背侧褶细（如竹叶蛙属）；背面多为绿色，有斑或无斑。在繁殖期间雄蛙胸部多有白色刺团，第一指基部粗大，婚垫发达。

本属世界现有 59 种，主要分布于亚洲亚热带和热带地区。我国目前已发现 38 种，主要分布于南方各省区。茂兰地区有 4 种，分种检索表如下。

茂兰地区种检索表

1 { 体背面纯绿色，与体侧颜色不同，界限分明 ································ 大绿臭蛙 *Odorrana graminea*
  { 体背面棕色，或以绿色为主，或多或少杂有黑色斑，体背面与侧面颜色不易分开 ···················· 2

2 { 雄蛙有 1 对咽侧下外声囊 ·· 花臭蛙 *Odorrana schmackeri*
  { 雄蛙无声囊 ··· 3

3 { 整个腹面满布极明显的深色花斑；鼓膜大，约为眼径的 4/5；雄性胸部未见刺群 ·········· 务川臭蛙 *Odorrana wuchuanensis*
  { 整个腹面无明显的深色花斑；鼓膜较小，约为眼径之半；雄性胸部有 1 个三角形细白刺群 ········ 绿臭蛙 *Odorrana margaretae*

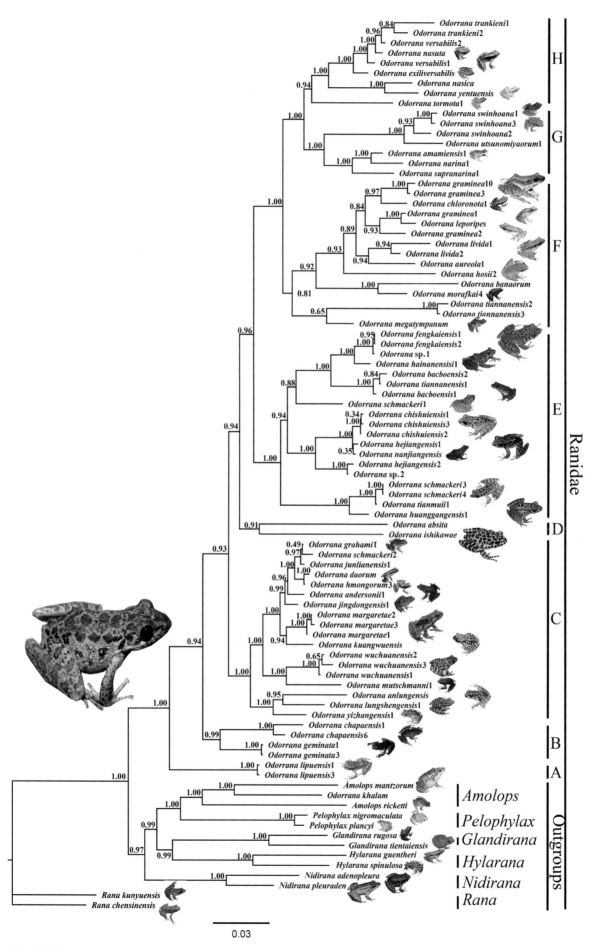

基于2个线粒体DNA（12SrRNA、16SrRNA）和3个核基因（Rhodopsin、RAG-1、Tyrosinase）构建的贝叶斯系统发育树

（22）大绿臭蛙 *Odorrana graminea* (Boulenger, 1899)

Rana graminea Boulenger, 1900 "1899", Proc. Zool. Soc., London, 1899: 958. Syntypes: BMNH (2 specimens); reported as BMNH 1947.2.27.96-97 by Bain, Lathrop, Murphy, Orlov and Ho, 2003, Am. Mus. Novit., 3417: 24. Type locality: Five-finger Mountains [= Mt. Wuzhi], Hainan Island, China.

Rana (*Hylorana*) *graminea* Boulenger, 1920, Rec. Indian Mus., 20: 127, 204.

Hylarana graminea Bourret, 1939, Annexe Bull. Gen. Instr. Publique, Hanoi, 1939: 46.

Rana (*Odorrana*) *graminea* Bain et al., 2003, Am. Mus. Novit., 3417: 24.

Huia graminea Frost et al., 2006, Bull. Am. Mus. Nat. Hist., 297: 368.

Odorrana graminea Che et al., 2007, Mol. Phylogenet. Evol., 43: 1-13.

Odorrana (*Odorrana*) *graminea* Fei et al., 2009, Fauna Sinica, Amph., 3: 1219.

[同物异名]　*Rana graminea*；*Hylarana graminea*；*Huia graminea*

[鉴别特征]　体背面纯绿色，背侧褶弱，背部有2～4个小黑斑，雌雄成体大小相差巨大，雄蛙有咽侧外声囊1对。

[形态描述]　依据茂兰地区标本描述　雄蛙头体长44.53～48.12mm，雌蛙头体长72.53～89.02mm；头扁平，头长大于头宽；吻端钝圆，略突出于下唇，吻棱明显，颊部向外侧倾斜，有深凹陷；鼻孔位于吻眼之间，鼻间距大于眼间距；鼓膜清晰；犁骨齿两端斜行；舌长，略呈梨形，后端缺刻深；声囊孔长裂形。前臂及手长近于体长之半；指细长，指端有宽的扁平吸盘，有腹侧沟，指关节下瘤明显，外侧3指有指基下瘤；掌突2个，内者大，外者小，均为椭圆形。后肢全长约为体长的1.8倍；后肢前伸贴体时胫跗关节超过吻端，左、右跟部重叠颇多；胫长大于体长的一半；足短于胫长；第五趾略长于第三趾，趾吸盘与指吸盘相同或略小；趾间全蹼，蹼均达趾端；第一、第五趾游离侧缘膜窄；外侧蹠间蹼达蹠基部；关节下瘤明显；内蹠突椭圆形，无外蹠突；无跗褶。皮肤光滑，背侧褶较弱而不明显，位于眼后角至胯部；眼下方有腺褶；腹面光滑。

生活时背面为鲜绿色；两眼前角间有一小白点；背部有2～4小黑斑；头侧、体侧及四肢浅棕色，四肢背面有深棕色横纹。趾蹼略带紫色；上唇缘腺褶及颌腺浅黄色；腹侧及股后有黄白色云斑。腹面白色。

[第二性征]　前臂较粗壮；第一指有灰白色婚垫，较大；有1对咽侧外声囊；无雄性线。

[生活习性]　该蛙生活于茂兰地区森林茂密的大中型山溪及其附近。溪流内裸露石头或岩壁较多，环境极为阴湿，石上长有苔藓等植物。雄蛙常攀附于水边灌草丛的枝条上，隐于绿叶间，白天难以发现，夜间出来捕食，鸣叫求偶。雌蛙白天多隐藏于溪流岸边石下或在附近的密林落叶间；夜间多蹲在溪内露出水面的石头上或溪旁岩石上捕食。

[地理分布]　茂兰地区分布于翁昂、洞塘等有中大型水流之地。国内主要分布于陕西、四川、云南、贵州、安徽、浙江、江西、湖北、湖南、福建、广东、香港、海南、广西等地。国外分布于越南。

[模式标本]　全模标本（BMNH 1947.2.27.96、BMNH 1947.2.27.97）保存于英国伦敦自然历史博物馆

[模式产地]　海南五指山

成体量度 （单位：mm）

编号	头体长	头长	头宽	前臂及手长	后肢全长	胫长	蹠足长
GZNU20170723009♀	89.02	28.90	22.70	42.29	206.49	59.97	86.04
GZNU20170723001♂	48.12	18.15	15.31	27.01	102.44	30.40	45.35
GZNU20170723011♀	72.53	23.92	18.36	35.07	154.33	49.10	65.83
GZNU20160726045♂	44.53	15.63	12.64	21.58	83.53	20.34	33.97
GZNU201607231002♂	47.50	18.90	12.71	26.00	94.94	28.41	54.77
GZNU20151124088♂	46.05	18.10	12.20	22.90	93.16	28.17	39.03

[保护级别] 三有保护动物

[IUCN 濒危等级] 无危（LC）

[讨论] 翟晓飞（2015）以中国 12 个省区 29 个地理种群 195 个大绿臭蛙样本为研究对象，使用 12S rRNA 和 16S rRNA 两个线粒体基因的部分片段，通过最大似然法和贝叶斯法对 55 个单倍型数据构建的中国大绿臭蛙复合体系统发育关系树，发现中国大绿臭蛙复合体明显分成 3 个不同的支系（即 A、B 和 C 3 个支系），每个支系都得到了高的支持（最大似然值 100、后验概率 1.0），3 个支系之间的遗传距离达到 4.5%～5.0%，每个支系地理分布格局明显。A 支系主要分布于横断山脉东南云贵高原周缘 6 个省区，含产自越南的大绿臭蛙 *Odorrana graminea*；B 支系分布于珠江中下游以南的广西东南部、广东西南部和海南，包括大绿臭蛙 *Odorrana graminea* 的模式产地海南五指山；C 支系分布于中国东部第三阶梯的江南丘陵、珠江以北南岭地区，包括大绿臭蛙 *Odorrana graminea* 的模式产地广东中部、北部所有种群。

中国大绿臭蛙复合体包括 *Odorrana graminea*、*Odorrana leporipes*，以及分类地位和分布界限有待确定的 *Odorrana sinica* 和 *Rana nebulosa*。贵州地区的大绿臭蛙是否是模式产地的大绿臭蛙，有待进一步研究，在问题没有得到合理的解决之前，本书将 *Odorrana graminea* 作为有效种名。

a～d 整体照
e 在茂兰地区的分布

（23）绿臭蛙 *Odorrana margaretae* (Liu, 1950)

Rana margaretae Liu, 1950, Fieldiana, Zool. Mem., 2: 303. Holotype: FMNH 49418, by original designation. Type locality:
Panlungshan [= Mt. Panlong], Kwanhsien [= Guanxian], Szechwan [= Sichuan], 3500 feet altitude, China.

Rana margaratae Liu and Hu, 1961, Tailless Amph. China: 204. Incorrect subsequent spelling.

Odorrana margaretae Fei, Ye and Huang, 1990, Key to Chinese Amph.: 147; Ye, Fei and Hu, 1993, Rare and Economic Amph.
China: 262; Fei, 1999, Atlas Amph. China: 192.

Rana (*Odorrana*) *margaretae* Dubois, 1992, Bull. Mens. Soc. Linn. Lyon, 61: 329.

Odorrana (*Odorrana*) *margaretae* Fei et al., 2005, in Fei et al. (eds.), Illust. Key Chinese Amph.: 126.

Odorrana margaretae Chen et al., 2005, Herpetol. J., 15: 239; Che et al., 2007, Mol. Phylogenet. Evol., 43: 1-13.

Huia margaretae Frost et al., 2006, Bull. Am. Mus. Nat. Hist., 297: 368.

[同物异名]　*Rana margaretae*；*Huia margaretae*

[鉴别特征]　本种体背部深绿色，且无斑点；雄蛙胸部只有1团小白刺，略呈"△"形，无声囊。

[形态描述]　依据茂兰地区标本描述　雄蛙头体长44.06～72.39mm，雌蛙头体长74.48～89.03mm；头部扁平，头长略大于头宽；吻端钝圆，突出于下唇；吻棱明显，颊部略向外侧倾斜，颊面凹陷深；鼻孔位于吻眼之间，恰在吻棱下方；两眼前角之间有很清晰的小白点；眼间距小于鼻间距，与上眼睑等宽；鼓膜为眼径之半，犁骨齿2斜列，发达，从内鼻孔内侧斜向后中线，不相遇；舌后端缺刻深。前肢较粗壮，前臂及手长近于体长之半；前臂发达；指较细长而略扁，末端膨大成较扁的吸盘，具腹侧沟，指端被分隔成背腹指面，第一指沟不清晰；指吸盘背面有半月形凹痕；指式为3＞4＞1＞2；关节下瘤明显，外侧3指的指基下瘤明显；内掌突椭圆形，外掌突略呈卵圆形。后肢长而发达，前伸贴体时胫跗关节达吻端，左、右跟部重叠；胫长超过体长之半；趾式为4＞5≈3＞2＞1；趾间全蹼，蹼达趾端，第一、第五趾的游离侧缘膜窄；外侧蹠间蹼达蹠基部；无跗褶。皮肤光滑，无背侧褶，背部没有极细致而弯曲的深浅线纹；上、下唇缘，颔部，上眼睑后部，背侧褶部位和体背后部及四肢背面有小白刺，但不同个体有多或少、疏或密的变异，有的不甚明显；体侧有扁平圆疣或无；肛下方及股部近端后下方扁平疣密集；颞褶短而清晰。腹面皮肤光滑，腹侧有扁平疣，有的疣上有小白刺。活体背部深绿色，背部近后端及体侧棕色，散有黑色麻斑；吻端至眼前角有细黑线，颌缘灰黄色，间有黑色纹；四肢浅棕色，间以黑色横纹4～5条，外侧的指、趾亦有横纹；上臂及胫跗关节部位有深绿色的斑块。腹面浅米黄色散有细黑点，有的咽、胸部呈紫褐色，股后深色大花斑或碎斑很明显，腹部及四肢腹面斑点较少而细小。

[第二性征]　雄蛙体较小，前臂较粗壮；第一指上婚垫发达；胸部中央只有1团小白刺，略排列成"△"形，无声囊；雄性线背面细，腹面无。

[生活习性]　该蛙生活于茂兰地区中大型溪流边。溪流内裸露或半裸露石头甚多，溪水清澈，水流湍急。溪两岸多为巨石和陡峭岩壁；乔木、灌丛和杂草繁茂。成蛙常栖于山涧湍急溪段，多蹲在长有苔藓、蕨类等植物的巨石或崖壁上，头迎向水面，稍有惊扰即跳入急流或深潭中。白天隐于岸边灌丛之下，其常蹲卧点常常被蹚出一个小土窝，受惊扰时会全身伏入小土窝内，背部皮肤上的斑纹与周围环境融为一体，以此躲避危险。夜晚出没于水边进行捕食。

[地理分布]　茂兰地区见于洞塘溪流之中。国内主要分布于甘肃、陕西、山西、四川、重庆、贵州、湖北、湖南、广西、广东、云南。国外分布于越南（北部）。

[模式标本]　正模标本（FMNH 49418）保存于芝加哥菲尔德自然历史博物馆（Field Museum, Chicago, USA, FMNH）

[模式产地]　四川盘龙山，1066.8m

[保护级别]　三有保护动物；贵州省重点保护野生动物

[IUCN 濒危等级]　无危（LC）

成体量度　　　　　　　　　　　　　　　　　　　　（单位：mm）

编号	头体长	头长	头宽	前臂及手长	后肢全长	胫长	跗足长
GZNU20160731023 ♂	44.06	27.26	21.09	35.84	156.51	48.03	93.23
GZNU20151124099 ♂	59.58	20.77	16.74	28.75	109.36	34.17	40.04
GZNU20160731087 ♂	72.39	25.33	20.41	32.57	137.42		
GZNU20160731089 ♀	74.48	25.87	21.46	39.08	161.64	48.43	93.56
GZNU20160731088 ♀	89.03	31.01	27.23	45.11	190.72	54.99	84.1
GZNU20160731022 ♀	85.44	25.78	21.73	37.85	152.82	44.65	63.29

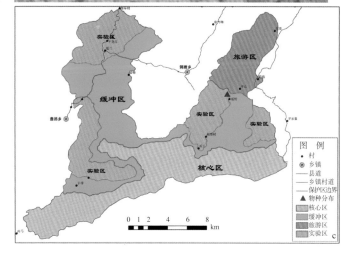

a、b　整体照
c　在茂兰地区的分布

（24）花臭蛙　*Odorrana schmackeri* (Boettger, 1892)

Rana schmackeri Boettger, 1892, Kat. Batr. Samml. Mus. Senckenb. Naturforsch. Ges.: 11. Holotype: SMF 6241 (formerly 1054.2a) according to Mertens, 1967, Senckenb. Biol., 48 (A): 46. Type locality: Kao-cha-hien [= Gaojiayan] bei Ichang [= Yichang], Hubei, Central-China.

Rana (*Hylorana*) *schmackeri* Boulenger, 1920, Rec. Indian Mus., 20: 126.

Rana melli Vogt, 1922, Arch. Naturgesch., Abt. A, 88: 144. Holotype: ZMB, by original designation. Type locality: aus der provinz Kuangtung die übringen sind im Yünnan gesammelt worden; rendered as Lienping, Kwangtung, China, by Liu, 1950, Fieldiana,

Zool. Mem., 2: 298. Tentative synonymy with *Rana andersonii* by Chang and Hsü, 1932, Contrib. Biol. Lab. Sci. Soc., China, Zool. Ser., 8: 161. Synonymy by Pope and Boring, 1940, Peking Nat. Hist. Bull., 15: 62-63; Bourret, 1942, Batr. Indochine: 358; Liu, 1950, Fieldiana, Zool. Mem., 2: 298.

Hylorana melli Deckert, 1938, Sitzungsber. Ges. Naturforsch. Freunde Berlin, 1938: 144.

Rana (*Hylarana*) *schmackeri* Bourret, 1942, Batr. Indochine: 357; Dubois, 1987 "1986", Alytes, 5: 42.

Odorrana schmackeri Fei, Ye and Huang, 1990, Key to Chinese Amph.: 151; Ye, Fei and Hu, 1993, Rare and Economic Amph. China: 266.

Rana (*Odorrana*) *schmackeri* Dubois, 1992, Bull. Mens. Soc. Linn. Lyon, 61: 329.

Odorrana (*Odorrana*) *schmackeri* Fei et al., 2005, in Fei et al. (eds.), Illust. Key Chinese Amph.: 130.

Odorrana schmackeri Chen et al., 2005, Herpetol. J., 15: 239; Che et al., 2007, Mol. Phylogenet. Evol., 43: 1-13.

Huia schmackeri Frost et al., 2006, Bull. Am. Mus. Nat. Hist., 297: 368.

[同物异名]　*Rana schmackeri*；*Rana melli*；*Hylorana melli*；*Huia schmackeri*

[鉴别特征]　本种鼓膜较大，约为第三指吸盘的两倍；上眼睑、后肢背面及背部均无小白刺；雄、雌成体体型大小差异较大。

[形态描述]　依据茂兰地区标本描述　雄蛙头体长44.57mm左右，雌蛙头体长82.36～83.94mm；头顶扁平，头长大于头宽；吻端钝圆而略尖，略突出于下唇，吻长于眼径；吻棱明显，眼至鼻孔处尤显；颊部微向外侧倾斜，颊面凹入颇深；鼻孔略近吻端，眼间距略小于鼻间距，与上眼睑几乎相等；鼓膜大而明显；犁骨齿2斜列，向后中线集中，二者一般相距较近，雄蛙的较弱，雌蛙的发达，末端在内鼻孔后方；舌呈长梨形，后端缺刻深。前臂及手长不到体长之半，前臂较粗；指较长，略扁平，指末端膨大成扁平吸盘，纵径大于横径，具腹侧沟，指腹侧沟将吸盘分隔成背、腹面，背面者宽大，有的标本第一指的沟不甚清晰，各指端背面有半月形或横置的凹痕；指式为3＞4≈1＞2；关节下瘤大，外侧3指指基下瘤较显或不显；内掌突椭圆形，位于第一指基部内侧，无外掌突。后肢全长大于体长的1.5倍；后肢前伸贴体时胫跗关节达眼与鼻孔之间或达鼻孔；左、右跟部重叠较多，胫长大于体长之半；第三、第五趾几乎等长，达第四趾的第二、第三关节下瘤之间；趾端与指端同；趾间全蹼，内外侧均达趾端，仅第四趾远端第二趾节两侧的蹼窄；外侧蹠间蹼达蹠基部；内蹠突卵圆形，无外蹠突；无跗褶。皮肤光滑，头体背面满布极细致而弯曲的深浅线纹，体侧有大小不一的扁平疣；两眼前角之间有一小白点；颞褶较细；口角后端有2～3粒浅黄色大腺粒，活体背部为绿色，间以大的棕褐色或褐黑色大斑点，多数斑点近圆形并镶以浅色边；颌缘及体侧黄绿色，有大小不一的黑棕色斑点；沿颞褶下方色深，而鼓膜色浅，上、下唇缘有棕褐色斑；四肢棕色或浅棕色，上有棕褐色横纹，较宽，股、胫部各有5～6条，股后方云斑状。腹面浅黄色或乳白色，咽喉部有浅棕色细点。

[第二性征]　与雌蛙相比，雄蛙体型明显较小；鼓膜较大；灰白色婚垫发达；有1对咽侧下外声囊，腹部无雄性线，背部雄性线显或略显；繁殖季节的雄蛙胸、腹部具白色刺群。

[生活习性]　该蛙生活于茂兰地区植被较为繁茂、环境潮湿、溪内石头甚多的大小山溪内。两岸岩壁长有苔藓，蛙常蹲在溪边岩石上，头朝向溪内，体背斑纹很像树叶投射到地面上的阴影，也与苔藓颜色相似。该蛙受惊扰后常跳入水中并潜入深水石间。雄蛙体型较小，常攀爬在近水边树枝上，夜晚开始集群鸣叫求偶、捕食。

[地理分布]　茂兰地区分布于翁昂、洞塘等地。国内分布于河南（南部）、四川、重庆、贵州、湖北、安徽（南部）、江苏、浙江、江西、湖南、广东、广西。国外分布于越南。

成体量度 （单位：mm）

编号	头体长	头长	头宽	前臂及手长	后肢全长	胫长	跗足长
GZNU20170510001♂	44.57	14.42	12.51	18.65	90.11	25.32	37.14
GZNU20170510002♀	82.36	25.75	20.77	30.78	158.42	48.81	63.47
GZNU20170510003♀	83.94	29.61	24.52	39.12	194.94	50.46	70.79

[模式标本] 正模标本（SMF 6241）保存于德国法森根堡自然博物馆（Forschungsinstitut und Natur-museum Senckenberg，Frankfurt，Germany）

[模式产地] 湖北宜昌高家堰

[保护级别] 三有保护动物

[IUCN 濒危等级] 无危（LC）

[讨论] 该种在我国的分布比较广泛，随着不断深入地野外考察及形态学比较和分子系统学等方面的研究，结果表明臭蛙属是一个自然类群，且并非单系（Chen et al.，2013）。随着人们对臭蛙类研究和关注度的提高，一些新种和隐存种陆续被认识和发现。刘承钊和胡淑琴（1961）认为安龙臭蛙 Odorrana anlungensis 和花臭蛙很相近。费梁等（2007）经过比较四川、重庆、贵州、湖北、湖南、安徽和福建花臭蛙的形态，发现分布于四川南江和万源的标本应为一新种——南江臭蛙 O. nanjingensis。陈晓虹等（2010a）经与花臭蛙模式产地标本湖北宜昌种群、花臭蛙浙江天目山种群及部分臭蛙类 12S mtDNA、16S mtDNA 序列相比，发现福建武夷山地区臭蛙标本的遗传分化已达种级水平，应为一新种——黄岗臭蛙 O. huanggangensis，浙江西天目山的花臭蛙与地模产地花臭蛙及花臭蛙 2 个新隐存种南江臭蛙和黄岗臭蛙之间存在显著差异，鉴定为一新种——天目臭蛙 O. tianmuii（陈晓红等，2010b）。然而，臭蛙复合体的物种组成和分布范围尚不明确。

陶娟（2010）基于线粒体 12S rRNA、16S rRNA 基因片段，对广义花臭蛙 12 个地理种群进行了研究，发现花臭蛙不同地理种群之间及种群内部都存在极其显著的分化；乔梁（2011）基于线粒体 12S rRNA、16S rRNA 基因片段对广义花臭蛙 11 个省区 15 个地理种群进行初步研究，结果显示，广义花臭蛙具有高的遗传多样性，种群间存在显著遗传分化。Chen 等（2013）通过广泛采样，获得较多的

a 侧面照
b 腹面照
c 背面照
d 在茂兰地区的分布

臭蛙属物种样品，基于 12S rRNA 与 16S rRNA 构建的系统发育树结果显示，广泛分布的花臭蛙物种并非单系，而是一个复合体，包括近缘种和隐存种。李永民（2015）基于联合 ND2 和 tRNAs 联合序列构建的树显示，来自贵州雷公山和梵净山的花臭蛙样品数据与来自武夷山的黄岗臭蛙 *O. huanggangensis* 聚为一支，贵州雷公山和梵净山可以看作黄岗臭蛙的新分布地；来自贵州冷水河和宽阔水的花臭蛙与来自湖北宜昌模式产地的花臭蛙，以及来自庐山-武功山的花臭蛙互为姐妹群，且支持率较高，种群间遗传距离较大。朱艳军（2016）基于 12S rRNA 与 16S rRNA 的联合序列研究表明，采自贵州印江的花臭蛙与来自湖北宜昌模式产地的花臭蛙聚为一支；来自贵州习水的花臭蛙与模式产地四川南江的合江臭蛙聚为一支；来自贵州三都、从江、雷山、江口、松桃的花臭蛙与福建武夷山的黄岗臭蛙聚为一支；同时提示贵州道真、务川等北部地区和三都、荔波等南部地区存在隐存种。

综合以上意见，贵州地区的花臭蛙分类地位并未得到很好的解决，关键在于没有使用统一的分子标记。因此，本书仍将茂兰地区作为花臭蛙的分布地。

（25）务川臭蛙 *Odorrana wuchuanensis* (Xu, 1983)

Rana wuchuanensis Xu, 1983, Acta Zool. Sinica, 29 (1): 66. Holotype: Dept. Biol., Zunyi Medical College 792238, by original designation. Type locality: Baicun, Wuchuan County, Guizhou, altitude 720m, China.

Odorrana wuchuanensis Fei, Ye and Huang, 1990, Key to Chinese Amph.: 150.

Rana (*Hylarana*) *wuchuanensis* Dubois, 1992, Bull. Mens. Soc. Linn. Lyon, 61: 329.

Odorrana wuchuanensis Ye, Fei and Hu, 1993, Rare and Economic Amph. China: 262; Fei, 1999, Atlas Amph. China: 192.

Odorrana (*Odorrana*) *wuchuanensis* Fei et al., 2005, in Fei et al. (eds.), Illust. Key Chinese Amph.: 126.

Huia wuchuanensis Frost et al., 2006, Bull. Am. Mus. Nat. Hist., 297: 368.

Odorrana wuchuanensis Che et al., 2007, Mol. Phylogenet. Evol., 43: 1-13.

[同物异名] *Rana wuchuanensis*；*Huia wuchuanensis*

[鉴别特征] 本种鼓膜大，约为眼径的 4/5，指、趾吸盘较大，趾间蹼缺刻较深，达第四趾第二关节下瘤，趾间蹼无大花斑，仅有少量细小花纹；体腹部满布深灰色和淡金黄色相间的大斑。

[形态描述] 依据茂兰地区标本描述 雄蛙头体长 50.57mm 左右，雌蛙头体长 70.68～101.12mm；头顶扁平，头长大于头宽；吻端钝圆，略突出于下唇，吻长大于眼径，吻棱明显，颊部微向外倾斜，颊面凹陷；鼻间距大于眼间距，鼻孔稍近吻端；眼间距较窄，约与鼓膜等宽；鼓膜大，约为眼径的 4/5，与眼睑宽几乎相等；犁骨齿强，左、右列在内鼻孔内侧斜向中央后方；舌梨形，后端缺刻深。前肢粗壮，前臂及手长约为体长之半；指式为 3 > 4 = 1 > 2，指端均有大而扁平的吸盘，除第一指外，吸盘前端较尖，指腹面两侧有沟，指端背面有横置的凹痕，吸盘横径约为鼓膜直径之半；关节下瘤明显，掌突不清晰。后肢长，向前伸贴体时胫跗关节达鼻孔，左、右跟部重叠，胫长超过体长之半，约与足等长；趾吸盘略小于指吸盘，吸盘两侧有沟，趾式为 4 > 5 = 3 > 2 > 1；趾间有蹼，但缺刻深，达第四趾第二关节下瘤的下方；关节下瘤发达；外侧蹠间有蹼；内蹠突呈长椭圆形，无外蹠突；无跗褶。头背部皮肤光滑或较粗糙。颞褶明显，自眼后角延伸到鼓膜背面与口角的白腺粒会合；两眼前角之间有一小白痣粒；前背部背侧褶处的皮肤有所增厚；在后背部、体侧及股、胫部背面有扁平疣粒。腹面皮肤光滑。活体背部绿色，上有分散的黑斑，有些黑斑周围镶以淡金黄色边，唇缘有深浅相间的斑纹；整个腹面具深灰色和淡金黄色相间的大斑块，交织成网状；四肢背面有深色宽带和浅色窄带相间的横纹，股、胫部各为 5～6 条。

[第二性征] 与雌蛙相比，雄蛙体型略小，背面皮肤较粗糙，在后背部、体侧及四肢背面均有分散的小白刺粒，后背部的小白刺粒尤为密集；第一指内侧有淡橘黄色（液浸标本为浅灰色）婚垫；无声囊及雄性线。

[生活习性] 该蛙生活于茂兰地区山区的溶洞内。洞内接近全黑，有暗河流出，水流缓慢。成蛙栖息于近洞口水塘周围的岩壁上。该蛙受惊扰后即跳入水中，并游到深水石下。

成体量度 （单位：mm）

编号	头体长	头长	头宽	前臂及手长	后肢全长	胫长	跗足长
20160728001 ♂	50.57	19.19	13.42	25.56	100.80	31.52	44.33
20160728002 ♀	78.41	28.01	21.85	41.24	147.01	44.80	66.55
20160729044 ♀	78.65	27.61	21.83	40.22	141.89	41.18	76.24
20160729042 ♀	71.25	28.47	22.25	38.21	141.60	44.18	62.64
20160729047 ♀	101.12	38.47	29.98	55.16	185.06	55.95	83.64
20160729041 ♀	70.68	26.65	19.04	37.43	134.34	41.66	88.50

[地理分布] 茂兰地区分布于翁昂、洞塘等地。国内见于贵州、湖北。国外无分布。

[模式标本] 正模标本（ZMC 792238）保存于遵义医科大学

[模式产地] 贵州务川

[保护级别] 三有保护动物

[IUCN 濒危等级] 极危（CR）

[讨论] 臭蛙属 Odorrana 是亚洲地区特有的一类两栖动物，目前世界已知 59 种广布于东亚、南亚和东南亚。我国有 38 种，广布于秦岭以南地区。由于其生活于山涧溪流、溪边乔木和灌丛的枝叶上，且指、趾端具有膨大而成的吸盘，早期被命名为树蛙（*Polyedates lividus* Blyth, 1856；*Buergeria ishikawae* Stejneger, 1901）、湍蛙（*Staurois hosii* Bourret, 1942）、水蛙（*Hylarana nasica* Bourret, 1939）或花臭蛙（*Rana schmackeri* Boettger, 1892）。费梁等（1990）以绿臭蛙 *Odorrana margaretae* 为属模建立臭蛙属，但是臭蛙属的分类地位在其后长时间内并未得到认可。例如，Dubois（1992）将臭蛙属作为广义蛙属的一个亚属，即 *Rana*（*Odorrana*）。

叶昌媛和费梁（2001）以 16 种臭蛙 29 种性状并以林蛙属物种为外群采用支序系统学进行了分析，研究表明，原始臭蛙可能起源于中国横断山区和云南高原西部，贵州高原可能是臭蛙物种的分化中心。

费梁等（2009b）基于形态学数据将原有臭蛙属 22 种划分为 2 亚属，即竹叶蛙亚属 *Odorrana*（*Bamburana*）和臭蛙亚属 *Odorrana*（*Odorrana*），并将臭蛙亚属划分为 6 个种组。Chen 等（2013）经过样本采集，在收集到包含地模和副模标本约 4/5 臭蛙属样本和序列的基础上，使用分子系统学方法对臭蛙属物种的系统发育和各物种的分歧时间做了较为全面的研究，结果表明，臭蛙属为一个单系，支持臭蛙属的有效性。臭蛙属的分类地位得到了 Amphibia Web（2017）、Frost（2017）及 Amphibia China（2017）的认可。

a、b 整体照

c 整体照
d 生境
e 在茂兰地区的分布

15）侧褶蛙属 *Pelophylax* Fitzinger, 1843

Pelophylax Fitzinger, 1843, Syst. Rept.: 31. Type species: *Rana esculenta* Linnaeus, 1758, by original designation.

Baliopygus Schulze, 1891, Jahresber. Abhandl. Naturwiss. Ver. Magdeburg, 1890: 177. Type species: *Rana ridibunda* Pallas, 1771, by subsequent designation of Dubois and Ohler, 1996 "1994", Zool. Polon., 39: 183. The earlier designation of *Rana esculenta* Linnaeus, 1758, by Stejneger, 1907, Bull. U.S. Natl. Mus., 58: 93, is in error inasmuch as *Rana esculenta* was not among the species listed by Schulze. Synonymy with *Rana* by Stejneger, 1907, Bull. U.S. Natl. Mus., 58: 93.

鼻骨较大，两内缘略分离或在前方相切，并与额顶骨相接；蝶筛骨或多或少显露；额顶骨窄长；前耳骨大；鳞骨颧枝长（约为耳枝的2倍）。肩胸骨基部不分叉；上胸软骨扇形，一般与剑胸软骨几乎等长（胫腺侧褶蛙的极小）；中胸骨较长，基部较粗；剑胸软骨后端缺刻较浅。舌角前突细短。第一掌骨正常；指骨节末端略膨大或尖。

指、趾末端钝尖或尖，腹侧无沟；无指基下瘤；趾间超过半蹼或近半蹼；外侧蹠间蹼发达，具1/2蹼或几乎达蹠基部。鼓膜大而明显；背侧褶宽厚；体背面以绿色为主。雄蛙婚垫位于第一指基部。

蝌蚪下唇乳突1排，完整或中央缺如；上唇齿1～2排；出水孔位于体左侧，无游离管；肛孔位于尾基部右侧。成体主要在静水水域（稻田、水塘）及其附近生活。卵和蝌蚪在静水内发育生长。

本属世界现有22种，广泛分布于古北界和东洋界。中国已知6种，除西藏和海南外，广泛分布。茂兰地区有1种，即黑斑侧褶蛙 *Pelophylax nigromaculatus*。

（26）黑斑侧褶蛙 *Pelophylax nigromaculatus* (Hallowell, 1860)

Rana chinensis Osbeck, 1771, Voy to China, London, 299. Type locality: Guangzhou (Canton), China.

Rana esculenta var. *japonica* Maak, 1859, Puteshestvie na Amur: 153. Type(s): not stated or known to exist. Type locality: left shore of Amur River below the Khingan range, Evrejskaja Autonomous Region, Russia. Nomen nudum. Based on specimens of *Rana nigromaculata*, according to Liu, 1950, Fieldiana, Zool. Mem., 2: 309. See discussion by Dubois and Ohler, 1996 "1994", Zool. Polon., 39: 162. Preoccupied by *Rana temporaria* var. *japonica* Günther, 1859 "1858", Cat. Batr. Sal. Coll. Brit. Mus.: 17. Synonymy (with *Rana esculenta chinensis*) by Boulenger, 1898, Tailless Batr. Eur., 2: 206; with *Rana nigromaculata* by Boulenger, 1882, Cat. Batr. Sal. Coll. Brit. Mus., ed. 2: 40; Nikolskii, 1918, Fauna Rossii, Zemnovodnye: 34.

Rana nigromaculata Hallowell, 1861 "1860", Proc. Acad. Nat. Sci. Philadelphia, 12: 500. Holotype: deposition not stated; presumably USNM or ANSP. Type locality: Japan; see discussion by Dubois and Ohler, 1996 "1994", Zool. Polon., 39: 163.

Hoplobatrachus reinhardtii Peters, 1867, Monatsber. Preuss. Akad. Wiss. Berlin, 1867: 711. Syntypes: ZMB 5900 (2 specimens) according to Bauer, Günther and Klipfel, 1995, in Bauer et al. (eds.), Herpetol. Contr. W.C.H. Peters: 49. Type locality: Malacca oder China; restricted to China by Liu, 1950, Fieldiana, Zool. Mem., 2: 209; see comments by Dubois and Ohler, 1996 "1994", Zool. Polon., 39: 164. Synonymy by Boulenger, 1882, Cat. Batr. Sal. Coll. Brit. Mus., ed. 2: 40; Boulenger, 1891, Proc. Zool. Soc., London, 1891: 376; Stejneger, 1907, Bull. U.S. Natl. Mus., 58: 94.

Hoplobatrachus davidi David, 1873 "1872", J.N. China Branch R. Asiat. Soc., N.S., 7: 227. Types: not designated or known to exist. Type localities: Pékin and Japon. Restricted to Beijing, China by Dubois and Ohler, 1996 "1994", Zool. Polon., 39: 165. Nomen nudum. Synonymy (with *Rana esculenta japonica*) by Boettger, 1888, Ber. Offenbach. Ver. Naturkd., 1888: 155; Dubois and Ohler, 1996 "1994", Zool. Polon., 39: 165.

Rana reinhardtii Möllendorff, 1877 "1876", J.N. China Branch R. Asiat. Soc., N.S., 11: 105.

Rana nigromaculata mongolia Schmidt, 1925, Am. Mus. Novit., 175: 1. Holotype: AMNH 18149, by original designation. Type locality: Mai Tai Chao [= Tumd Zuoqi], northern Shansi [now part of Nei Mongol], China.

Pelophylax nigromaculata Fei, Ye and Huang, 1990, Key to Chinese Amph.: 133-134; Ye, Fei and Hu, 1993, Rare and Economic Amph. China: 224.

[同物异名] *Rana nigromaculata*；*Rana esculenta* var. *japonica*；*Rana marmorata*；*Hoplobatrachus reinhardtii*；*Hoplobatrachus davidi*；*Rana reinhardtii*；*Rana chinensis*；*Hylarana nigromaculata*

[鉴别特征] 本种背侧褶间有数行长短不一的纵肤棱，雄蛙有1对颈侧外声囊，肩上方无扁平腺体。

[形态描述] 依据茂兰地区标本描述 雄蛙头体长43.79～63.38mm，雌蛙头体长56.47～81.83mm；头长大于头宽；吻部略尖，吻端钝圆，突出于下唇；吻棱不明显，颊部向外倾斜；鼻孔在吻眼中间，鼻间距等于眼睑宽，眼大而突出，眼间距窄，小于鼻间距及上眼睑宽；鼓膜大而明显，近圆形；犁骨齿两小团，突出于内鼻孔之间；舌宽厚，后端缺刻深。前肢短，前臂及手长小于体长之半；指末端钝尖，指式为3＞1＞2＞4；指侧缘膜不明显；关节下瘤小而明显。后肢较短而肥硕，前伸贴体时胫跗关节达鼓膜和眼之间，左、右跟部不相遇，胫长小于体长之半；趾末端钝尖；背面皮肤较粗糙，背侧褶明显，褶间有多行长短不一的纵肤棱，后背、肛周及股后下方有圆疣和痣粒；体侧有长疣或痣粒；鼓膜上缘有细颗褶，口角后的颌腺窄长；胫背面有多条由痣粒连缀成的纵肤棱；无蹠跗褶。腹面光滑。活体背面颜色多样，有淡绿色、黄绿色、深绿色、灰褐色等，杂有许多大小不一的黑斑纹，背侧褶金黄色、浅棕色或黄绿色；沿背侧褶下方有黑纹，自吻端沿吻棱至颞褶处有1条黑纹；四肢背面浅棕色，前臂常有棕黑横纹3条，股、胫部各有4条，股后侧有酱色云斑。腹面为一致的乳白色或带微红色。唇缘有斑纹；鼓膜灰褐色或浅黄色；颌腺棕黄色或淡黄色，关节下瘤米黄色。雄蛙外声囊浅灰色，第一指内侧的婚垫浅灰色。

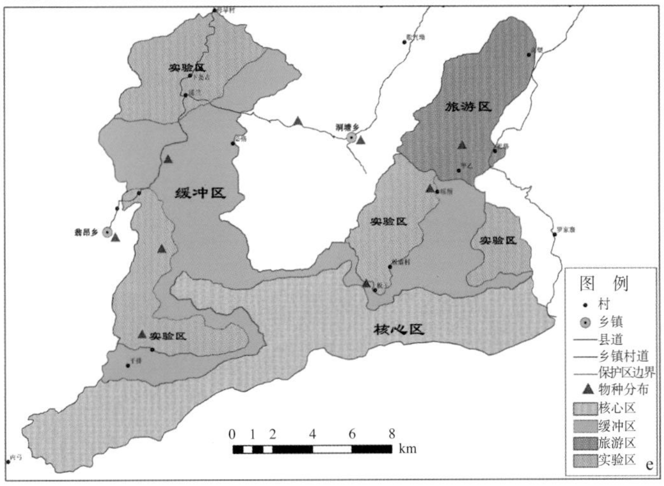

a～d 整体照
e 在茂兰地区的分布

成体量度 （单位：mm）

编号	头体长	头长	头宽	前臂及手长	后肢全长	胫长	跗足长
GZNU20170513001 ♀	56.47	21.49	16.47	24.92	100.48	27.50	47.65
GZNU20170513003 ♂	63.38	22.56	17.38	29.17	111.72	31.97	51.72
GZNU20170509001 ♀	70.90	22.21	16.46	28.78	111.93	32.77	52.09
GZNU20170513006 ♀	79.31	26.63	20.93	29.97	133.00	38.34	63.94
GZNU20170513007 ♀	81.83	29.75	23.58	37.41	140.12	41.21	66.09
GZNU20170513004 ♂	43.79	16.06	12.96	19.82	75.73	21.89	36.09

[第二性征]　雄蛙体较小；前臂较粗壮，第一指内侧的婚垫发达；有1对颈侧外声囊；背侧及腹侧都有雄性线，背侧者较粗。

[生活习性]　该蛙广泛生活于茂兰地区大型河流、水塘、鱼塘、农田、沼泽等多种类型生境中，适应能力强，多集中生活在农田、沼泽环境中。白天常隐于田埂杂草之下，夜晚出没于水田中捕食、鸣叫求偶。

[地理分布]　茂兰地区全区分布。国内除青海、台湾、海南外，广布于全国各地。国外分布于俄罗斯、日本、朝鲜半岛。

[模式标本]　模式标本保存于美国费城自然博物馆或美国国家自然历史博物馆

[模式产地]　日本

[保护级别]　三有保护动物

[IUCN 濒危等级]　近危（NT）

8. 树蛙科 Rhacophoridae Hoffman, 1932 (1858)

Polypedatidae Günther, 1858, Proc. Zool. Soc., London, 1858: 346. Type genus: *Polypedates* Tschudi, 1838.

Polypedatinae Mivart, 1869, Proc. Zool. Soc., London, 1869: 292.

Polypedatinae Boulenger, 1888, Proc. Zool. Soc., London, 1888: 205.

Polypedatidae Noble, 1927, Ann. New York Acad. Sci., 30: 105.

Rhacophoridae Hoffman, 1932, S. Afr. J. Sci., 29: 581. Type genus: *Rhacophorus* Kuhl and van Hasselt, 1822.

Rhacophorinae Laurent, 1943, Bull. Mus. R. Hist. Nat. Belg., 19: 16.

Racophoridae Hellmich, 1957, Veröff. Zool. Staatssamml. München, 5: 28.

Philautinae Dubois, 1981, Monit. Zool. Ital., N.S., Suppl., 15: 258. Type genus: *Philautus* Gistel, 1848. Synonymy by Channing, 1989, S. Afr. J. Zool., 24: 116-131.

Philautini Dubois, 1987 "1986", Alytes, 5: 34, 69; Dubois, 1992, Bull. Mens. Soc. Linn. Lyon, 61: 335.

Buergeriinae Channing, 1989, S. Afr. J. Zool., 24: 127. Type genus: *Buergeria* Tschudi, 1838.

Buergeriini Dubois, 1992, Bull. Mens. Soc. Linn. Lyon, 61: 335.

Rhacophorini Dubois, 1992, Bull. Mens. Soc. Linn. Lyon, 61: 336.

Nyctixalini Grosjean et al., 2008, J. Zool. Syst. Evol. Res., 46: 174. Type genus: *Nyctixalus* Boulenger, 1882, by original designation. Tribe of Rhacophorinae formulated originally to contain the monophyletic group of *Theloderma* + *Nyctixalus*.

Rhacophorini Grosjean et al., 2008, J. Zool. Syst. Evol. Res., 46: 174. Reformulation of tribe to contain *Aquixalus*, *Chiromantis* (including *Feihyla* in their sense), *Kurixalus*, *Philautus*, *Polypedates* and *Rhacophorus*.

骨骼方面的主要特征与蛙科相同。椎体参差型或前凹型；指、趾末端两节间有介间软骨；指、趾骨末节呈"Y"形或"T"形；指、趾末端膨大成显著的吸盘和边缘沟，吸盘的背面一般无横凹痕，腹面呈肉垫状。多树栖，外形与生活习性与雨蛙颇近似。

树蛙科物种分布很广，覆盖非洲、南亚、东亚和东南亚。树蛙科目前已收录19属，分为2亚科，即溪树蛙亚科 Buergeriinae（1属5种，即溪树蛙属 *Buergeria*）和树蛙亚科 Rhacophorinae（18属412种）。中国已知12属75种，即溪树蛙属 *Buergeria*、螳臂树蛙属 *Chiromantis*、费树蛙属 *Feihyla*、纤树蛙属 *Gracixalus*、原指树蛙属 *Kurixalus*、刘树蛙属 *Liuixalus*、棱鼻树蛙属 *Nasutixalus*、小树蛙属 *Philautus*、泛树蛙属 *Polypedates*、灌树蛙属 *Raorchestes*、树蛙属 *Rhacophorus*、棱皮树蛙属 *Theloderma*，分布于秦岭以南地区。

中国所有的物种上颌均有齿，一般舌后端缺刻深，瞳孔大多横置；多有筑泡沫卵巢的习性，卵粒盛于卵泡内或胶质团内，卵泡或被树叶包裹。蝌蚪（如树蛙属 *Rhacophorus* 和跳树蛙属 *Chirixalus* 物种）生活于静水水域内；有的种类产卵于树洞或陆地上，有短暂的非摄食性的蝌蚪阶段，从卵直接发育成幼蛙（如小树蛙属 *Philautus*、棱皮树蛙属 *Theloderma* 和夜跳蛙属 *Nyctixalus* 物种）；有的种类雌蛙在有积水的竹桩内或树洞内壁多次产卵，后产的卵群供先产出的卵孵化出的蝌蚪摄食，雌蛙有保护蝌蚪发育的习性（如原指树蛙属 *Kurixalus* 物种）；有的种类生活于溪流（如溪树蛙属 *Buergeria* 物种），卵贴附于溪边石下或水生植物上或呈小块状浮于水面，不呈泡沫状。蝌蚪属于有唇齿左孔型。

本科物种在茂兰地区有 3 属 5 种和 1 待定种，分属检索表如下。

茂兰地区属检索表

1. 指间无蹼；多数种类背面黄褐色或灰褐色，一般有褐黑色"X"形斑或纵列条纹，股后有褐黑色网状斑 ············ 泛树蛙属 *Polypedates*
   指间有蹼或微具蹼；多数种类背面为绿色、棕黑色或两者交织，股后无褐黑色网状斑 ············ 2
2. 四肢和背面无疣粒，蝌蚪口部位于吻端腹面，眼部位于头的背侧 ············ 树蛙属 *Rhacophorus*
   四肢和体背面有疣粒，蝌蚪口部位于吻前端，眼部位于头的顶部 ············ 原指树蛙属 *Kurixalus*

16）原指树蛙属 *Kurixalus* Ye, Fei and Dubois, 1999

Kurixalus Ye, Fei and Dubois, 1999, in Fei (ed.), Atlas Amph. China: 383. Type species: *Rana eiffingeri* Boettger, 1895, by original designation. Delorme, Dubois, Grosjean and Ohler, 2005, Bull. Mens. Soc. Linn. Lyon, 74: 166. Type species: *Philautus odontotarsus* Ye and Fei, 1993, by original designation. Synonymy with *Kurixalus* by Li, Che, Bain, Zhao and Zhang, 2008, Mol. Phylogenet. Evol., 48: 31; Yu, Rao, Zhang and Yang, 2009, Mol. Phylogenet. Evol., 50: 571-579.

Aquixalus Delorme et al., 2005, Bull. Mens. Soc. Linn. Lyon, 74: 166. Type species: *Philautus odontotarsus* Ye and Fei, 1993, by original designation. Synonymy with *Kurixalus* by Li et al., 2008, Mol. Phylogenet. Evol., 48: 31; Yu et al., 2009, Mol. Phylogenet. Evol., 50: 571-579.

本属蛙类体型小，鼓膜明显；有犁骨齿或弱小；第一、第二指与第三、第四指不呈对指握物状，原拇指发达；指、趾端具吸盘，背面无"Y"形骨迹，有马蹄形边缘沟；指间几乎无蹼，趾间蹼较发达；胫跗关节外侧有 1 个大的白色疣粒，沿跗、蹠后缘有若干颗粒状疣粒。

蝌蚪口部位于吻前端略向上方，角质颌很强，上唇中部无唇乳突，上唇两侧及下唇部均有大的乳突，唇齿式为 Ⅱ / Ⅱ，眼位于头的顶部。

成蛙生活在林区，雌蛙将卵群产在有积水的竹桩内或树洞内壁内，每年产卵多次，每次产卵 33～129 粒，卵呈单粒状。雌蛙有保护蝌蚪发育的习性，第二次以后产的卵群多供蝌蚪捕食。

该属物种世界现有 17 种，国外分布于柬埔寨（南部）、越南（中部）、马来半岛、印度尼西亚（苏门答腊岛）、加里曼丹岛和菲律宾等地。中国有 10 种，分布于中国南部。茂兰地区有 1 种，即锯腿原指树蛙 *Kurixalus odontotarsus*。

（27）锯腿原指树蛙 *Kurixalus odontotarsus* (Ye and Fei, 1993)

Philautus odontotarsus Ye and Fei, 1993, in Ye, Fei and Hu (eds.), Rare and Economic Amph. China: 318, 320. Holotype: CIB 57311, by original designation. Type locality: Laiyanghe, Mengla County, altitude 1000m, Yunnan, China.

Philautus (*Philautus*) *odontotarsus* Bossuyt and Dubois, 2001, Zeylanica, 6: 59.

Aquixalus (*Aquixalus*) *odontotarsus* Delorme et al., 2005, Bull. Mens. Soc. Linn. Lyon, 74: 166.

Kurixalus odontotarsus Li et al., 2008, Mol. Phylogenet. Evol., 48: 310.

Aquixalus odontotarsus Fei et al., 2009, Fauna Sinica, Amph., 2: 678.

[同物异名]　*Philautus odontotarsus*（锯腿小树蛙）；*Aquixalus odontotarsus*（锯腿水树蛙）

[鉴别特征]　前臂及跗、蹠部外侧有锯齿状肤突；鼓膜较大，约为眼径的2/3；体背面皮肤较粗糙，具许多小疣粒；雄蛙具单咽下内声囊。

[形态描述]　依据茂兰地区标本描述　雄性头体长33.45～47.34mm；头长大于头宽；吻棱明显，自鼻孔向前下方倾斜，左、右向中线集中，吻略尖，显著突出于下唇；颊面内陷，鼻孔近吻端，鼻孔周围明显突起；瞳孔平置椭圆；眼间距小于上眼睑宽；鼓膜约为眼径的2/3，紧接眼的后下方；犁骨齿列短，起自内鼻孔内侧前部，呈倒"八"字形排列，相距甚宽；舌上有分散的乳突，舌后端缺刻深。前肢长，前臂及手长约为体长之半；指端均有吸盘及边缘沟，第一指吸盘甚小，第三指吸盘比鼓膜略小；指侧有缘膜，基部微具蹼；关节下瘤大，近掌部的较小；掌部有小疣，有内、外掌突，外掌突几乎一分为二。后肢细长，前伸贴体时胫跗关节达眼或鼻眼之间，左、右跟部重叠；胫长约为体长之半，足比胫短；趾端均有吸盘及边缘沟，内蹠突扁平，无外蹠突。皮肤粗糙。头上背部、颞部及四肢背面都有大小、形状不同的疣粒，上眼睑上疣粒颇多，胫跗关节上方常有明显的肤突；沿前臂外侧及跗、蹠部至第五趾外侧有排列成行的浅色锯齿状肤突；颞褶较平置；肛孔下方有1团分散的圆锥形疣。咽喉、胸、腹各部及股部下方均密布扁平圆疣，腹部的疣特别明显。

活体背面浅褐色或绿褐色，两眼间常有1条深色横纹，背部有"Y"形或排列不规则的深色斑；体侧为灰绿色，具黑褐色斑点；四肢背面有黑褐色横纹，股前、后侧呈橘红色。腹面灰红色，胸部、腹部及四肢腹面散有不明显的浅灰色蠕虫状斑纹。

[第二性征]　雄蛙第一指基部背面有乳白色婚垫；有单咽下内声囊；有雄性线。

[生活习性]　该蛙生活于茂兰地区灌木林地带。晚上成蛙栖息在灌木枝叶上或藤本植物及杂草的叶片上，受惊扰后停止鸣叫，静止不动，因体型小，极不易被发现。

[地理分布]　茂兰地区分布于翁昂、洞塘等地。国内分布于云南、贵州、广西、广东、海南。国外分布于越南（北部）、老挝（北部）。

[模式标本]　正模标本（CIB 57311）保存于中国科学院成都生物研究所

[模式产地]　云南勐腊县莱阳河，1000m

[保护级别]　三有保护动物

[IUCN 濒危等级]　无危（LC）

[讨论]　锯腿原指树蛙 *Kurixalus odontotarsus* 的分类地位在 Frost（2017）、Amphibia Web（2017）、Amphibia China（2017）中与费梁等（2012）有所不同。主要分歧是将本种归为原指树蛙属 *Kurixalus* 还是水树蛙属 *Aquixalus*。通过对采集于本种模式产地样品的分子系统学研究，表明锯腿原指树蛙与原指树蛙属物种，包括原指树蛙属模式物种 *Kurixalus eiffingeri* 聚在一起（Yu et al., 2010）。

成体量度　　　　　　　　　　　　　　　　　　（单位：mm）

编号	头体长	头长	头宽	前臂及手长	后肢全长	胫长	跗足长
GZNU20170509036 ♂	33.45	10.11	11.58	15.91	51.53	16.16	19.72
GZNU20170509034 ♂	33.63	11.62	10.36	14.73	50.74	15.62	21.01
GZNU20170509035 ♂	47.34	18.75	12.26	19.43	79.36	21.87	36.47

Yu 等（2010）使用两个线粒体（12S rRNA 和 16S rRNA）基因构建了锯腿原指树蛙 Kurixalus odontotarsus 种组（复合体）的系统发育，试图划定中国境内锯腿原指树蛙种内水平的物种边界。根据较高的支持率获得了 3 个主要分支，并且所有的系统发育分析都否定了锯腿原指树蛙的单系性。从而建议将锯腿原指树蛙、泰北原指树蛙 Kurixalus bisacculus 和多疣原指树蛙 Kurixalus verrucosus 作为 3 个独立的物种。锯腿原指树蛙种组的西藏谱系不属于锯腿原指树蛙，并暂时将其放入多疣原指树蛙中。海南原指树蛙 Kurixalus hainanus 被认为是泰北原指树蛙的同物异名。泰北原指树蛙的分布范围应扩大到华南大部分地区，而在中国，锯腿原指树蛙的分布应局限于其模式产地及其附近地区。

基于此，虽然在 Yu 等（2010）的研究中使用了来自贵州荔波的分子数据，并指出荔波应该是泰北原指树蛙的分布地，但由于未注明贵州荔波分子数据的本源标本是否来自茂兰地区，因此，本书基于目前的研究结果和参考相关意见，将本种纳入原指树蛙属，记为锯腿原指树蛙。

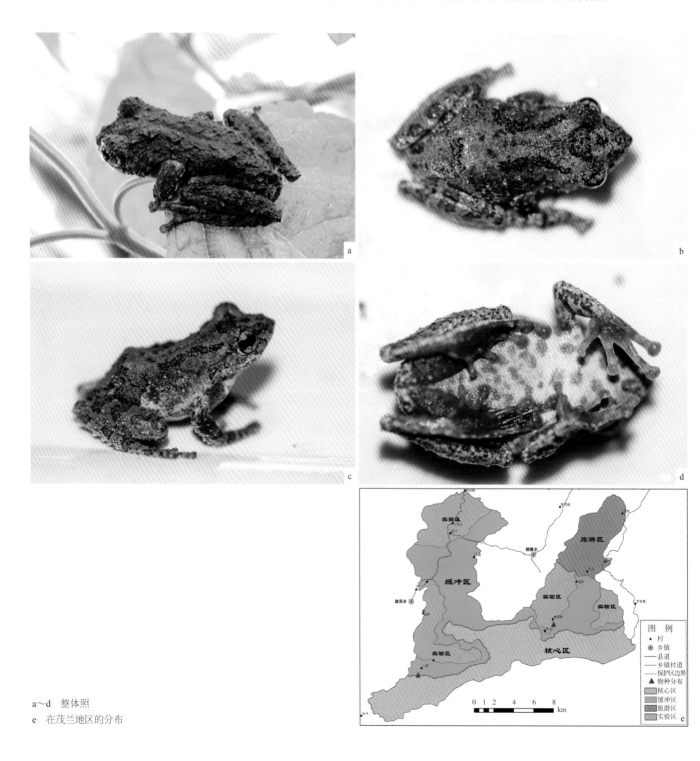

a～d 整体照
e 在茂兰地区的分布

17）泛树蛙属 *Polypedates* Tschudi, 1838

Polypedates Tschudi, 1838, Classif. Batr.: 75. Type species: *Hyla leucomystax* Gravenhorst, 1829, by subsequent designation of Fitzinger, 1843, Syst. Rept.: 31.

Polypedotes Tschudi, 1838, Classif. Batr.: 34. Alternative original spelling.

Trachyhyas Fitzinger, 1843, Syst. Rept.: 31. Type species: *Polypedates rugosus* Duméril and Bibron, 1841, by original designation. Synonymy with *Polypedates* by Günther, 1859 "1858", Cat. Batr. Sal. Coll. Brit. Mus.: 78; by implication; by Stejneger, 1907, Bull. U.S. Natl. Mus., 58: 143.

鼻骨小，蝶筛骨完全显露；额顶骨宽短；有的种有次生真皮，覆盖在额顶骨和鳞骨侧边，一般头部皮肤紧贴于头骨。舌骨无舌角前突，有冀状突；喉器的环状软骨有侧突；食管突呈锥状突起或无二锥体前凹型或参差型；肩胸骨在基部略分叉，中胸骨细长，长于咏骨；指、趾骨末节呈"Y"形，介间软骨呈心形；第二掌骨远端有结节。

体型中等大小；背面皮肤光滑或具小痣粒；前臂、跟部和肛上方有明显的皮肤褶，雄蛙吻端斜尖；舌后部缺刻深；鼓膜明显；指、趾吸盘大，背面可见到"Y"形迹；指与指不形成对指握物状，指间无蹼或仅有蹼迹，趾间约为半蹼；一般外侧跖间蹼不发达。成体背面多为黄褐色，头顶皮肤多与骨骼紧贴。

蝌蚪体型中等大小，口部位于吻端腹面，眼位于头的两极侧；尾鳍高而薄，尾末段细尖。一般为陆栖或树栖，卵粒小，数量多，卵群形成泡沫状；卵泡多产在水塘、沼泽和稻田边泥窝内。

本属世界已知有24种，主要分布于斯里兰卡、印度到东南亚热带地区、中国（南部）及琉球群岛。

我国现有4种，主要分布于秦岭以南热带和亚热带地区，向北达陕西和甘肃（南部）。茂兰地区有2种，即斑腿泛树蛙 *Polypedates megacephalus* 和无声囊泛树蛙 *Polypedates mutus*，分种检索表如下。

茂兰地区种检索表

雄蛙无声囊；多数个体背面有6条深色纵纹·· 无声囊泛树蛙 *Polypedates mutus*

雄蛙有声囊；体背面有"X"形斑或呈纵条纹状·· 斑腿泛树蛙 *Polypedates megacephalus*

（28）斑腿泛树蛙 *Polypedates megacephalus* Hallowell, 1860

Hyla leucomystax Gravenhorst, 1829, Delic. Mus. Zool. Vart., I: 26. Type locality: Indonesia.

Polypedates megacephalus Hallowell, 1861 "1860", Proc. Acad. Nat. Sci. Philadelphia, 12: 507. Holotype: Presumably ANSP or USNM but likely now lost. Type locality: Hong Kong, China.

Polypedates maculatus var. *unicolor* Müller, 1878, Verh. Naturforsch. Ges. Basel, 6: 585. Types: NHMB. Type locality: China. Nomen nudum. Synonymy by Stejneger, 1925, Proc. U.S. Natl. Mus., 66: 30; Zhao and Adler, 1993, Herpetol. China: 156.

Rhacophorus leucomystax megacephalus Stejneger, 1925, Proc. U.S. Natl. Mus., 66: 30; Pope, 1931, Bull. Am. Mus. Nat. Hist., 61: 574; Inger, 1966, Fieldiana, Zool., 52: 308.

Polypedates megacephalus Matsui, Seto and Utsunomiya, 1986, J. Herpetol., 20: 483-489.

Rhacophorus (*Rhacophorus*) *leucomystax megacephalus* Dubois, 1987 "1986", Alytes, 5: 81.

[同物异名]　*Hyla leucomystax*（斑腿树蛙）；*Rhacophorus mutus*

[鉴别特征]　背前部多有"X"形深色斑，股后有网状斑；外侧指间无蹼；雄蛙有1对咽侧下内声囊。

[形态描述]　依据茂兰地区标本描述　体扁而窄长，雄蛙头体长41.41～45.33mm，雌蛙头体长48.27～62.48mm；头长大于头宽；吻长，吻端钝尖或钝圆，突出于下唇，呈倾斜状；吻棱明显。颊面内陷；鼻孔近吻端，鼻间距小于眼间距；鼓膜明显，约为眼径的2/3；犁骨齿强，由内鼻孔前内角向内侧中线斜伸至内鼻孔后角，左、右间距约与内鼻孔宽度同；舌后端缺刻深。前肢细长，前臂及手长超过体长之半；指式为3＞4＞2＞1；指端均有吸盘，指腹面有马蹄形边缘沟，指间微蹼。指侧均有一缘膜，第四指缘膜延至掌部；关节下瘤均很发达，掌突3个，不甚明显。后肢细长，前伸贴体时胫跗关节达眼与鼻孔之间，左、右跟部重叠，胫长约为体长之半，足短于胫。趾吸盘略小于指吸盘，指、趾吸盘背面可见到"Y"形迹；趾间具半蹼，外侧3趾间蹼达第二关节下瘤，内侧2趾蹼达第一关节下瘤，外侧蹠间蹼不甚发达；关节下瘤发达，内蹠突扁平，外蹠突甚小。皮肤平滑，背面有很小的痣粒；颞褶平直而长，达肩上方；头顶皮肤不紧贴头骨。腹面有扁平疣，咽、胸部的较小，腹部的大而稠密。活体颜色常随栖息环境而改变，背面为浅棕色，耳上面有黑褐色斑纹，斑纹形状变异大，有的个体有4条或6条深色纵纹，有的前背部有"X"形斑或仅有不规则的点状斑。四肢背面有黑色斑纹；肛部至股后方有黄棕色及乳白色交织成的网状斑。腹面乳白色。

[第二性征]　雄蛙体较小；第一、第二指基部内侧背面有乳白色婚垫，有时第三指上也有不甚明显的婚垫；有1对咽侧下内声囊；雄性线明显。

[生活习性]　该蛙生活于茂兰地区的山区，常栖息在稻田、草丛或泥窝内，或在田埂石缝及附近的灌木、草丛中。行动较缓，跳跃力不强。白天隐于石缝中，夜晚出没于稻田或水塘附近捕食、鸣叫求偶、抱对产卵等。

[地理分布]　茂兰地区全区分布。国内分布于香港、广东、广西、海南、湖南、贵州、云南、西藏。国外分布于泰国、柬埔寨、老挝、越南（北部）、缅甸等地。

成体量度　　　　　　　　　　　　　　　　（单位：mm）

编号	头体长	头长	头宽	前臂及手长	后肢全长	胫长	跗足长
GZNU20170423099 ♂	45.33	16.92	13.51	20.94	75.48	23.14	33.03
GZNU20170423017 ♂	41.41	14.17	12.56	21.91	73.60	21.60	39.05
GZNU20170423055 ♀	56.81	19.68	16.50	25.56	90.90	27.49	54.26
GZNU20170423089 ♀	62.48	20.74	18.54	31.95	111.78	33.61	65.12
GZNU20170423090 ♀	48.27	17.74	15.81	27.14	89.22	26.22	48.03
GZNU20170423083 ♀	56.61	18.93	16.02	26.11	85.50	26.95	49.23

a、b　整体照

［模式标本］ 可能丢失 ［保护级别］ 三有保护动物
［模式产地］ 香港 ［IUCN 濒危等级］ 无危（LC）

c、d 整体照
e 在茂兰地区的分布

（29）无声囊泛树蛙 *Polypedates mutus* Smith, 1940

Rhacophorus mutus Smith, 1940, Rec. Indian Mus., 42: 473. Syntypes: BMNH 1940.6.1.3-4, by original designation. Type locality: N'Chang Yang, Northern Myanmar.

Polypedates mutus Inger, 1985, in Frost (ed.), Amph. Species World: 541; Fei, Ye and Huang, 1990, Key to Chinese Amph.: 184; Ye, Fei and Hu, 1993, Rare and Economic Amph. China: 326.

Rhacophorus mutus Tian et al., 1986, Handb. Chinese Amph. Rept.: 64.

Rhacophorus (*Rhacophorus*) *mutus* Dubois, 1987 "1986", Alytes, 5: 77.

［同物异名］ *Rhacophorus mutus*（无声囊树蛙）

［鉴别特征］ 股后有网状斑；外侧指间无蹼，第四趾外侧蹼达第二、第三关节下瘤之间，与斑腿泛树蛙极相似，但本种背面一般有 6 条深色纵纹；雄蛙无声囊。

［形态描述］ 依据茂兰地区标本描述 雄蛙头体长 44.15～58.89mm，雌蛙头体长 62.99～65.44mm；体窄长而扁平；头顶扁宽，头长大于头宽；吻端较尖，突出于下唇；鼻孔极近吻端，鼻间距小于眼间距；吻棱平直达鼻孔，颊部

略向外侧倾斜；眼颇大，眼间距略小于或几乎等于眼径；鼓膜大而清晰，距眼后角近；犁骨齿列在内鼻孔内前缘，平置，左、右列之间相距较远；舌后端缺刻深。前肢细长，前臂及手长约为体长之半；指端有大的吸盘，指腹面边缘有马蹄形沟，第一指吸盘较小；指式为 3 > 4 > 2 > 1；指间无蹼，指侧均有缘膜；关节下瘤小而明显，指基下瘤明显；掌突 3 个，有的不显。后肢细长，约为体长的 1.7 倍，后肢前伸贴体时胫跗关节达吻端或鼻孔，左、右跟部重叠，足比胫短；趾端与指端相同，但吸盘略小，指、趾吸盘背面可见到"Y"形迹；第三趾略短于第五趾；趾间具蹼，缺刻深，第四趾外侧蹼达第二、第三关节下瘤之间，其余各趾以缘膜达趾端；外侧蹠间蹼不甚发达；关节下瘤很发达，跟部有成行小疣；内蹠突细窄，外蹠突小。皮肤较光滑，头顶部皮肤与头骨相连；背面满布均匀很小的痣粒；颞褶平直，达肩部后逐渐消失。胸部、腹部、体侧及股下方密布扁平圆疣。背部为棕灰色，有 6 条明显的深色纵纹，一对始自吻端，一对始自上眼睑前端，最外侧的一对沿颞褶平向后，有的断续成点斑，排列相缀成行；四肢横纹清晰，股后方有与斑腿泛树蛙相同的网状斑；上唇缘有白色的细线纹。腹面白色，咽喉部具棕色小点，少数标本腹面也有小斑点。

[第二性征]　雄蛙体较小；第一、第二指背面基部有白色婚垫；无声囊；有雄性线。

[生活习性]　该蛙生活于茂兰地区的山区，多栖息于水塘边、稻田埂边的草丛或泥窝内，以及污水池、粪坑边的石缝内。

[地理分布]　茂兰地区分布于翁昂、洞塘等地。国内分布于云南、贵州、海南、广西、重庆。国外分布于缅甸、老挝、越南、泰国。

[模式标本]　模式标本（BMNH 1940.6.1.3-4）保存于英国伦敦自然历史博物馆

[模式产地]　缅甸

[保护级别]　三有保护动物；贵州省重点保护野生动物

[IUCN 濒危等级]　无危（LC）

成体量度　　　　　　　　　　　　　　　　　　　　（单位：mm）

编号	头体长	头长	头宽	前臂及手长	后肢全长	胫长	跗足长
GZNU20170508009 ♂	58.89	19.01	16.17	27.09	89.72	26.21	36.84
GZNU20170508010 ♂	47.23	18.62	16.20	25.96	88.13	25.83	51.18
GZNU20170508006 ♂	52.72	18.77	15.33	25.37	90.09	26.45	36.51
GZNU20170508007 ♂	44.15	15.90	13.01	22.54	74.49	22.20	31.30
GZNU20170508008 ♀	62.99	22.76	19.88	33.99	117.66	34.83	69.59
GZNU20170508005 ♀	65.44	23.60	21.12	33.06	123.86	37.02	73.48

a、b　整体照

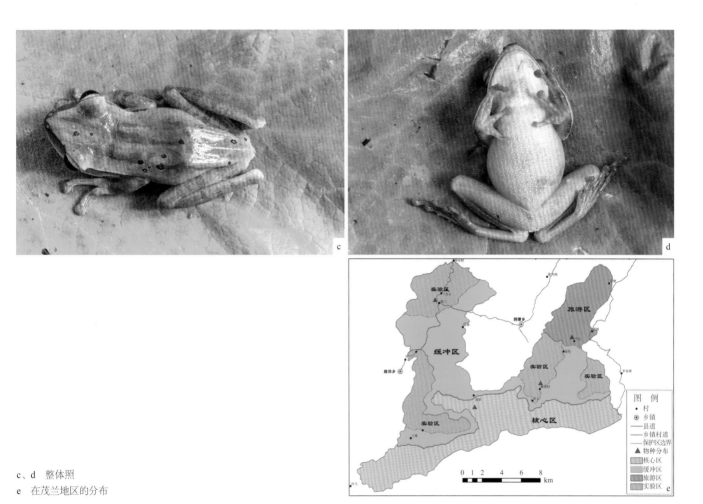

c、d 整体照
e 在茂兰地区的分布

18）树蛙属 *Rhacophorus* Kuhl and van Hasselt, 1822

Rhacophorus Kuhl and van Hasselt, 1822, Algemeene Konsten Letter-Bode, 7: 104. Type species: *Rhacophorus moschatus* Kuhl and van Hasselt, 1822 (= *Hyla reinwardtii* Schlegel, 1840), by monotypy (see comments by Dubois, 1989 "1988", Alytes, 7: 101-104; Ohler and Dubois, 2006, Alytes, 23: 123-132).

Racophorus Schlegel, 1826, Bull. Sci. Nat. Geol., Paris, Ser. 2, 9: 239. Incorrect subsequent spelling.

Leptomantis Peters, 1867, Monatsber. Preuss. Akad. Wiss. Berlin, 1867: 32. Type species: *Leptomantis bimaculata* Peters, 1867, by monotypy. Synonymy by Ahl, 1931, Das Tierreich, 55: 52; Harvey, Pemberton and Smith, 2002, Herpetol. Monogr., 16: 48.

Rhacoforus Palacky, 1898, Verh. Zool. Bot. Ges. Wien, 48: 374.

Huangixalus Fei, Ye and Jiang, 2012, Colored Atlas Chinese Amph. Distr.: 598. Type species: *Rhacophorus translineatus* Wu, 1977, by original designation. Provisionally retained in this synonymy because its recognition would render *Rhacophorus* paraphyletic (DRF).

鼻骨小，蝶筛骨显露，额顶骨宽或宽短，有的种类有次生真皮板，覆盖在额顶骨或鳞骨侧面；舌骨无舌角前突，有翼状突；喉器环状软骨有或无侧突；椎体前凹型或参差型；肩胸骨基部分叉，长于喙骨；指、趾末端骨节呈"Y"形，介间软骨菱形或心形或翼状；第三掌骨远端有结节；吴氏管卷曲；无瓶状储精囊。

体型中等或大，一般雄蛙吻端斜尖，显然与雌性有别；舌后端缺刻深，犁骨齿发达，鼓膜明显；指间有蹼或发达。

趾间近全蹼或满蹼，外侧蹠间蹼较发达；部分种前臂及跟部或肛上方有皮肤褶；第一、第二指与第三、第四指不相对，不形成握物状；指、趾末端吸盘及边缘沟明显，吸盘背面可见"Y"形迹。

蝌蚪体型中等大小；口部位于吻端腹面；眼位于头的背侧；尾鳍多低平，一般末端钝尖；大多数种类树栖性；雌蛙产出的卵粒较小，数量较多，卵粒多为乳黄色。卵团呈泡沫状，附着于水塘边植物上或泥窝内，或者悬挂在水塘上空的枝叶上。蝌蚪生活在静水水域内。

本属种类较多，有92种，分布在亚洲东部和南部亚热带及热带地区。中国现有34种，分布于秦岭以南各省区，以热带和亚热带种类较多。茂兰地区有2种，分种检索表如下。

茂兰地区种检索表

指间全蹼，第三、第四指间半蹼；体背面绿色，有棕黄色斑点·····································大树蛙 Rhacophorus dennysi
指间半蹼；背面有绿色与棕色斑纹交织成的网状斑·····································峨眉树蛙 Rhacophorus omeimontis

（30）峨眉树蛙 *Rhacophorus omeimontis* Stejneger, 1924

Polypedates omeimontis Stejneger, 1924, Occas. Pap. Boston Soc. Nat. Hist., 5: 119. Holotype: USNM 66548, by original designation. Type locality: Shin-kai-si, Mount Omei, Szechwan [= Sichuan], China.

Rhacophorus schlegelii omeimontis Wolf, 1936, Bull. Raffles Mus., 12: 195.

Rhacophorus davidi Pope and Boring, 1940, Pek. Nat. Hist. Bull., 15 (1): 70.

Rhacophorus omeimontis Liu, 1950, Fieldiana, Zool. Mem., 2: 379; Tian, Jiang, Wu, Hu, Zhao and Huang, 1986, Handb. Chinese Amph. Rept.: 65.

Polypedates omeimontis Liem, 1970, Fieldiana, Zool., 57: 98; Fei, Ye and Huang, 1990, Key to Chinese Amph.: 185.

Rhacophorus (Rhacophorus) omeimontis Dubois, 1987 "1986", Alytes, 5: 77.

Polypedates pingbianensis Kou, Hu and Gao, 2001, A new species of the genccs *Polyhedates* from Yunnan, China, 26 (2): 229-233.

Rhacophorus omeimontis Rao, Wilkinson and Liu, 2006, Zootaxa, 1258: 17.

［同物异名］ *Rhacophorus davidi*；*Polypedates pingbianensis*（屏边泛树蛙）；*Polypedates omeimontis*

［鉴别特征］ 体窄长而扁平；指间半蹼；趾蹼发达，除第四趾以缘膜达趾端外，其余均为全蹼；皮肤粗糙，满布小刺疣；背面多为绿色与棕色斑纹交织成的网状斑。

［形态描述］ 体窄长而扁平；头宽略大于头长；雄蛙吻端斜尖，明显突出于下唇，雌蛙吻端较圆而高，略突出于下唇；吻棱明显；鼻孔略近吻端，鼻间距与上眼睑几乎等宽，而小于眼间距；鼓膜大而圆；犁骨齿粗壮，左、右两列位于内鼻孔内侧上方，几乎平置，间距宽；舌后端缺刻深。皮肤粗糙，液浸标本整个背面布满细痣粒或呈白色小刺粒；颞褶平直，向后延至前肢基部上方。腹面及股部下方密布扁平疣。体色变异很大，生活时背面（包括四肢）多为草绿色，与不规则的棕色斑纹交织成网状斑；有的个体背面呈棕色，斑纹为绿色，有的个体体色为棕黑色。腹面乳白色，有的个体在咽胸部、腹侧及四肢腹面有大小不等的黑点。液浸标本的色斑为紫灰色与棕色构成的不规则斑纹（费梁等，2012）。

［第二性征］ 雄性吻端斜尖；第一、第二指基部背面内侧有乳白色婚垫；雄性有1对咽侧下内声囊，声囊孔长裂形；有雄性线。

［地理分布］ 该蛙是1984年李德俊等在茂兰地区调查时记录到的，在后期个别文献中也有提到，本次调查未曾发现，故不做详细描述。国内分布于云南、贵州、广西、四川、湖北、湖南等地。

［模式标本］ 正模标本（USNM 66548）保存于美国国家自然历史博物馆

［模式产地］ 四川峨眉山新开寺

[保护级别]　三有保护动物　　　　　　　　　　　　[IUCN 濒危等级]　无危（LC）

（31）大树蛙 *Rhacophorus dennysi* Blanford, 1881

Rhacophorus dennysi Blanford, 1881, Proc. Zool. Soc., London, 1881: 224. Holotype: ZRC (according to the original publication); RMNH (according to Stejneger, 1925, Proc. U.S. Natl. Mus., 66: 31, apparently in error). Type locality: China; but the precise locality is unknown.

Rhacophorus exiguus Boettger, 1894, Ber. Senckenb. Naturforsch. Ges., 1894: 148. Types: not stated; SMF 6987 is holotype according to Mertens, 1967, Senckenb. Biol., 48 (A): 49. Type locality: Chinhai bei Ningpo, China. Considered a subspecies of *Rhacophorus schlegelii* by Wolf, 1936, Bull. Raffles Mus., 12: 194, but this is currently untenable. Synonymy by Liu and Hu, 1961, Tailless Amph. China: 258; Zhao and Adler, 1993, Herpetol. China: 155.

Polypedates dennysi: Stejneger, 1925, Proc. U.S. Natl. Mus., 66: 31; Fei, Ye and Huang, 1990, Key to Chinese Amph.: 184.

Hyla albotaeniata Vogt, 1927, Zool. Anz., 69: 287. Holotype: ZMB 24118 (according to Pope, 1931, Bull. Am. Mus. Nat. Hist., 61: 462). Type locality: Südchina. Probably Yunnan Province, according to Zhao and Adler, 1993, Herpetol. China: 155. Synonymy by Liu and Hu, 1961, Tailless Amph. China: 258.

Polypedates feyi Chen, 1929, China J. Sci. Arts, Shanghai, 10: 198. Holotype: Sun Yat-sen Univ. Collection 1205, by original designation. Type locality: Yaoshan [= Mt. Dayao], Kwangsi [= Guangxi]", China. Synonymy by Pope, 1931, Bull. Am. Mus. Nat. Hist., 61: 563; Liu and Hu, 1961, Tailless Amph. China: 258.

Rhacophorus (*Rhacophorus*) *exiguus* Ahl, 1931, Das Tierreich, 55: 113.

Rhacophorus (*Rhacophorus*) *dennysi* Ahl, 1931, Das Tierreich, 55: 162; Dubois, 1987 "1986", Alytes, 5: 77.

Rhacophorus nigropalmatus exiguus Wolf, 1936, Bull. Raffles Mus., 12: 201.

Rhacophorus nigropalmatus dennysi Wolf, 1936, Bull. Raffles Mus., 12: 201.

Rhacophorus dennysi Inger, 1966, Fieldiana, Zool., 52: 321; Wilkinson, Drewes and Tatum, 2002, Mol. Phylogenet. Evol., 24: 272.

Polypedates dennysi Liem, 1970, Fieldiana, Zool., 57: 98.

Rhacophorus (*Rhacophorus*) *dennysi* Dubois, 1987 "1986", Alytes, 5: 77.

[同物异名]　*Rhacophorus exiguous*；*Polypedates dennysi*（大泛树蛙）；*Hyla albotaeniata*；*Polypedates feyi*

[鉴别特征]　体型大，雄蛙平均体长81mm，雌蛙体长99mm左右；第三、第四指间半蹼；背面绿色，其上一般散有不规则的少数棕黄色斑点，体侧多有成行的乳白色斑点或连成乳白色纵纹；前臂后侧及跗部后侧均有1条较宽的白色纵线纹，分别延伸至第四指和第五趾外侧缘。

[形态描述]　依据茂兰地区标本描述　雄蛙头体长60.67～68.08mm；体略扁平；鼓膜大而圆；头长略大于头宽，吻端斜尖，明显地突出于下唇；吻棱棱角状，鼻孔近吻端，鼻间距小于眼间距而略大于上眼睑宽；瞳孔呈横椭圆形；鼓膜大而圆；犁骨齿强壮，位于内鼻孔内前方，左、右列几乎平置，相距颇宽；舌宽大，后端缺刻深。前肢粗壮，前臂及手长略大于体长之半；指端均具吸盘，指腹面边缘有马蹄形沟，背部可见到"Y"形迹，腹面有清晰的肉质垫，第一指吸盘略小；指式为 3 > 4 > 2 > 1；指间全蹼。第三、第四指间为半蹼，蹼厚而色深，有网状纹，第一、第四指游离侧缘膜明显；关节下瘤发达，有单个或成行的指基下瘤；内掌突椭圆形，外掌突小或不显。后肢长，前伸贴体时胫跗关节达眼或眼前方，左、右跟部不相遇或仅相遇，胫长小于体长之半；趾端与指端同，但吸盘较小；趾间全蹼，蹼厚而色深，上有网状纹；第一、第五趾游离侧具缘膜；关节下瘤发达，蹠底部有成行的小疣；内蹠突小，无外蹠突。背面皮肤较粗糙有小刺粒；腹部和后肢股部密布较大扁平疣；体色和斑纹有变异，生活时多数个体背面绿色，体背部有镶浅色线纹的棕黄色斑点；沿体侧一般有成行的白色大斑点或连成乳白色纵纹，下颌及咽喉部为紫罗蓝色；腹面其余部位灰白色。前臂后侧及跗部后侧均有1条较宽的白色纵线纹，

成体量度 （单位：mm）

编号	头体长	头长	头宽	前臂及手长	后肢全长	胫长	跗足长
GZNU20170504097 ♂	63.63	26.69	24.04	32.50	105.32	31.02	63.66
GZNU20170504098 ♂	68.08	24.19	22.86	37.00	128.41	38.65	47.55
GZNU20170504099 ♂	60.67	22.67	18.95	30.16	100.81	32.85	38.95

a～d 整体照
e 在茂兰地区的分布

分别延伸至第四指和第五趾外侧缘。

［第二性征］ 雄蛙体略小；吻长而斜尖；第一、第二指基部内侧背面有浅灰色婚垫，具单咽下内声囊，声囊孔长裂形；有雄性线。

［生活习性］ 该蛙生活于茂兰地区山区的树林里或附近的田边、灌木及草丛中。常在水边高大乔木的枝干上，白天隐于树冠枝叶密集层，夜间则会爬至近地面鸣叫求偶。

［地理分布］ 茂兰地区全区分布。国内分布于广东、广西、海南、福建、湖南、江西、浙江、湖北、上海、安徽、河南、贵州、重庆。国外分布于缅甸、越南。

［模式标本］ 标本保存于美国佛罗里达州发现和科学博物馆（Museum of Discovery and Science, Fort Lauderdale, Florida, USA）

［模式产地］ 具体地点不详，可能位于中国南方

［保护级别］ 三有保护动物

［IUCN濒危等级］ 无危（LC）

（32）树蛙未定种 *Rhacophorus* sp.

本次采集到树蛙属1未定种，该种形似侏树蛙或白线树蛙，但与侏树蛙及白线树蛙在形态上有多处不同，目前尚未确定其种名，由于标本使用福尔马林保存，没有提取出足量的DNA，因此没有进行系统发育的遗传分析，故将该物种先列为未定种。

［物种描述］ 标本均为成体，平均头体长37.13mm。头长均值11.44mm，头宽略大于头长；吻棱明显，呈棱角状，微突出于下唇；鼻孔位于吻眼之间，略近于吻端，鼻间距小于眼间距；颊褶弯曲，鼓膜圆而明显，距眼后角较近；舌后端缺刻较浅。

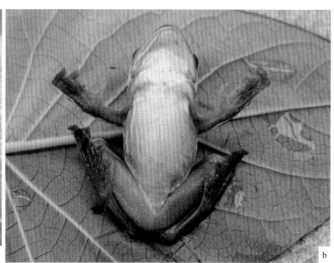

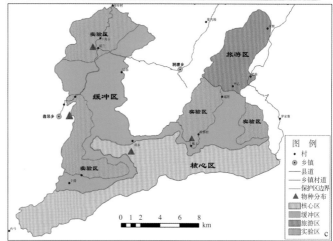

a 侧面照（成体）
b 腹面照（成体）
c 在茂兰地区的分布

测量性状数据

（单位：mm）

项目		标本1	标本2	标本3	标本4	标本5	平均值（占头体长的比例）
头体长		40.28	37.85	36.54	30.89	40.09	37.13（100%）
头长		13.04	11.24	11.12	9.71	12.08	11.44（30.81%）
头宽		13.86	12.57	13.63	10.02	13.29	12.67（34.12%）
吻长		5.46	5.05	4.87	3.91	4.87	4.83（13.01%）
鼻间距		3.98	3.81	3.60	3.51	3.88	3.76（10.13%）
眼间距		5.13	4.22	4.12	4.56	5.65	4.74（12.77%）
眼径		4.35	4.07	4.11	4.06	5.03	4.32（11.63%）
眼鼻距		3.63	3.38	3.21	3.51	4.42	3.63（9.78%）
上眼睑宽		2.29	2.71	3.02	2.69	3.31	2.80（7.54%）
鼓膜径		2.56	2.36	2.37	2.17	2.94	2.48（6.68%）
前臂及手长		17.41	15.79	16.69	13.45	17.42	16.15（43.50%）
前臂宽		3.79	3.03	2.88	2.29	2.84	2.97（8.00%）
后肢全长		56.44	50.54	49.72	41.85	53.24	50.36（135.63%）
胫长		19.12	17.78	16.88	13.87	19.50	17.43（46.94%）
胫宽		4.04	2.93	2.80	2.75	3.38	3.18（8.56%）
足长		16.17	13.29	13.26	10.68	16.35	13.95（37.57%）
第三指吸盘	横径	2.57	1.28	1.58	1.17	2.02	1.72（4.63%）
	纵径	1.42	1.65	1.43	1.13	1.52	1.43（3.85%）
第四趾吸盘	横径	1.68	1.11	1.13	0.70	1.71	1.27（3.42%）
	吸盘	1.20	0.94	1.29	0.71	1.20	1.07（2.88%）

前肢较短，指端具吸盘，指腹面有边缘沟，第三指吸盘小于鼓膜；吸盘背面可见明显"Y"形迹，外侧3指具微蹼；关节下瘤较发达，有指基下瘤，掌部有小瘤，内掌突卵圆形，外掌突不显。后肢较短，前伸贴体时胫跗关节达眼的后方，左、右跟部不相遇，胫长不到体长之半，足略长于胫长；趾端与指端同，趾吸盘略小，趾吸盘背面可见"Y"形迹；第三、第五趾等长，达第四趾的第三关节下瘤；趾间约为1/3蹼，蹼缘缺刻深，缘膜达趾端，外侧蹠间几无蹼；关节下瘤明显，蹠部有小疣，内蹠突椭圆形，无外蹠突。

皮肤较光滑，上眼睑、颌后部和前臂外侧有小疣，咽胸部有小疣或不明显，腹部和股部腹面布满扁平疣，股后部疣粒较大，肛上方略显肤棱。体和四肢背面呈浅绿色，散有褐色细点；从上唇经颞部下方和体侧至胯部有1条黄白色带纹；四肢内外侧、手和跗足的外侧均为黄白色；四肢被遮盖部位及指端、趾端均为肉黄色；腹面肉黄色或肉红色，无斑或有褐色斑。

［地理分布］ 茂兰地区分布于永康、翁昂、洞塘等地。

［模式标本］ 标本保存于贵州师范大学生命科学学院生物学实验室

［模式产地］ 荔波茂兰

第三章
爬行动物术语及分类描述

一、爬行纲分类检索名词术语

（一）龟鳖目特征及分类检索常用术语

龟鳖目种类体被背腹甲，是现存爬行动物中特化的一类。许多淡水龟与陆龟的背甲与扁平的腹甲构成甲壳，遇敌或受惊时，头、尾和四肢可缩入甲壳内。龟鳖目动物上、下颌无齿，无胸骨，脊椎和肋骨多与背甲的骨质板愈合，仅颈椎与尾椎骨游离，椎体有前凹、后凹及两凹型。龟类硬甲的背甲以甲桥和腹甲相连，盾片缝和骨板缝不重叠，比较坚固。鳖类的背甲和腹甲以韧带相连，没有盾片，骨板外被有革质皮肤；背甲无缘板，肋骨向两侧突出于肋板之外。肩带由肩胛骨、前肩胛骨及喙骨构成，锁骨和间锁骨分别与腹甲的内板及上板合并。

世界现存龟鳖目种类 13 科 89 属 270 多种，中国有 6 科 22 属 38 种，贵州现有 2 科 4 属 5 种，茂兰地区有 1 种，即中华鳖 Pelodiscus sinensis。

龟鳖目的鉴定主要依据骨骼结构、皮肤及头部鳞片、四肢形状、背甲和腹甲特征、骨板和盾片特征等。

1. 龟壳背甲的骨板

颈板（proneural）：背甲腹面中央最前面的 1 块骨板，即颈盾变为的骨板。

椎板（neural）：颈板之后中央的 1 列骨板，与背椎的椎弓相愈合，一般为 8 块。

肋板（pleural）：椎板两侧，常左、右各 8 块，分别与一肋骨相接。

缘板（peripheral）：肋板外侧，背甲边缘的骨板，常左、右各 11 块。鳖类的背甲无缘板。

上臀板（epipygal）：椎板之后，常有 1~3 块，由前至后分别称第一上臀板和第二上臀板。

臀板（pygal）：上臀板之后，1 枚。

2. 龟壳背甲的盾片

颈盾（pronentral）：背甲背面前部（椎盾前方），嵌于左、右 2 枚缘盾间的 1 枚小盾片。

椎盾（central）：颈盾之后中央的 1 列盾片，常为 5 枚。

肋盾（lateral）：位于椎盾两侧，左、右各 4 枚。

缘盾（marginal）：背甲边缘两列较小的盾片，通常左、右各 12 枚。

上缘盾（supramar scute）：肋盾左、右两侧与左、右缘盾之间的两列细长的盾片，通常左、右各 7~8 枚。

3. 龟壳腹甲的骨板

腹甲骨板由 11 枚骨板组成，除内板为 1 枚外其余 10 枚均成对。由前向后依次为上板、内板、舌板、间下板、下板及剑板。

上板（epiplastron）：腹甲最前缘 1 对骨板。

内板（entoplastron）：单枚，介于上板与舌板中央，形状与位置变化较大，或缺失。

舌板（hyoplastron）：又称中腹板，1 对，位于上板、内板和中下板之间。

间下板（mesoplaston）：舌板和下板之间的 1 对盾片。

下板（hyoplaston）：间下板和剑板之间的 1 对盾片。

剑板（xiphiplaston）：腹甲最后 1 对盾片。

4. 龟壳腹甲的盾片

腹甲的盾片由 6 对左右对称的盾片和 1 枚间喉盾组成，由前至后依次为间喉盾、喉盾、肱盾、胸盾、腹盾、股盾及肛盾。左、右喉盾相连的沟称为喉盾沟，喉盾与肱盾的沟称为喉肱沟。其余依次类推。

间喉盾（intergular）：腹甲最前缘正中央 1 枚盾片。

喉盾（gular）：间喉盾和肱盾之间的 1 对盾片。

肱盾（humeral）：喉盾和胸盾之间的 1 对盾片。

胸盾（pectoral）：肱盾和腹盾之间的 1 对盾片。

腹盾（abdominal）：胸盾和股盾之间的 1 对盾片。

股盾（femoral）：腹盾和肛盾之间的 1 对盾片。

肛盾（anal）：腹甲后部的 1 对盾片。

5. 龟壳的甲桥

甲桥为腹甲舌板及下板向两侧延伸的部分，以骨缝或韧带与背甲相连接。此处外层的盾片有以下 3 种。

腋盾（axillary）：靠近腋凹的盾片，左、右各 1 枚。

胯盾（inguinal）：靠近胯凹的盾片，左、右各 1 枚。

下缘盾（inframarginal）：腹甲的胸盾、腹盾与背甲的缘质之间的几枚小盾片。

（二）有鳞目特征及分类检索常用术语

Ⅰ. 蜥蜴亚目

蜥蜴通体覆盖的鳞片（scale、shield、scute），鳞的大小、形状和数目总称鳞被（lepidosis）。鳞被特征是分类检索的重要依据。此外，外部形态的其他特征，如舌的形态，指、趾形状及结构等，也是分类的检索依据。本书主要介绍各部位鳞片及有关结构的名称。

1. 头部的鳞片及相关结构

1）头背的鳞片及相关结构

吻鳞（rostral）：吻端口缘正中的 1 枚鳞片，一般较其两侧的上唇鳞宽、高，如较高时，从吻背可以看到。

吻后鳞（postrostral）：与吻鳞后（上）缘相切的 1 或数枚小鳞。

鼻间鳞（internasal）：左右鼻鳞之间的单枚、成对或一团不规则的鳞片。除上鼻鳞（如有的话）外，凡位于左右鼻鳞之间的鳞片都称为鼻间鳞；也可能有其他鳞片次生地伸入此范围。

上鼻鳞（supranasal）：与鼻鳞上缘相切的 1 或数枚小鳞，形态与其他鼻间鳞有别。上鼻鳞也可能是很大的 1 枚鳞片，参与围成鼻孔，且左右上鼻鳞在吻背相切（胎生蜥蜴）。

额鼻鳞（frontonasal）：介于鼻间鳞（前）、前额鳞（后）与颊鳞（两侧）之间的鳞片，如前述某种鳞片缺如，则可根据位置判定。额鼻鳞可能是一团小鳞，也可能是 1 横排 3 枚，或者只有单枚。

前额鳞（prefrontal）：通常是额鳞前方的 1 对鳞片，彼此相切或间隔 1 枚小鳞，或由于额鳞与额鼻鳞相接而被分开不相切。

额鳞（frontal）：位于左右眶背之间，通常是 1 枚较大的鳞片，五边形或六边形（龟甲形），也有成一团不规则的小鳞。

额顶鳞（frontoparietal）：介于额鳞与顶鳞之间的鳞片，通常成对。

顶间鳞（interparietal）：额顶鳞之后正中的 1 枚鳞片，有的由于左右额顶鳞相切而将额鳞与顶间鳞分开；顶"眼"即位于顶间鳞上，呈一小白点，有的种类不明显。

顶鳞（parietal）：额顶鳞之后、顶间鳞两侧的鳞片，通常是头背最大的 1 对鳞片，由于顶间鳞的存在，左右顶鳞互相不切或在顶间鳞之后以一短的鳞沟相切；如无顶间鳞则左右顶鳞在中线相切。

枕鳞（occipital）：顶间鳞正后方的较小鳞片，如有时，顶鳞在顶间鳞之后也不相切。

眶上鳞（supraocular）：眼眶背面、额鳞两侧的较大鳞片，一般每侧 1～4 枚（少数 5 枚），前后 2 枚较小，中间 2 枚较大。蜥蜴科许多种类眶上鳞周围镶以粒鳞。

上睫鳞（superciliari）：眼眶上方、眶上鳞外侧的 1 列小鳞，构成头背眶上方两侧的上睫脊。上睫鳞或为较小的平砌鳞片（石龙子科），或为覆瓦状排列或扭成麻花状的较大鳞片（鬣蜥科岩蜥属）。有的种类上睫脊发达，棱出，甚至略翘向上方；有的向前延伸几达鼻鳞。

颈鳞（nuchal）：蜥蜴科头背顶鳞之后多为较小鳞片，石龙子科顶鳞之后往往有左右交错排列的 0～4 对较大鳞片，称为颈鳞。颈鳞也可能两侧不对称（数目不一致）。

2）头侧的鳞片及相关结构

鼻鳞（nasal）：鼻孔开口于其上的鳞片，通常为完整的 1 枚，有的有局部鳞沟。有的鼻鳞退化为小鳞，与其他鳞片共同围成鼻孔。

前鼻鳞（prenasal）：鼻鳞前方的 1 枚小鳞。鼻鳞周围的小鳞视情况可分别称为前鼻鳞、后鼻鳞、上鼻鳞和下鼻鳞。前鼻鳞也可能是前述的吻后鳞。

后鼻鳞（postnasal）：与鼻鳞后缘相切的 1～2 枚小鳞。

颊鳞（loreal）：吻部两侧、眶前鳞与鼻鳞或后鼻鳞（如有的话）之间，吻棱下方与上唇鳞之间的 1 列数枚小鳞。

眶周鳞：除前面已提及的眶上鳞位于眶背外，眼眶周围细小粒鳞以外的若干稍大的小鳞片。视其位置分别称为眶前鳞（preocular）、眶后鳞（postocular）与眶下鳞

（subocular）。

眼睑鳞（palpebral）：被覆眼睑的鳞片。可分为上眼睑鳞（upper palpebral）与下眼睑鳞（lower palpebral）。滑蜥属物种的部分下眼睑鳞愈合为一大的半透明的睑窗（pallpebral disc）。壁虎科物种没有活动的眼睑，亦无眼睑鳞。

睑缘鳞（ciliarie）：上下眼睑游离缘的1行矩形小鳞。可分为上睑缘鳞（upper ciliarie）与下睑缘鳞（lower ciliarie）。凡有眼睑的种类都有睑缘鳞；沙蜥属物种的睑缘鳞呈明显的锯齿状。

眼鳞（ocular）：罩于眼外的1枚大鳞，退化穴居的种类具有。

上唇鳞（supralabial）：上颌口缘除前端吻鳞以外的鳞片。

上唇后鳞（upper postlabial）：上唇鳞之后，口角后方与上唇鳞在一条线上的若干鳞片。石龙子属、滑蜥属与蛇蜥属物种有此鳞片。

颞鳞（temporal）：眼与耳孔（鼓膜）间、顶鳞与上唇鳞之间颞部的鳞片。如排列有序，可分为前列的初级颞鳞（primary ternporal），后列的次级颞鳞（secondary temporal），有的还可区分出三级颞鳞（tertiary temporal）。

上颞鳞（supratemporal）：颞鳞较小而排列不规则的种类颞区上部的鳞片较大，可称为上颞鳞。如鬣蜥科岩蜥属物种就有较大的上颞鳞，其下方为粒鳞。

鼓膜（tympaunrn）：在耳孔上的1层薄膜，有的裸露透明，有的被鳞，有的种类鼓膜位于表面，有的种类鼓膜下陷。

外耳道（external auditory meatus）：鼓膜下陷的种类耳孔内形成外耳道。

耳孔瓣（auricular lobule）：有的种类耳孔前缘数枚鳞片饰变成突出的瓣突。

3）头腹的鳞片及相关结构

颏鳞（mental）：下颌口缘前端的1枚鳞片，一般大于其两侧的下唇鳞，略呈倒三角形或馒头形。

后颏鳞（submental）：颏鳞之后沿腹中线不成对或成对的鳞片。

下唇鳞（infralabial）：下颌口缘除颏鳞以外的鳞片。

下唇下鳞（sublabial）：下唇鳞内侧与之平行的几行窄长鳞片。

颏片（chin shield）：颏鳞或后颏鳞（如有的话）后方成对的大鳞片，前部者常左右相切，后部者常被喉鳞分开，左右不相切。也有称颏片的。

喉鳞（gular）：头腹中央许多较小的鳞片。

喉褶（gular fold）：横跨颈腹，恰在前肢前方的皮肤褶皱，褶缘被细鳞，如蜡皮蜥。

喉囊（gular pouch 或 gular sac）：鬣蜥科一些种类雄性喉部皮肤延伸形成的囊状结构，常有鲜艳颜色，可由动物自身控制伸长或缩小。

颈侧囊（lateral flap）：颈侧皮肤形成的囊状构造，如斑飞蜥。

领围（collar）：颈侧皮肤褶，褶缘具1排大鳞，如蜥蜴属与麻蜥属物种。

肩褶（fold in front of the shoulder）：肩前形成的皮肤皱褶，如鬣蜥科某些属种。

2. 躯干部的鳞片

背鳞（dorsal）：覆于躯干背部的鳞片。背鳞数目可以反映鳞片的大小，在分类检索时有一定的意义，计算方法为前端始自顶间鳞后，沿背脊中线，到两后肢基部前缘连线的中点。背鳞又可分为脊鳞与脊侧鳞。

脊鳞（vertebral）：沿背中线一行或若干行略为扩大的鳞片。

脊侧鳞（paravertebral）：脊鳞两外侧的背鳞，位于脊鳞与侧鳞之间。

侧鳞（latera）：躯干两侧主要是腋胯之间的鳞片。

腹鳞（ventral）：覆于躯干腹面的鳞片。腹鳞数目具有分类检索的意义。计算腹鳞的方法是前端始自喉褶（如无喉褶则取两前肢前缘连线的中点），沿腹中线数至肛孔前。

胸鳞（pectoral）：躯干腹面胸部的鳞片。

股间鳞（interfemoral）：两后肢间的腹鳞。

肛前鳞（preanal）：泄殖肛孔前方的数枚较大鳞片，往往有2枚、4枚或更多，两侧常有略小的鳞片数枚。

胼胝鳞（callosity）：鬣蜥科部分种类雄性肛前或腹部中央有成团增厚的鳞片，相应称为肛前胼胝鳞（preanal callosity）与腹部胼胝鳞（abdominal callosity）。胼胝鳞也称板鳞。

环体一周鳞（scales around mid-body）：有分类检索意义。计算方法为沿前后肢间身体中部自背至腹再回到背部起始处一周的鳞片数目。

3. 四肢的鳞片及相关结构

四肢各部鳞片都有专门术语，现简介如下。前肢上臂的鳞片称为上臂鳞（brachial），按其位置分背、腹（内侧）、前、后，分别称为上臂背鳞（suprabrachial）、上臂腹鳞或上臂内侧鳞（infrabrachial）、上臂前鳞（prebrachial）和上臂后鳞（postbrachial）。前臂的鳞片称为前臂鳞（antebrachial），后肢股部（大腿）的鳞片称为股鳞（femoral），胫部（小腿）的鳞片称为胫鳞（fibial），都可以参照上臂鳞分为4个部位，分别冠以supra-、infra-、pre-与post-，如前臂背鳞（supraantebrachial）、前臂前鳞（preantebrachial）、股后鳞（postfemoral）与胫内侧鳞（infratibial）等。现将四肢其他鳞片分别介绍如下。

掌背鳞（supracarpal）：掌部（不包括指）背面的鳞片。

掌下鳞（infracarpal）：掌部（不包括指）腹面的鳞片。

蹠背鳞（supratarsal）：蹠部（不包括趾）背面的鳞片。

蹠下鳞（infratarsal）：蹠部（不包括趾）腹面的鳞片。

指（趾）下瓣（subdigital lamellae）：指（趾）腹面的鳞片，分类检索常用到第四指（趾）下瓣。指（趾）背面的鳞片称为指（趾）背鳞（supradigital lamellae），不常用。

翼膜（wing-like membrane）：体侧前后肢间由伸长的肋骨支持的皮肤膜，如飞蜥属物种。

肛前孔（preanal pore）：泄殖肛孔前方鳞片上的小孔，常呈一横排，如壁虎科物种。

鼠蹊孔（inguinal pore）：鼠蹊部鳞片上的小孔，1～5个，如地蜥属与草蜥属物种。

股孔（femoral pore）：股部腹内侧鳞片上的小孔，一列有数个至20余个，如鬣蜥科某些属种。

4. 尾部的鳞片

尾鳞（caudal）：覆于尾部鳞片的通称。

尾下鳞（subcaudal）：尾部腹面的鳞片，一般横向扩大，只有1行。

尾环（caudal whorl）：尾部鳞片整齐排列成环，如岩蜥属物种。

节（segment）：呈环状排列的尾鳞往往数环组成一节，如新疆岩蜥每四环组成一节，吴氏岩蜥每三环组成一节。

按鳞片的形态及排列方式，有以下各种情况。

圆鳞（cycloid）：较大的鳞片，游离缘呈弧形，一般呈覆瓦状排列，如石龙子科物种。

方鳞（pavimentous）：较大而呈方形的鳞片，如蜥蜴科物种的腹鳞。

粒鳞（granular）：较小而表面微凸的鳞片。

统鳞（tubercle）：杂于粒鳞间略大的圆球状鳞片。

棱鳞（carinate）：表面起棱呈纵脊的鳞片。

平鳞（smooth）：表面平滑无棱的鳞片。

尖鳞（mucronate）：棱鳞的棱在鳞片游离缘向后尖出者。

刺鳞（spinous）：扁平尖出，上翘如刺的鳞片。

锥鳞（conical）：尖出耸立如圆锥状的鳞片。

鬣鳞（crestal）：组成鬣蜥科物种颈鬣或背鬣的一列尖出的刺鳞或锥鳞。

粗鳞（rugose）：表面粗糙不光滑，或凹凸不平的鳞片。

纹鳞（striated）：表面有饰纹的鳞片。

缺凹（notched）：鳞片游离缘向内凹入呈缺刻的鳞片。

锯齿状鳞（denticulate）：指某部分游离缘的鳞片排列如锯齿状，如一般的睑缘鳞和草原蜥的上下唇鳞。

缨脊或栉状缘（fringe）：身体某部位边缘的一列鳞片尖出如刺，形似缨或栉状，如沙蜥属物种指（趾）侧缘的鳞片。

5. 蜥蜴各部分的量度

头体长（吻肛长，snout-vent length，SVL）：吻端到泄殖肛孔前缘的长度。

尾长（tail length，TL）：完整尾部从泄殖肛孔前缘到尾尖的长度，如断尾与再生尾应注明。

体全长（total length，TOL）：吻端到尾尖的长度，完整的尾部才可使用体全长的量度。

头长（head length，HL）：吻端到耳孔（鼓膜）后缘的长度，应取吻端到左右耳孔后缘连线的中点测量。如无耳孔或鼓膜隐蔽，则另定一位置（如颌角、顶鳞后缘等）并加以注明，以便重复。

头宽（head width，HW）：可取若干位置为标准，如颞部宽度或眼眶处宽度，一般可取头部最宽处的直线长度。

头高（head depth，HD）：一般取头部最高处的直线长度。

前肢长（length of foreleg）：腋下到最长指端（不包括爪）的长度。

后肢长（length of hind leg）：鼠蹊部（后肢基部）到最长趾端（不包括爪）的长度。

腋胯距（axilla-groin）：前肢后缘基部到后肢前缘基部之间的直线距离。

前后肢贴体相向（limbs adpressed）：前肢贴体后伸，同侧后肢贴体前伸时，指（趾）端（不包括爪）的距离，或指（趾）重叠的长度。后肢贴体前伸时，趾端所达位置（如腋后、肩前、颞角、耳孔、眼耳间、眼前角、吻端等）也可作为描述项目。

Ⅱ. 蛇亚目

蛇类具有以下特征：①没有肩带及附肢；②没有活动眼睑；③没有外耳及鼓膜；④没有上颞弓，鳞骨消失；⑤头骨简化，泪骨、轭骨、方轭骨、上翼骨及后额骨消失；⑥由于额骨及顶骨侧突下延而脑颅完全封闭；⑦脊椎有前、后关节突；⑧尾椎的脉突不在腹面会合。此外，尚有以下一些特征：下颌左右两半以韧带相连（盲蛇类除外）；枕大孔由外枕骨与基枕骨围成；左主动脉弓大于右主动脉弓（蟒类有些种除外）；眼无睫状肌。有些特征为蛇类与某些蜥蜴所共有，例如，没有四肢，某几类蜥蜴也无四肢，但蛇类绝无前肢与肩带，而无四肢的蜥蜴也有肩带；长而深分叉的前舌，并可缩入后舌，与巨蜥类共有此特征。

蛇通体覆盖鳞片，鳞片的排列和性状总称"鳞被"，鳞被的特征是分类检索常用的依据。除此之外，其他特征如上颌齿的数目、大小及排列方式，椎体下突是否存在及其位置等，也常被用作分类检索的重要依据。在此，主要对蛇类各部分鳞片做如下介绍。

1. 头部的鳞片

头部的鳞片如图 3-1 所示。除盲蛇科、蟒科、蚺科、瘰鳞蛇科、蝰科［圆斑蝰属、蝰属和烙铁头属（广义）］蛇类头部的鳞片或者较特殊或者都是一些小鳞片外，其他蛇类头部的鳞片一般可作如下的区分和命名。

1）头背的鳞片

头背的鳞片由前至后依次如下。

鼻间鳞（internasal）：位于头背最前端的 1 对鳞片，介于左右两枚鼻鳞之间。正常 1 对，有的种类没有（两头蛇属、海蛇亚科等的蛇类），有的只有 1 枚（黄腹杆蛇等），游蛇科水蛇属蛇类只有 1 枚且位于彼此相切的 1 对鼻鳞之后。

前额鳞（prefrontal）：位于鼻间鳞正后方的大鳞。正常 1 对，有的种类只有单枚（后棱蛇属蛇类、黄腹杆蛇等），有的种类纵分为两枚以上（滇西蛇等）。

额鳞（frontal）：位于前额鳞正后方的单枚大鳞，介于左右两枚眶上鳞之间，略呈六角形或龟甲形。

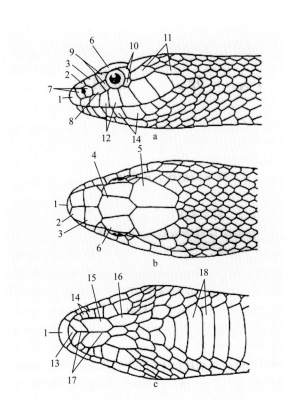

图3-1 蛇头部鳞片示意图（引自赵尔宓，2006）
a. 侧面观；b. 背面观；c. 腹面观。1. 吻鳞；2. 鼻间鳞；3. 前额鳞；4. 额鳞；5. 顶鳞；6. 眶上鳞；7. 鼻鳞；8. 颊鳞；9. 眶前鳞；10. 眶后鳞；11. 颞鳞；12. 上唇鳞；13. 颔鳞；14. 下唇鳞；15. 前颌片；16. 后颌片；17. 颌沟；18. 腹鳞

顶鳞（parietal）：位于额鳞正后方的大鳞，正常 1 对。闪鳞蛇属 *Xenopeltis* 蛇类为前后两对（图 3-2）。

顶间鳞（interparietal）：闪鳞蛇科蛇类 4 枚顶鳞中央围绕的单枚鳞片称顶间鳞（图 3-2）。

枕鳞（occipital）：眼镜王蛇 *Ophiophagus hannah* 顶鳞正后方 1 对大鳞片，称枕鳞（图 3-3）。

眶上鳞（supraocular）：位于额鳞两侧、围成眼眶上缘的 1 对大鳞片。

2）头侧的鳞片

正常头侧的鳞片是左右两侧对称排列，也有因变异而左右不完全对称者。每侧由前至后依次如下。

鼻鳞（nasal）：鼻孔周围的鳞片。有的种类鼻鳞为完整的 1 枚，有的种类鼻孔上方或（和）下方有鳞沟，将鼻鳞局部或完全分为前后两半。在没有鼻间鳞（如海蛇亚科蛇类）或虽有 1 枚鼻间鳞（游蛇科水蛇属 *Enhydris* 蛇类）

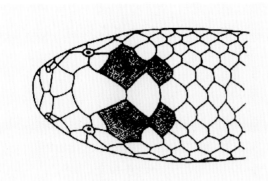

图3-2　闪鳞蛇属头部背视（引自田婉淑和江耀明，1986）
示前后2对顶鳞及其间的顶间鳞1枚

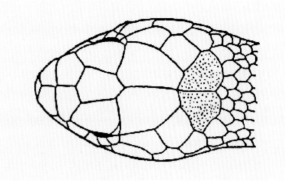

图3-3　眼镜王蛇头部背视（引自田婉淑和江耀明，1986）
示1对顶鳞正后1对较大的枕鳞（加网点者）

的情况下，左右鼻鳞在头背相切（图3-4）。

颊鳞（loreal frenal）：介于鼻鳞与眶前鳞之间的较小鳞片，通常1枚。有的没有（如两头蛇属、眼镜蛇科等的蛇类），有的多于1枚（如鼠蛇属物种）。有的种类没有眶前鳞或眶前鳞较小，颊鳞后伸，参与围成眼眶或入眶（entering eye）。

眶前鳞（preocular, praeocular）：位于眼眶前缘，1至数枚。如没有或较小时，可能由颊鳞和（或）前额鳞参与围成眼眶。

眶后鳞（postocular）：位于眼眶后缘，1至数枚。如没有，则由颞鳞参与围成眼眶。

眶下鳞（subocular）：多数种类没有，由部分上唇鳞参与围成眼眶下缘。如有眶下鳞，或者呈一长条完全围成眼眶下缘（钝头蛇属大多如此），或者较小，靠近眼前下方[眶前下鳞（presubocular）]或眼后下方[眶后下鳞（postsubocular）]。

颞鳞（temporal）：眼眶之后，介于顶鳞与上唇鳞之间，一般可分为前后2列或3列。其数目可以式表示，如1+2、2+2+3，分别表示前颞鳞（anterior ternporal）及后颞鳞（pvsterior temporal）或前中后2列的数目。

吻鳞（rostral）：位于吻端正中的一枚鳞片，其下缘（口缘）一般有缺凹，口闭合时，细长而分叉的舌可经此缺凹伸出。从背面一般只能看见吻鳞的上缘，如颈槽蛇属 *Rhabdophis*（图3-5），但如小头蛇属 *Oligodon* 的吻鳞甚高，且弯向吻背，从背面可看到的部分较多（图3-6）。

上唇鳞（upper labial, supralabial, superior labial）：位于吻鳞两侧之后上唇边缘的鳞片，其数目有鉴别意义，一般两侧对称，也有因变异而不对称者。有的中间几枚上唇鳞参与围成眼眶。如两侧上唇鳞均为9，第4、第5两枚入眶，其式可写为3-2-4；如上唇鳞左8右9，可写为3-2-3/4，

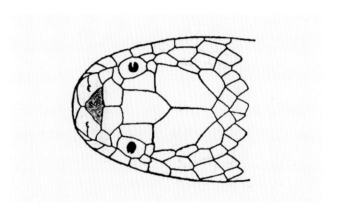

图3-4　水蛇属头部背视（引自田婉淑和江耀明，1986）
示单枚鼻间鳞（图中颜色较深区域），左右鼻鳞在吻背相遇

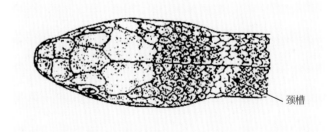

图3-5　颈槽蛇属头颈部背视（引自赵尔宓和鹰岩，1993）
示颈背正中的颈槽及前端吻鳞

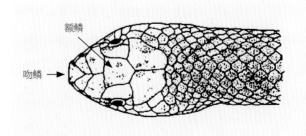

图3-6　小头蛇属背视（引自赵尔宓和鹰岩，1993）
示吻鳞向上弯，头部自吻背可见甚多

其余类推。

颏鳞(mental)：下颌前缘正中的一枚鳞片，略呈三角形。其位置恰与吻鳞相对应。

下唇鳞（lower labial, infralabial, inferior labial, sublabial）：位于颏鳞之后，下颌两侧下唇边缘的鳞片都称下唇鳞，两侧对称或不对称。大多数蛇类的第一对下唇鳞在颏鳞之后彼此相切，将颏鳞与前颏片分开；少数种类（如颈斑蛇属蛇类、美姑脊蛇和多种海蛇）的第一对下唇鳞左右不相切，故前颏片与颏鳞相切。下唇鳞的数目及其切前颏片的鳞数有鉴别意义，如两侧均为下唇鳞10，前5枚切前颏片，可写为10（5）；如下唇鳞左10右9，可写为10/9（5）；如下唇鳞左10右9，分别有5枚及4枚切前颏片，可写为10（5）/9（4），其余类推。

3）头腹的鳞片

颏片（chin-shield, sublingual）：位于颏鳞之后，左右下唇鳞之间的成对窄长鳞片。一般为两对，分别称为前颏片（anterior chin-shield）和后颏片（posterior chin-shield）。前颏片左右两枚常彼此相切，后颏片左右两枚则常有小鳞片将其分开。左右颏片之间形成的鳞沟，称为颏沟（mental groove, submental groove, symphysial groove）。钝头蛇属蛇类一般有3对颏片，左右镶嵌排列，无颏沟。瘰鳞蛇也无颏沟。蝰科蛇类一般只有1对颏片，其后往往还有几对较小呈对称排列的鳞片（图3-7）。

喉鳞（gular）：位于头部腹面除最前端中央的颏鳞外，下唇鳞、颏片及腹鳞之间的全部小鳞片，统称喉鳞，即头腹喉部的鳞片。一般在分类检索时很少用到它们的特征。

2. 躯干部的鳞片

腹鳞（ventral, abdominal, gastrostege, scutum）：位于躯干腹面、肛鳞之前、正中的一行较宽大的鳞片，统称腹鳞。腹鳞在陆栖蛇类的运动中起着重要的作用。树栖蛇类的腹鳞中央较平，甚至凹入，两侧略向背侧翘起，形成侧棱（lateral keel）；有的种类在相当于侧棱的腹鳞游离缘处尚形成缺凹（notch）。半水栖蛇类的腹鳞较不发达，长期适应水生生活的海蛇类腹鳞有程度不等的退化，甚至缺失；有的腹鳞中央有纵沟将其纵分为二。较低等的或穴居的种类腹鳞或者没有分化出来（如盲蛇科蛇类，周身被覆大小一致的鳞片）或者较窄，仅略宽于相邻的背鳞（如闪鳞蛇属、蟒科与蚺科蛇类）（图3-8）。腹鳞的大小和数目，以及前述特殊构造，有分类鉴别意义。

肛鳞（anal, preanal, postahdominal）：最末一枚腹鳞之后、紧覆于泄殖肛孔上方的鳞片，称为肛鳞。它或是完整的一片，称肛鳞完整（entire）；或是有裂沟纵分为二，称肛鳞二分（divided）。

背鳞（dorsal, costarl）：位于被覆躯干部的鳞片，除腹鳞和肛鳞外，统称背鳞。背鳞排列前后略成纵列，可以计算行数。计数一般取颈部（头后2个头长处）、中段（吻端到肛孔之间的中点处）及肛前（肛孔前2个头长处）3个数据。可以式表示，如写作21-19-17，表示背鳞在颈部21行、中段19行、肛前17行。如果只写背鳞19行，一

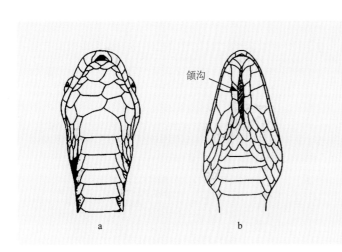

图3-7　钝头蛇属 *Pareas*（a）和紫沙蛇 *Psammodynastes pulverulentus*（b）头部腹视（引自赵尔宓和鹰岩，1993）

a. 示3对颏片交错排列，其间没有颏沟；b. 示3对颏片对称排列，其间形成颏沟

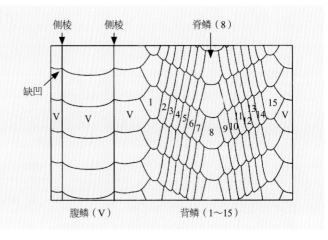

图3-8　过树蛇属 *Dendrelaphis* 躯干部自腹鳞剪断展开（引自赵尔宓和鹰岩，1993）

示背鳞（1～15）、脊鳞扩大（8）、腹鳞（V）的侧棱及其游离缘的缺凹

一般多指中段行数。背鳞行数一般都是奇数（odd），唯乌梢蛇属蛇类的背鳞行数为偶数（even）。除行数外，背鳞的形状如菱形（rhomboid）、披针形（lanceolate）、平砌或并列（juxtaposed）、起棱（keeled）或平滑（smooth）及起棱的强或弱等，有鉴别意义。有的种类背鳞排列成斜行（oblique，如斜鳞蛇属蛇类）。若需表示背鳞的行数，一般都从两外侧紧邻腹鳞的鳞片开始计数，如自外侧数第一行背鳞写为 D_1，第 5 行写为 D_5，如需表示第 4～6 行，可以写为 D_{4-6}，其余类推。此外，有的蛇种的背鳞近游离端在放大镜或解剖镜观察下可看到成对的小窝——端窝，也有分类鉴别的参考意义。

脊鳞（vertebral）：背鳞在蛇背正中的一行称为脊鳞。有的种类脊鳞扩大（enlarge）呈六角形，如过树蛇属 Dendrelaphis 和链蛇属 Lycodon 蛇类。

3. 尾部的鳞片

尾下鳞（subcaudal, caudal, urostege, scutellum）：尾部的鳞片，只有尾腹面尾下鳞的计数及单行（成单）或双行（成对）才有分类鉴别意义（图3-9）。尾末端（尾尖）往往是一枚小圆锥棒状角质构造，尾下鳞成单的，可将此构造加入计数；尾下鳞成对的，如该蛇有 98 对尾下鳞，其完整的表示法为 98/98+1。有一些具成对尾下鳞的蛇，可能有少数几枚成单，可以提到第几、第几成单；反之，如以成对的为主，就可以提到第几、第几成对。如成对和成单的鳞片都较普遍，可以笼统地用"单双不定"来说明。有的树栖蛇类不仅腹鳞两侧具棱，有的尾下鳞也有侧棱。

4. 鳞片的排列方式及其结构

在描述蛇类鳞片时，常使用到以下一些名词术语。

覆瓦状排列（imbricate）：正如前面介绍背鳞时提到，按鳞片排列的方式，如前一枚鳞片的后缘覆盖在后一枚鳞片的基部，就如旧式房屋盖瓦的方式一样。

平砌排列或并列（juxtaposed）：与覆瓦状排列不同，两两鳞片彼此并列，就如房屋地砖或墙砖的排列方式一样。

相接（contact）：在平砌排列的情况下，两枚相邻鳞片以整个边缘或局部彼此相接的情况（过去曾描述为"相切"）。

鳞沟（sulcus, groove）：平砌排列的鳞片彼此相接的地方形成一条缝称为鳞沟。以一般蛇类头部背面的鳞片举例：成对鼻间鳞彼此相接的鳞沟，称鼻间鳞沟；成对顶鳞彼此相接的鳞沟，称为顶鳞沟；顶鳞与其前方额鳞相接的沟，称为额顶鳞沟；余以此类推。

颔沟（mental grove）：许多蛇类头部腹面有成对排列的颔片，其间形成的鳞沟称颔沟。但如游蛇科钝头蛇属的蛇，其颔片不呈对称排列，而是交错排列，因而没有颔沟。

端窝（scale pit, apical pit, apical pore, scale fossa）：有的蛇种背鳞近游离端在放大镜或解剖镜观察下可看到成对的小窝，有分类鉴别的参考意义。

5. 牙齿的类型

牙齿也是分类检索常用到的重要依据。

上颌齿（maxillary teeth）：游蛇科属种的检索，尤其是属的鉴别时，常用到上颌齿的特征。上颌齿着生于上颌骨上，其数目、大小、排列方式等在分类鉴别上有参考意义。

齿间隙（diastema）：上颌齿列中，两组牙齿之间有 1 个或宽或窄的间隙称为齿间隙。

后沟牙（opisthoglyphic tooth）：游蛇科后沟牙类毒蛇着生在上颌骨后端、表面有沟的毒牙。

前沟牙（proteroglyphic tooth）：眼镜蛇科前沟牙类毒蛇着生在上颌骨前端、表面有沟的毒牙。视不同属、种，前沟牙之后没有或有少数几枚普通上颌齿。

管牙（solenoglyphic tooth）：蝰科管牙类毒蛇上颌骨甚短而高，只着生较长而略弯曲、中空有管的毒牙。在有功能的毒牙之后，还有几枚发育中准备替补的预备牙。此外，上颌骨没有其他上颌齿。

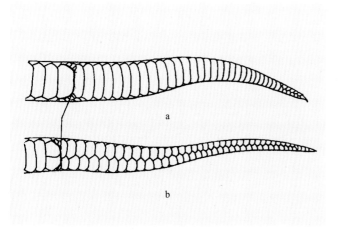

图3-9　蛇尾部腹视（引自赵尔宓和鹰岩，1993）
示肛鳞完整及尾下鳞成单或单行（a）和肛鳞二分及尾下鳞成对或双行（b）

6. 其他与分类检索有关的名词术语

以下几种构造也是分类检索常用到的名词术语。

椎体下突（hypapophysis，复数为 hypapophyses）：椎体下突在整个躯椎上都具有，或是前部躯椎才具有，或是后部躯椎上才具有，有分类鉴别意义。

半阴茎（hemipenis）：蛇类的交接器是成对器官，每侧的交接器称半阴茎。由于它是雄蛇的交接器官，在生殖隔离上有一定的作用，因此有人强调它在种的鉴别上的意义。其实半阴茎的构造也有个体变异，不宜将细微的差异或外部形态很相似而仅半阴茎有少许差别的个别标本视为不同的种。半阴茎在收缩态的长度（相当于第几枚尾下鳞位置），外翻态的构造（如分叉与否、分叉点的位置）、表面的构造（如刺、萼的形状、大小与分布范围）、精沟有无、走行方向、分叉与否及分叉位置等，都是描述半阴茎的内容。

7. 斑纹的描述

顶斑（parietal spot）：游蛇科部分蛇类在靠近顶鳞沟中部两侧各有1个镶深色边的浅色小斑点，称为顶斑，在分类鉴别上有参考意义。因为它们成对存在，所以在英文中用复数。

腹链纹（ventral with chain pattern）：游蛇科腹链蛇属 Amphiesma、东亚腹链蛇属 Hebius 和喜山腹链蛇属 Herpetoreas 的蛇类大多数的腹鳞两外侧各有一黑褐色斑点，斑点可大可小，起始或前或后，这些斑点前后缀连成两条链纹，称为腹链纹。许多种蛇的腹链纹在尾下鳞也有，贯通全身，也是识别蛇种的标志之一。

此外，在科、属的分类鉴别中还用到一些骨骼特征。此处不做详细介绍。

二、茂兰地区爬行动物物种分类描述

（一）龟鳖目 TESTUDINES

Testudines Batsch, 1788, Anleit. Kennt. Theire. Mineral., Ⅰ: 437.
Chejouia Macartney, 1802, in Ross's Transl. Cuvier's Lect, Comp. Anat., 1: tab. Ⅲ.
Testudinata Oppel, 1811, Ordn. Rept.: 3.
Testudoformes Chang（张孟闻），1957, Science, 33 (1): 50.

龟鳖类为爬行动物中较为奇特的一类，全身包裹在硬甲中，头、尾、四肢外露。头骨上没有颞孔，也无顶孔。上下颌没有齿，但颌缘被有角质硬鞘，用以切割食物。舌着生于口腔后部，伸缩性很小。鼻孔位于吻的前端，眼具有眼睑及瞬膜，瞳孔圆形，鼓膜位于眼后，没有外耳。嗅觉及触觉较发达。交接器单个，着生在肛道的腹壁上，泄殖肛孔圆形或纵裂。均以肺呼吸。但许多水栖种类在泄殖肛孔两侧具有1对副膀胱，上面密布微血管，此外，咽壁上的隐窝亦富有微血管网。这两者均能在水中交换气体，帮助呼吸。趾式常为 2-3-3-3-3。

龟鳖动物陆栖或水栖。草食、肉食或杂食，耐饿能力很强，长期不吃食物也能生存。均为卵生。卵圆形或椭圆形，卵壳多为钙质，少数为革质软壳（海龟）。雌雄差异不明显，但多数雄龟尾部较长，泄殖肛孔的位置较靠后，有的种类雄龟的腹甲中央稍凹入，以此与雌龟相区别。也有一些种类雌、雄的色斑不同。

1. 鳖科 Trionychidae Bell, 1828

Trionychidae Bell, 1828, Zool. Journ., Ⅲ: 515.

体扁平，呈椭圆形或卵圆形；吻长，吻端较尖，称为吻突，鼻孔位于吻突前端；鼓膜不显；颈较长，头与颈全部能缩入壳内，背甲、腹甲均有骨板而无角质盾片，外覆以柔软的革质皮肤；背甲没有缘板；腹甲各骨板退化缩小，彼此多不相连；椎骨后凹型或两凹型；四肢扁平，指、趾间蹼大，内侧三指、趾具有爪；尾较短。生活于湖沼河流中，肉食性。

1）鳖属 *Pelodiscus* Fitzinger, 1835

Trionyx Geoffroy, 1809, Ann. Mus. Hist. Nat. Paris., XIV: 4, 20. Type species: *T. aegyptiacus* (Geoffroy, 1809).

Pelodiscus Fitzinger, 1835, Entwurf einer systematischen Anordnung der Schildkröten nach den Grundsätzen der natürlichen Methode. Annalen des Wiener Museums der Naturgeschichte, 1: 105-128 (*Pelodiscus*, new genus, p. 120).

Pelodiscus Rhodin, van Dijk, Iverson and Shaffer, 2010, Turtles of the World 2010 Update: Annotated Checklist of Taxonomy, Synonymy, Distribution and Conservation Status. Archived from the original.

吻突较长，约等于眼径，上、下颌粗壮；眼后弓较眼窝为窄；翼骨后端游离，无高出的突起（即上突）；腹甲胼胝体不多于5个；尾很短；四肢外露。

本属有15种，分布于亚洲、非洲及北美洲。贵州有2种，茂兰地区仅有中华鳖1种。

（1）中华鳖 *Pelodiscus sinensis* (Wiegmann, 1835)

Trionyx (*Aspidonectes*) *sinensis* Wiegmann, 1835, Nova. Acta. Acad. Leop. Carol., XVII: 189.

Pelodiscus sinensis Fitzinger, 1835, Ann. Wien Mus., I: 127.

Tyrse peroellata Gray, 1844, Cat. Tort, Croe. Amphisb. Brit. Mus.: 48.

Dogania subplana Gray, 1862, Proc. Zool. Soe., London: 265.

Landemania irrorata Gray, 1869, Proc. Zool. Soc., London: 216, fig. 18.

Gymnopus perocellatus David, 1872, Nouv. Arch. Mus. Hist. Nat., Paris. VIII, Bull: 37.

Oscaria swinhoei Gray, 1873, Ann. Mag. Nat. Hist., (4) XII: 157, Pl. V.

Yuen leprosus Heude, 1880, Mem. Hist. Nat. Emp., Chinois, I: 20.

Amyda sinensis Stejneger, 1907, Herp. Japan: 524.

Pelodiscus sinensis Stuckas and Fritz, 2011, Identity of *Pelodiscus sinensis* revealed by DNA sequences of an approximately 180-year-old type specimen and a taxonomic reappraisal of *Pelodiscus* species (Testudines: Trionychidae). Journal of Zoological Systematics and Evolutionary Research, 49 (4): 335-339.

该物种是李德俊等于1984年在荔波首次调查到（贵州省林业厅，1987），分布于王蒙和城关地区，后期科考未再进行过系统的调查，2016～2017年调查时未发现该物种实体，仅访问到在洞腮河有分布，故其简介描述引自《中国动物志》。

[同物异名] *Trionyx sinensis*；*Tyrse peroellata*；*Dogania subplana*；*Landemania irrorata*；*Gymnopus perocellatus*；*Oscaria swinhoe*；*Yuen leprosus*；*Amyda sinensis*

[俗名] 甲鱼、团鱼、水鱼、王八

[鉴别特征] 淡水生。通体被柔软的革质皮肤，无角质盾片。体色基本一致，无鲜明淡色斑点。吻端具肉质吻突，吻突长，约与眼径相等；颈基两侧及背甲前缘均无明显的

瘰粒或大疣。腹部有 7 个胼胝体。

[形态描述] 体中等。成体背盘长 192～345mm，宽 138.8～256.0mm。头中等大，前端瘦削。吻长，形成肉质吻突，鼻孔位于吻突端。眼小，瞳孔圆形。吻突长于或等于眼间距，等于或略短于眼径。耳孔不显。两颌有肉质唇及宽厚的唇褶，唇褶分别朝上下翻褶。颈长，颈背有横行皱褶而无显著瘰粒。

背盘卵圆形，后缘圆，其上无角质盾片而被覆柔软的革质皮肤。背盘前缘向后翻褶，光滑而有断痕，呈 1 列扁平疣状，正对颈项中线，并列 2 枚平疣粒。背盘中央有脊棱，脊侧略凹，呈浅沟状。盘面有由小瘰粒组成的纵棱，每侧 7～10 条，近脊部略与体轴平行。近外侧者呈弧形，与盘缘走向一致。骨质背板后的软甲部分有大而扁平的棘状疣，疣之末端尖出，游离。

腹甲平坦光滑，具 7 块胼胝体，分别在上腹板、内腹板、舌腹板与下腹板联体及剑板上。腹甲后叶短小。

四肢较扁。第五指、趾外侧缘膜发达，向上伸展至肘、膝部，形成一侧游离的肤褶。其宽可达 10mm。前臂前缘有 4 条横向扩大的扁长条角质肤褶，宽 10～22mm，排列略呈 "品" 字形。胫跗后缘亦有一横向扩大的角质肤褶。指、趾均具 3 爪，满蹼。

体背青灰色、黄橄榄色或橄榄色。腹乳白色或灰白色，有灰黑色排列规则的斑块。幼体裙边有黑色具浅色镶边的圆斑。腹部有对称的淡灰色斑点，腭与头侧有青白间杂的虫样饰纹。幼体背部隆起较高，脊棱明显。雌鳖尾较短，不能自然伸出裙边，体型较厚。雄鳖尾长，尾基粗，能自然伸出裙边，体型较薄。

[生活习性] 多生活于江河、湖沼、池塘、水库等水流平缓、鱼虾繁生的淡水水域。也常出没于大山溪中。在安静、清洁、阳光充足的水岸边活动较频繁。喜晒太阳或乘凉风。卵生。每年 4～8 月为繁殖期。

[地理分布] 茂兰地区分布于瑶山、城关，茂兰保护区内仅访问到洞塘有分布。除宁夏、新疆、青海及西藏未见报道外，我国其他各省区均有分布。国外分布于日本、朝鲜、越南。

[模式标本] 模式标本（ZMB38、ZMB39、ZMB37 已丢失）保存于德国柏林洪堡大学自然历史博物馆

[模式产地] 澳门

[保护级别] 三有保护动物；贵州省重点保护野生动物

[IUCN 濒危等级] 濒危（EN）

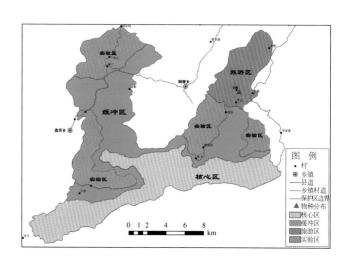

在茂兰地区的分布

（二）有鳞目 SQUAMATA

I. 蜥蜴亚目 Lacertilia

蜥蜴亚目是爬行纲有鳞目下一亚目。体表被以角质鳞片；绝大多数蜥蜴类有四肢，少数种类四肢退化，但肢带残存。茂兰地区蜥蜴亚目共计 6 科 8 属 12 种，分科检索表如下。

茂兰地区科检索表

```
1 { 头背无对称排列的大鳞 ·································································································································· 2
    头背有对称排列的大鳞 ·································································································································· 3

2 { 无活动眼睑 ········································································································································· 壁虎科 Gekkonidae
    有活动眼睑 ·············································································································································· 4

3 { 无四肢，体呈蛇形，体侧有纵沟 ··················································································································· 蛇蜥科 Anguidae
    有四肢或退化 ·············································································································································· 5
```

4 { 背部一般有鬣鳞，无环状花纹，尾部无明显膨大 ·· 鬣蜥科 Agamidae
　 背部无鬣鳞，具有鲜艳的环状花纹，尾部明显膨大 ··· 睑虎科 Eublepharidae

5 { 腹鳞近方形，有股孔或鼠蹊孔 ··· 蜥蜴科 Lacertidae
　 腹鳞近圆形，无股孔或鼠蹊孔 ··· 石龙子科 Scincidae

2. 睑虎科 Eublepharidae Boulenger, 1883

Eublepharidae Boulenger, 1883, Ann. Mag. Nat. Hist., 12 (5): 308.

体通常纵扁。头背无对称排列的大鳞。具眼睑，瞳孔垂直。椎骨前凹型。顶骨合并。有四肢，指、趾细直，无攀瓣。爪通常被部分覆盖或可收缩。具肛后囊和肛后骨。背部具有漂亮的环状花纹，与乳白色腹部呈强烈对比。正常的个体具有与身体一样粗壮的膨大尾部。

本科目前有6属32种，在东南亚、南亚、澳大利亚、西非、东非、北美洲、中美洲分布。在中国仅有睑虎属物种分布。

2）睑虎属 *Goniurosaurus* Barbour, 1908

Goniurosaurus Barbour, 1908, Bull. Mus. Comp. Zool. Coll., Cambridge, 51: 316. Type species: *Goniurosaurus hainanenszs* Barbour, 1908 (= *Eublepharis lichrenfelderi* Mocquard, 1897), by monotypy.

体被六角形粒鳞及疣鳞，上、下眼睑均发达，可活动。瞳孔垂直。指、趾细直，指、趾下有1列横鳞。爪部分夹藏在2枚侧鳞及1枚上鳞间。尾短于头体长。雄性具肛前孔和股孔。

本属已知有17种，分布于南亚和东南亚。中国有10种。茂兰地区有2种，即里氏睑虎 *Goniurosaurus lichtenfelderi* 和荔波睑虎 *Goniurosaurus liboensis*。

茂兰地区种检索表

体型较小，体背3条环状纹 ··· 里氏睑虎 *Goniurosaurus lichtenfelderi*
体型较大，体背4条环状纹 ··· 荔波睑虎 *Goniurosaurus liboensis*

（2）里氏睑虎 *Goniurosaurus lichtenfelderi* (Mocquard, 1897)

Eublepharis lichtenfelderi Mocquard, 1897, Bull. Mus. Hilt. Nat., Paris, 3: 213.
Goniurosaurus hainanensis Barbour, 1908, Bull. Mus. Comp. Zool., Harvard Coll., Cambridge, 51: 316.
Goniurosaurus lichtenfelderi Borner, 1981, Misc. Art. Saurnl. (privately printed), Cologne, 9: 1.
Gorciurosaurus lichtenfelderi hainanensu Grismer, 1987, Acta Herpetol. Sinica, Beijing, 6: 45.
Goniurosaurus murphyi Orlov and Darevshy, 1999, Russ. J. Herpetol., 6 (1): 72-78.
Goniurosaurus lichtenfelderi Seufer et al., 2005, Die Lidgeckos. Kirschner und Seufer Verlag: 238.

［同物异名］ *Eublepharis lichtenfelder*；*Goniurosaurus hainanensis*；*Goniurosaurus murphyi*

［俗名］ 睑虎

［鉴别特征］ 头及颈部长，躯干及尾较短。上、下眼睑均发达，可活动。爪大部分夹藏在2枚侧鳞及1枚上鳞间，基部的腹面被1枚小鳞覆盖。

［形态描述］ 头长，呈三角形；颈部明显；躯干相对较短；四肢细长；尾较短。吻部圆锥形，吻长约为眼径的2倍，等于或略大于眼至耳孔之距。耳孔大而显著，其长径约为眼径之半。吻鳞大，宽大于高，略呈五角形，上缘有中裂，约占吻鳞高的一半。吻鳞、第一上唇鳞不与鼻孔相接。大的上鼻鳞1对，被2枚前后排列的小鳞隔开。上唇鳞9～11，下唇鳞8～11。颏鳞大，亚三角形，一对大的颏片位于下侧第一下唇鳞的后角。

头部背面被粒鳞，顶部粒鳞间有小疣鳞，自眼后经耳孔至枕部散布大疣鳞。颈背及躯干背面粒鳞间均匀散布圆形或锥形的大疣鳞，约25个不规则纵列，大疣鳞直径小于疣鳞的间隔。头部腹面具粒鳞，喉后渐渐转为覆瓦状鳞。躯干腹面被覆瓦状鳞，过体中部处约23列。四肢背侧被粒鳞，指、趾下有1列横鳞。尾短于头体长，基部膨大，末端尖细，切面呈圆形，上被粒鳞。尾基部分节，8～9横排的粒鳞成一节，每节有1横列锥状疣鳞。雄性具肛前孔和股孔28～30个。

［生活习性］ 常生活在山洞中，也会出现于树林间开阔地的石头上。

［地理分布］ 茂兰地区见于洞塘。国内分布于海南、广西、贵州。国外分布于越南。

［模式标本］ 正模标本（ROM 32456）保存于皇家安大略博物馆。副模标本（MNHN 97-91、MNHN 97-92）保存于法国国家自然历史博物馆（Muséum National d'Histoire Naturelle，MNHN）

［模式产地］ 越南21°12′48″N，106°28′38″E，海拔250m

［保护级别］ 三有保护动物；贵州省重点保护野生动物

［IUCN 濒危等级］ 易危（VU）

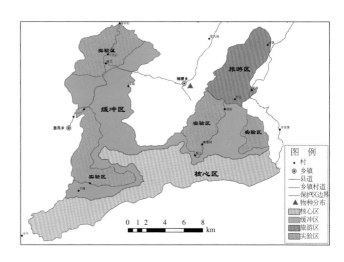

在茂兰地区的分布

（3）荔波睑虎 *Goniurosaurus liboensis* Wang, Yang and Grismer, 2013

Goniurosaurus liboensis Wang, Yang and Grismer, 2013, A New Species of *Goniurosaurus* (Squamata: Eublepharidae) from Libo, Guizhou Province, China. Herpetologica, 69 (2): 214-226.

［同物异名］ 无

［俗名］ 睑虎

［鉴别特征］ 头长大于头宽，体背有4条白色环带，尾粗短，基部膨大，与臀部连接处有1环形尖疣鳞。

［形态描述］ 依据茂兰地区标本描述 头长大于头宽，雄性头体长97.86～111.28mm，后肢略长于前肢，尾长51.98～60.47mm，头、颈区分明显；耳孔大而显著，体背有4条白色环带，第1条位于前肢后侧，第2条位于躯干中部，第3条位于后肢略前，第4条位于肛部之上；每条环带边均有黄色小疣鳞，头部腹面密布细粒鳞，延伸至颈部；前胸部至肛门前部密布乳白色覆瓦状鳞片，尾基部有1环状尖疣鳞。吻鳞、第一上唇鳞不与鼻孔相接。大的上鼻鳞1对，被2枚前后排列的小鳞隔开。上唇鳞10，下唇鳞9或8。颏鳞大，略呈三角形，一对大的颏片位于下侧第一下唇鳞的后角。头部背面被粒鳞，顶部粒鳞间有小疣鳞，自眼后经耳孔至枕部散布大疣鳞。颈背及躯干背面粒鳞间均匀散布圆形或锥形的大疣鳞。头部腹面粒鳞至喉后，转为覆瓦状鳞。躯干腹面被覆瓦状鳞至大腿腹面，前胸鳞片小于腹部鳞片。四肢背面被粒鳞，腹面覆瓦状鳞片。指、趾下有1列横鳞。尾短于头体长，基部膨大，

成体量度

编号	产地	头长（mm）	头宽（mm）	头高（mm）	头体长（mm）	尾长（mm）	上唇鳞（枚）	下唇鳞（枚）
GZNU20160727 ♂	荔波茂兰	35.23	20.66	12.94	111.28	60.47	10	9
GZNU20170512005 ♂	荔波茂兰	34.59	18.62	11.18	97.86	51.98	8	8

a 背面照
b 腹面照
c 侧面照
d 整体照
e 在茂兰地区的分布

末端尖细，上被粒鳞，尾部波浪状环斑3道，与臀部连接处有环形尖疣鳞。

[生活习性] 常栖息于岩壁或巨石的缝隙中，夜间出没于岩壁或巨石表面捕食昆虫。

[地理分布] 茂兰地区分布于永康、洞塘。国内分布于贵州、广西。国外无分布。

[模式标本] 正模标本（SYS r000218，成年雄性），副模标本（SYS r000216，成年雌性，450m）、（SYS r000217，成年女性）、（SYS r000219，成年雄性）保存于中山大学生物博物馆

[模式产地] 茂兰保护区，25°15′37.73″N，108°5′45.74″E，海拔 660m 广西环江县，25°12′50.0″N，107°59′56.04″E

[保护级别] 贵州省重点保护野生动物

[IUCN 濒危等级] 濒危（EN）

3. 壁虎科 Gekkonidae Gary, 1825

Ascalabotes Cuvier, 1817, Regne Anim., 2: 44.
Gekkonidae Gray, 1825, Ann. Philos., 10 (2): 198.
Platyglossae Wagler, 1830, Syst. Amphib.: 141.
Eublepharidae Boulenger, 1883, Ann. Mag. Nat. Hist., 12 (5): 308.
Uroplatidae Boulenger, 1884, Ann. Mag. Nat. Hist., 14 (5): 119-120.
Gekkonidae Smith, 1935, Rept. Amphi. F.B.I., 2: 21.

体大多扁平，头顶无对称排列的大鳞。无眼睑，瞳孔大多垂直，少数圆形。垂直的瞳孔有直弧型和分叶型两类。皮肤柔软。头顶无对称排列的大鳞。体背面通常被粒鳞或疣鳞，少数种具圆形或六角形的覆瓦状鳞。体腹面被圆形或六角形的覆瓦状鳞。无眼睑，眼在透明膜下转动自如。鼓膜大多裸露内陷，外耳道明显。侧生齿，齿小而数多。舌中等长而宽，前端微缺。舌面被绒毛状乳突。舌能伸出，但本身不能伸缩。四肢发达，具五指、趾或第Ⅰ指、趾退化成痕迹状。指、趾的形状及构造变化很大，具爪或无爪，有些科的爪能伸缩。许多种指、趾扩展，腹面有攀瓣，上具微毛垫，可吸着在光滑表面上。尾呈圆柱形、纵扁圆形、侧扁圆形或其他形状，易断，具缠绕性的极少。

本科共有 56 属 996 种，主要分布在中南半岛、印度、中国（南部）、日本和澳大利亚等地。茂兰地区有 1 属 2 种。

3）壁虎属 *Gekko* Laurenti, 1768

Gekko Laurenti (= Laurent), 1768, Synops, Rept., Vienna: "34" (43). Type species: *Gakko cillatus* Laurent, 1768 (= *Laccerta gecko* Linnaeus, 1758), by absolute tautonomy.
Platydactylus Goldfuss, 1820, Handb. Dierk., 2: 342.
Scelotretus Fitzinger, 1843, Syn. Pept., 99: 101.

背面被均一的粒鳞或混杂有较大的疣鳞。瞳孔分叶型。指、趾扩展，攀瓣不对分，上具微毛垫，远端的攀瓣前缘直或浅凹，指、趾间无蹼或部分具蹼。第Ⅰ指、趾发达，无爪。其余各指、趾具爪。雄性具肛前孔、肛股孔或股孔。

本属已命名的约 51 种，分布于中国、韩国、日本、东南亚及南亚。中国有 16 种，贵州有 5 种，茂兰地区有 2 种，即多疣壁虎 *Gekko japonicus* 和荔波壁虎 *Gekko liboensis*。

茂兰地区种检索表

上鼻鳞相接，体背面疣鳞扁平···荔波壁虎 *Gekko liboensis*
上鼻鳞不相接，体背两侧及后肢疣鳞中等大小·····································多疣壁虎 *Gekko japonicus*

（4）多疣壁虎 Gekko japonicus (Duméril and Bibron, 1836)

Platydactyus japonicas Duméril and Bibron, 1836, Erpetol, Gen., Paris, 3: 337.

Platydactylus jamori Temminck and Schlegel, 1838: 103. In Ph. Fr. de Siebold: 85-144. Fauna Japonica auctore. L. Batavorum.

Hemidactylus nanus Cantor, 1842b, Ann. Mag. Nat. Hist., London, Ser. 1, 9: 482.

Gecko japonicus Boulenger, 1885: 188. Catalogue of the Lizards in the British Museum (Nat. Hist.) I. Geckonidae, Eublepharidae, Uroplatidae, Pygopodidae, Agamidae, London.

Gecko japonicus Günther, 1888, Ann. Mag. Nat. Hist., (1) 6: 165-172.

Platydactylus yamori Fritze, 1891, Mitteilungen der Deutschen Gesellschaft für Natur- Völkerkunde Ostasiens in Tokyo, 5: 235-248.

Gekko japonicus Stejneger, 1907, Bull. U. S. Natl. Mus., 58: xx, 1-577.

Gekko japonicus Schmidt, 1927, Bull. Amer. Mus. Nat. Hist., New York, 54: 477.

Luperosaurus amissus Taylor, 1962, Univ. Kansas Sci. Bull., 43: 209-263.

Gekko japonicus Ota, 1989, in Matsui et al. (eds.), Current Herpetology in East Asia: Proceedings of the Second Japan-China Herpetological Symposium Kyoto: 222-261.

Gekko japonicus Rösler, 2000, Gekkota, 2: 28-153.

Gekko japonicus Luu et al., 2014, Zootaxa, 3895 (1): 73-88.

[同物异名] *Platydactylus japonicus*；*Platydactylus jamori*；*Hemidactylus nanus*；*Gecko japonicas*；*Gecko japonicas*；*Platydactylus yamori*；*Gekko japonicus*；*Luperosaurus amissus*；*Gekko japonicas*；*Gekko japonicas*；*Gekko japonicas*；*Gekko japonicus*

[俗名] 壁虎、四脚蛇

[鉴别特征] 指、趾间具蹼迹。体背粒鳞较小，疣鳞显著大于粒鳞。前臂和小腿有疣鳞。尾基部肛孔多数每侧3个。

[形态描述] 身体扁平，体全长99～149mm，头体长为尾长的0.87～1.10倍。吻长稍大于眼径的2倍。耳孔直径0.55～1.5mm，为眼径的15%～43%。吻鳞长方形，宽约为高的2倍，上缘中央无缺刻。鼻孔位于吻鳞、第一上唇鳞、上鼻鳞及1～3枚后鼻鳞间。两上鼻鳞被1枚圆形小鳞隔开。上唇鳞9～13枚，下唇鳞8～13枚。颏鳞五角形。颏片弧形排列，内侧1对较大，呈长六角形，长大于宽。外侧1对较小。

体背被粒鳞。吻部粒鳞扩大，自鼻孔至眼的纵列鳞约15枚。眶间部横列鳞32～35枚。体背疣鳞显著大于粒鳞，呈圆锥状，颞部、枕部、颈背及荐部疣鳞甚多。过体中部处有12～14枚不规则粒鳞。体腹面被覆瓦状鳞，过体中部处42～46列。四肢背面被小粒鳞，前臂粒鳞间有少量疣鳞，小腿粒鳞间的疣鳞较多。四肢腹面被覆瓦状鳞。指、趾间具蹼迹。后足第Ⅰ～第Ⅴ趾扩展部的攀瓣Ⅰ6～9、Ⅱ6～9、Ⅲ6～10、Ⅳ6～10、Ⅴ6～10。雄性具肛前孔4～8个，多数6个。

尾稍纵扁，基部每侧大多有3个肛疣，有些标本在肛疣之下有3～6枚疣鳞。尾背面被小覆瓦状鳞，每7～9行成一节。尾腹面的覆瓦状鳞较大，中央具1列横向扩大的鳞板。

体背面灰棕色。多数从吻端经眼至耳孔有1黑色纵纹。头及躯干背面有深褐色斑，并在颈及躯干背面形成5～7条横斑。有些个体褐斑不明显。四肢及尾背面亦具褐色横斑，尾背的横斑9～13条。体腹面淡肉色。

个别标本两上鼻鳞间有2枚以上的小鳞相隔。颏片的变异亦多，内侧1对颏片有一分为二的、有在上方中间隔一小鳞的、有特长或特短的、有与外侧一对等大的。

[生活习性] 栖息在建筑物的缝隙及岩缝、石下、树下或草堆柴堆内。夜晚常在有灯光照射处捕食，常数只在同一墙上出现，有时为争食而互相争斗。食物主要有蛾类、蚊类等。该种在5～7月繁殖，5月中旬到6月中旬为产卵旺季。

[地理分布] 茂兰地区分布于洞塘。国内分布于安徽、江苏、上海、浙江、福建、江西、湖北、广西、贵州和四川等地。国外分布于日本和韩国。

[模式标本] 正模标本（BMNH 1946.8.26.9-10）保

存于英国伦敦自然历史博物馆

[模式产地] 日本

[保护级别] 三有保护动物；贵州省重点保护野生动物

[IUCN 濒危等级] 无危（LC）

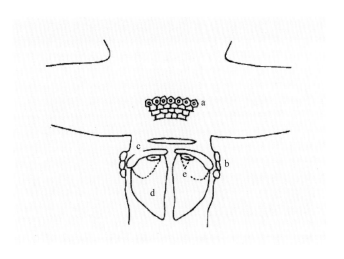

肛前部和尾基部腹面观（引自《中国动物志》）
a. 肛前孔；b. 肛疣；c. 肛后骨；d. 肛后囊；e. 肛后囊孔

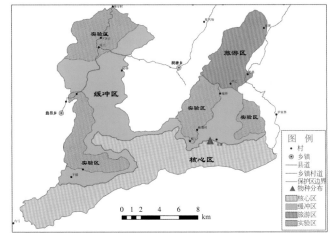

在茂兰地区的分布

（5）荔波壁虎 *Gekko liboensis* Zhou and Li, 1982

Gekko liboensis Zhou and Li, 1982, Acta Zootaxon. Sinica, 7: 438-447.

Goniurosaurus liboensis Yang, 2015, Zootaxa, 3936 (2): 287-295.

Goniurosaurus liboensis Jono et al., 2015, Asian Herpetological Research, 6 (3): 229-236.

[同物异名] 无

[俗名] 壁虎

[鉴别特征] 两上鼻鳞大且在中线相接。体背粒鳞间扁圆疣鳞约10列。前后肢均不具疣鳞。尾基部每侧肛疣1个。前足第Ⅰ～第Ⅲ指间有蹼迹，第Ⅲ～第Ⅴ指间微蹼。后足第Ⅰ～第Ⅳ趾间有蹼迹。

[形态描述] 形态根据贵州荔波一雌性个体描述。个体甚大，全长122mm，头体长85mm，再生尾长为37mm。吻长为眼径的1.8倍，明显大于眼至耳孔之距。耳孔直径2mm，为眼径的40%，吻鳞宽大于高，上缘中央略凹。鼻孔位于吻鳞、第一上唇鳞、上鼻鳞及2枚后鼻鳞间。两上鼻鳞大且在中线相接。上唇鳞12，下唇鳞11。颏鳞三角形。内侧1对颏片的长为宽2倍，后接1对多角形小颏片。自鼻孔至眼的纵列鳞18枚左右，眼眶间的横列鳞约40枚。体背面自头顶枕部至尾基部有扁圆疣鳞，稀布在均匀粒鳞间，在躯干背面的疣鳞约10列，前肢及后肢均不具疣鳞。体腹面自颈部以后被覆瓦状鳞，肛前有扩大的鳞10枚。指扩展部攀瓣Ⅰ 8、Ⅱ 8、Ⅲ 9、Ⅳ 9、Ⅴ 8～9，第Ⅰ～第Ⅲ指间有蹼迹，第Ⅲ～第Ⅴ指间微蹼，蹼缘着于指1/3部。后肢发达，长占腋胯距95%。趾扩展部攀瓣Ⅰ 8、Ⅱ 7～8、Ⅲ 9、Ⅳ 9、Ⅴ 8～9；第Ⅰ～第Ⅳ趾间有蹼迹。尾基部每侧有1个肛疣。再生的尾甚短。

液浸标本体背肉灰色。1褐纹沿眼眶下缘后行至近耳孔处。颈及躯干背面有9条褐色横斑。四肢的背面也有褐色横斑。体腹面淡肉色。

[生活习性] 生活于喀斯特森林地区，栖息于农舍墙上。常在夜间活动，极罕见，其生物学信息较少。

[地理分布] 仅在贵州荔波发现，茂兰地区分布于洞塘。

[模式标本] 正模标本（ZMC 791669）保存于遵义医科大学

[模式产地] 贵州荔波

[保护级别] 贵州省重点保护野生动物

[IUCN 濒危等级] 无危（LC）

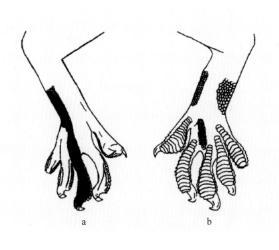
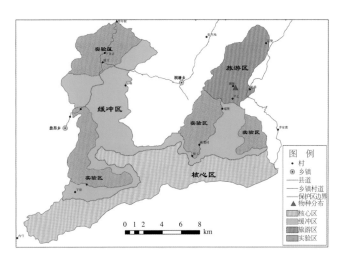

后足（引自《中国动物志》）　　　　　　　在茂兰地区的分布
a. 背面观；b. 腹面观

4. 石龙子科 Scincidae Gary, 1825

Scincidae Gray, 1825, Ann. Phil., 26: 201 (in part), and 1827, Zool. Journ., 3: 130.

体型一般中等大小，头顶具有对称排列的大鳞，通体被以覆瓦状排列的圆鳞；眼较小，大多数具活动眼睑，下眼睑有睑窗或被鳞，亦有个别种因下眼睑中央显著扩大和变薄（如 Riopa albopunctata）而常被误认是睑窗；瞳孔圆形；鼓膜深陷或被鳞；舌中等长，前端微缺，覆盖鳞状乳突；四肢发达或退化；四肢退化的类群几乎都是体相应延长，脊椎数目增加，体脊椎数一般 18～25 枚；尾较粗，横切面圆形，易断，断后能再生，再生部分无脊椎；无股孔或鼠蹊孔。

前颌骨成对，或前端联合；眶后骨弓完整，除退化的类群外；颞弓完整；头骨具有真皮骨板，头骨与真皮骨板联合时，能盖上顶孔。常有翼骨齿；侧生齿圆锥形或钩状，齿冠侧扁或球形；新牙通常是从旧牙基部空洞中长出。锁骨近端常扩大，并有小孔；四肢退化的类群，体内尚有肩带或腰带的残余；腹肋主要存在于穴居类群；体、四肢、尾内亦有真皮骨板保护。

多数种类陆栖，白天活动；穴居或地下生活的种类在黄昏或夜间活动；有半水栖或树栖生活的种类。多数种类完全吃昆虫及其幼虫；体型较大者也吃小型脊椎动物，少数种类兼吃植物。卵生（如石龙子属所有种）或卵胎生（如棱蜥属所有种）。

本科有 1582 种，隶属于 149 属。广泛分布于各大洲，以澳大利亚、西太平洋诸岛、南亚、东南亚及非洲分布最多，美洲种类较少。我国现知 10 属 38 种。茂兰地区有 2 属 3 种，分属检索表如下。

茂兰地区种检索表

下眼睑被扩大的鳞片··石龙子属 Plestiodon
下眼睑被小鳞··蜓蜥属 Sphenomorphus

4）蜓蜥属 Sphenomorphus Fitzinger, 1843

Sphenomorphus Fitzinger, 1843, Syst., Vienna, 1: 23. Type species: Lygosoma melanopgon Dmneril and Bibron, 1839, of the Celebes and New Guinea, by original designation.

腭骨在中线相遇，次生腭凹缺前端没有达到两眼中心的水平；翼骨在前端相接，翼骨齿细小或无；上颌齿圆锥形。眼睑发达，下眼睑被小鳞。无上鼻鳞，鼻孔位于单枚鼻鳞上；前额鳞彼此相接或不相接，额顶鳞愈合，间顶鳞显著；耳孔小而显著。大多数种鳞片平滑；四肢发达，五趾型；大型肛前鳞1对；尾巴腹面及背部的鳞片大小大致相同。

分布于越南、泰国、柬埔寨、马来西亚、印度尼西亚、菲律宾、斯里兰卡、印度、巴布亚新几内亚、澳大利亚、琉球群岛。我国产6种，茂兰地区有1种，即铜蜓蜥 Sphenomorphus indicus。

（6）铜蜓蜥 Sphenomorphus indicus (Gray, 1853)

Sphenomorphus indica Gray, 1853, Ann. Mag. Nat. Hist., London, 12 (2): 388. Type locality: Sikkim, Himalaya.

Sphenomar indicus Andeson, 1871, Proc. Zool. Soc., London: 158.

Lygosoma zebratum Boulenger, 1887, Ann. Mus. Civ. Stor. Nat. Genova, 2, Ser. 5: 474-486.

Lygosoma indicum Boulenger, 1893, Ann. Mus. Civ. Stor. Nat. Genova, 2, Ser. 8: 303-347.

Lygosoma cacharense Annandale, 1905, J. Asiat. Soc. Bengal, (1) 2: 139-151.

Sphenomorphus indicus Stejneger, 1907, Bull. U. S. Natl. Mus., 58: xx, 1-577.

Sphenomorphus indicus Schmidt, 1928, Copeia, 1928: 77-80.

Lygosoma (Hinulia) indicum taeniatum Werner, 1922, Mathemat.-Naturwissensch., Vienna, 59 (24-25): 220-222.

Lygosoma (Hinulia) indicum multilineatum Werner, 1922, Mathemat.-Naturwissensch., Vienna, 59 (24-25): 220-222.

Lygosoma indicum indicum Smith, 1935, Rept. Amphib. F.B.I., 2: 281.

Sphenomorphus indicus indicus Taylor, 1963, Univ. Kansas Sci. Bull., 44: 687-1077.

Sphenomorphus formosensis Greer and Parker, 1967, Breviora: 275.

Sphenomorphus indicus indicus Grandison, 1972, Bull. Br. Mus. Nat. Hist. (Zool.), London, 23: 45-101.

Sphenomorphus indicus Manthey and Grossmann, 1997, Natur und Tier Verlag (Münster): 512.

Sphenomorphus indicus Cox et al., 1998: 119. A Photographic Guide to Snakes and Other Reptiles of Peninsular Malaysia, Singapore and Thailand. Ralph Curtis Publishing: 144.

Sphenomorphus indicus Ziegler, 2002, Natur und Tier Verlag (Münster): 342.

Lygosoma indicum Li et al., 2003, Acta Zool. Sinica, 49 (4): 547-550.

Sphenomorphus indicus Grismer, 2006, Herpetological Natural History, 9 (2): 151-162.

Sphenomorphus indicus Nur-Amalina et al., 2017, Malays. Appl. Biol., 46 (4): 119-129.

[同物异名] Lygosoma indicum；Sphenomorphus indica；Lygosoma zebratum；Lygosoma cacharense；Sphenomorphus indicus；Sphenomorphus formosensis

[俗名] 铜楔蜥、铜石龙子、山龙子、石蜴、四脚蛇

[鉴别特征] 背面古铜色，背脊有1条黑脊纹，体两侧各有1黑色纵带，纵带上不间杂白色斑点或点斑，纵带上缘镶以浅色窄纵纹；环体中段鳞一般34～38行，第Ⅳ趾趾下瓣16～22枚。

[形态描述] 雄性头体长55～90mm，尾长97～158mm；雌性头体长63～95.6mm，尾长101～150mm。吻短而钝，吻长约等于眼耳间距；吻鳞突出，宽大于高，后缘与单枚额鼻鳞相接较宽；吻鳞加上额鼻鳞与额鳞长度的比值平均约0.6。鼻孔位于单枚鼻鳞中部；无上鼻鳞和后鼻鳞；前额鳞一般不相接；前额鳞与额鼻鳞的比值大于1/3；额鳞长等于或长于额顶鳞与间顶鳞之和，前缘与额鼻鳞相接；2枚额顶鳞相接较宽，1枚间顶鳞中型大，顶鳞后端彼此相接。颊鳞2枚，前枚略小；无颈鳞。眶上鳞4枚，第1、第2枚较大；上睫鳞8～10枚，第1枚最大；鼓膜小而下陷；耳孔卵圆形，与眼径等长或略大，无瓣突。颏鳞2枚，较大，重叠排列，下枚颏鳞楔形。上唇鳞7枚，第五、第6枚最大，位于眼下方；下唇鳞多数7枚；颏和后颏鳞各1枚，颏片3对。体型中等大小，体表被覆圆

鳞，覆瓦状排列，平滑无棱；背鳞几等大，环体中段鳞行多数34～38行；肛前鳞4枚，中间1对显著大于两侧2枚。尾长为头体长的1.5～2倍，尾基至尾尖渐缩小成圆锥形，尾背鳞片几等大，尾腹面正中一行鳞略扩大，尾易断，断后能再生。腋胯间距相当于吻至前肢距离的2倍。四肢较弱，前后肢贴体相向时，相遇或达腕部、掌部和肘关节；指、趾略侧扁，具爪，第Ⅳ趾趾下瓣16～20枚，起棱。

生活时背面古铜色，背脊部常有1条断断续续的黑脊纹，背脊两侧的褐色或黑色斑点缀连成行，从头、体侧至尾部两侧各有1条占3～4鳞行宽的黑色纵带，不间杂白色斑点，黑纵带上缘有1条明显的窄的浅色线纹，下缘浅色纵纹宽，略呈灰棕色，杂有细黑点。腹面浅色无斑，四肢背面黄棕色，间杂黑色和浅色小点。唇缘浅色，具黑色纵纹。

[生活习性]　一般生活于平原及山地阴湿草丛中，荒石堆或有裂隙的石壁处。雨后天晴活动较多，深秋中午活动觅食。

[地理分布]　茂兰地区分布于永康、洞塘。国内广泛分布于南方。国外分布于印度（大吉岭、锡金）、缅甸、泰国。

[模式标本]　正模标本（ZSI 12076）保存于爱尔兰动物学协会，同模标本（BMNH 1946.8.15.49、BMNH 1946.8.19.27）保存于英国伦敦自然历史博物馆

[模式产地]　喜马拉雅山

[保护级别]　三有保护动物；贵州省重点保护野生动物

[IUCN 濒危等级]　无危（LC）

a　背面照
b　腹面照
c　整体照
d　在茂兰地区的分布

5）石龙子属 *Plestiodon* Wiegmann, 1834

Eumzces Wiegmann, 1834, Herpetol. Mex., Berlin, 1: 36. Type species: *Scincus pavimentatus* I. Geoffroy Saint-Hilaire, 1827
　　(= *Scincus schneideri* Daudin, 1802), of Nor Africa and the Middle East, by subsequent designation (A. F. A. Wiegmann, 1835,

Arch Naturgesch., Berlin, 1 (2): 288.

Plestiodon Duméril and Bibron, 1839, Erp. Gen., Ⅴ: 697.

Plestiodon Duméril and Bibron, 1843, Syst. Rept.: 22.

Pariocela Fitzinger, 1843, Syst. Rept.: 22.

Lamprosaurus Hallowell, 1852, Proc. Acad. Philad., Ⅵ: 206.

Eurylepis Blyth, 1854, J. Asiat. Soc. Beng., XXⅢ: 739.

腭骨不在腭部的中线相遇；有翼骨齿；侧生齿圆锥形或具球形齿冠。头顶有对称排列的大鳞，上鼻鳞、前额鳞、额顶鳞和间顶鳞显著。眼较小，眼睑发达，瞳孔圆形，下眼睑被鳞。鼻孔位于鼻鳞上。体圆柱形，通体被覆圆鳞，覆瓦状排列，鳞下层有来源于真皮的骨板。四肢发达，五趾型。

本属世界已知37种，分布于中美洲、北美洲、亚洲（东南部和西南部）、非洲（北部）。我国有8种，茂兰地区有2种，即蓝尾石龙子 *Plestiodon elegans* 和中华石龙子 *Plestiodon chinensis*。

茂兰地区种检索表

后颏鳞1枚；有后鼻鳞，股后面有1团具棱大鳞···蓝尾石龙子 *Plestiodon elegans*

后颏鳞2枚；无后鼻鳞，股后无具棱大鳞···中华石龙子 *Plestiodon chinensis*

（7）蓝尾石龙子 *Plestiodon elegans* (Boulenger, 1887)

Mabouia chinensis Swinhoe, 1863, Ann. Mag. Nat. Hist., (12) 3: 219-226.

Eumeces pulchra Bocourt, 1879, Etudes sur les reptiles, I-XⅣ: 1-1012.

Eumeces elegans Boulenger, 1887, Catal. Liz. Brit. Mus., London, 3: 371. Type locality: Ningpo [= Ningbo], Formosa [= Taiwan] Pscadore [= Penghu] Islands, and Kiukiang [= Jiujiang] China.

Eumeces elegans Stejneger, 1907, Bull. U.S. Natl. Mus., 58: xx, 1-577.

Eumeces elegans Taylor, 1936, Univ. Kansas Sci. Bull., 23 (14): 1-643 [1935].

Eumeces elegans Zhao and Adler, 1993, Herpetology of China. SSAR, Oxford/Ohio: 1-522.

Eumeces elegans Griffith, Ngo and Murphy, 2000, Russ. J. Herpetol., 7 (1): 1-16.

Plestiodon elegans Schmitz et al., 2004, Hamadryad, 28 (1-2): 73-89.

[同物异名] *Eumeces elegans*；*Mabouia chinensi*；*Eumeces pulchra*

[俗名] 四脚蛇、蓝尾四脚蛇

[鉴别特征] 有上鼻鳞；有后鼻鳞；后颏鳞1枚；顶鳞1对；股后有1团大鳞；成体褐色侧纵纹显著；幼体背面5条浅铜褐色纵纹，尾末端蓝色。

[形态描述] 雄性头体长62.7～91mm，尾长81～132mm；雌性头体长60.5～82mm，尾长87～153mm。本次调查捕获雄性标本1号，其头体长63.79mm，尾长114.50mm。

吻高，从背面观稍窄，能见到大部分；吻长与眼耳间距相等；上鼻鳞中等大，在吻鳞之后彼此相接；额鼻鳞宽大，不与吻鳞相接，一般与额鳞和颊鳞相接；前额鳞比额鼻鳞小，彼此分离（个别个体相接），与两枚颊鳞相接，和额鼻鳞相接构成较宽的缝沟；额鳞长度大于额顶鳞与间顶鳞之和，并显著长于它到吻端的距离，前部略宽，与3枚眶上鳞相接；额顶鳞长大于宽，或长宽相等，构成的中缝沟等于其长度的一半；间顶鳞比额顶鳞小，后部窄，将顶鳞分隔，直接与颈鳞相接；顶鳞大，最宽处约为长度的3/4。多数颈鳞1对，有个体变异现象。

鼻孔位于单枚鼻鳞的前部；有后鼻鳞；前颊鳞宽不大于高的2倍，略高于后颊鳞，一般高为长的3/4，与上鼻鳞、额鼻鳞和两枚唇鳞相接；后颊鳞五边形，长宽约相等，与第二和第3枚唇鳞相接；上睫鳞6～8枚，前部鳞比后部鳞大3倍，中间4枚眼睑鳞直接与上睫鳞相接；眶上鳞4枚，第2枚最大，前3枚与额鳞相接；1枚小的眶前鳞；2枚眶后鳞；眶前下鳞2枚；眶后下鳞4枚；颞鳞1+2，第1枚颞鳞大，长方形，边缘与两枚第二列颞鳞相接；第二列上颞鳞大，近

成体量度

编号	产地	头长（mm）	头宽（mm）	头高（mm）	头体长（mm）	尾长（mm）	上唇鳞（枚）	下唇鳞（枚）
GZNU2016072937♂	荔波茂兰	19.94	9.69	9.28	63.79	114.50	8	7

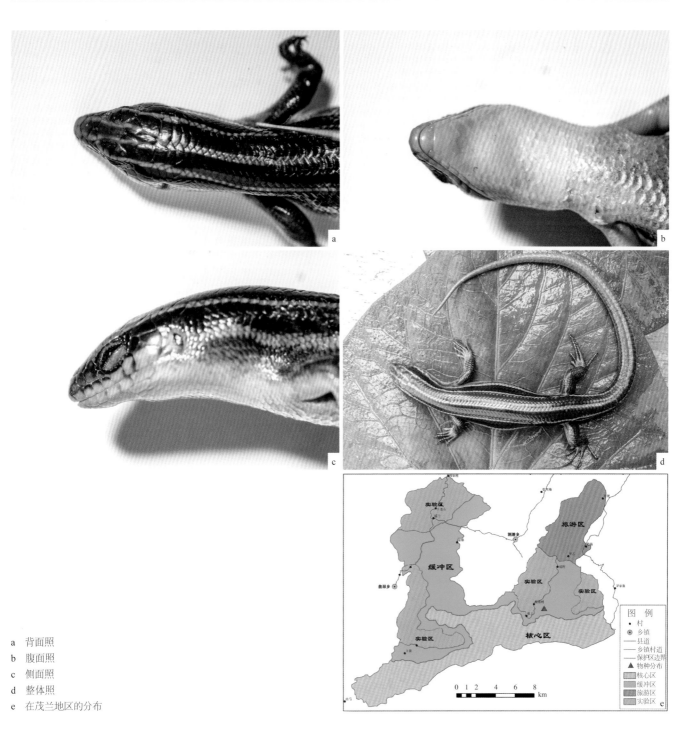

a 背面照
b 腹面照
c 侧面照
d 整体照
e 在茂兰地区的分布

于三角形，顶端向前，其后两枚大小相等的鳞片垂直排列，与颈部大鳞相接；第二列下颚鳞长方形，上下缘几乎平行。上唇鳞8枚，第1枚比它紧接的3枚鳞片略高大，第七枚唇鳞最大，和耳孔间隔1行小鳞；下唇鳞7枚，第6枚最大；颏鳞大，显著大于吻鳞；后颏鳞1枚；颏片3对（贵州雷山雄雌各一，右侧2枚，1雌左侧2枚），前对最小，第三对最大，后接1枚窄长鳞片。耳孔卵圆形，宽约为眼径的1/2，前缘有2或3枚瓣突，耳孔周围约有鳞20枚；鼓膜深陷；耳后颈部一周有鳞32～36枚。体鳞平滑，覆瓦状排列，背中段鳞行略大于相邻侧鳞；环体中段鳞23～26行；肛前鳞8枚，中间一对特大，外侧鳞小，部分重叠于内侧；肛后鳞强烈起棱；肛部两侧各有1棱鳞，雄性尤为明显。尾长不到头体长的1.5倍，尾基部一周有鳞数枚，尾下面正中1行鳞横向扩大，约有105枚。四肢贴体相向时，指、趾不遇，或恰相遇，或超越；前肢前伸时，指端可达眼；前肢基部一周约有鳞15枚，腕外侧粒鳞显著，紧邻2或3枚小鳞片；掌部有4枚大小不等的粒鳞；每指基部皮瓣扩大加厚；后肢基部一周有鳞18枚；股部后方有一团不规则排列的3～4行大鳞；膝部有两对垫状粒鳞；指、趾侧扁，具爪，基部鳞片密集，第Ⅳ趾趾下瓣一般为16～18枚。

生活时成体雄性背面棕黑色，有5条浅黄色纵纹。雌性成体背面色深暗，5条纵纹尤为显著；正中1条浅纵线纹，在间顶鳞部位分叉向前沿额鳞两侧到吻部，向后延伸到尾背的1/2处；两条背侧浅纵线，从前额鳞或上睫鳞起，自眼上方，经顶鳞外缘沿背侧第三行鳞片向后延至尾部2/3以上；体侧两条浅纵线从第七枚唇鳞起经耳孔上方沿体侧第六、第七行鳞延伸至胯部；体侧线下缘有1较宽的深褐色带状纹，向腹部渐浅，融入腹面浅灰色。雌雄老年个体背面橄榄棕色，浅色纵纹甚浅或消失，特别是雄性，但褐色侧纹一般很明显。面颊红色斑与体侧红色隐约贯穿，形成2条红色纵线。雄性体侧及泄殖腔部分有不显著的分散紫红色斑，繁殖期体较长；雌性无此斑，繁殖期腹部膨大。

幼体背面棕黑色，具5条浅铜褐色纵线纹，有金属光泽，背面3条向后达尾1/3处，逐渐不清晰，到后部1/2处消失。体侧浅铜褐色纵线纹延至胯部。躯体腹面灰白色到黄白色。尾自基部向末端背面棕黑色渐深而鲜明，后端几呈纯蓝色。四肢背面黑褐色，腹面灰白色。液浸标本无金属光泽，蓝色消失，但浅纵线亦清晰。

[生活习性] 一般栖息于山区路旁草丛、石缝或树林下溪边乱石堆杂草中。多见于有阳光照射的山坡，受惊扰后立即进入草丛、土洞或石缝。食性较广。

[地理分布] 茂兰地区分布于洞塘。国内广泛分布于南方。

[模式标本] 全模标本（BMNH 54.2.10.1-4）保存于英国伦敦自然历史博物馆

[模式产地] 宁波

[保护级别] 三有保护动物；贵州省重点保护野生动物

[IUCN 濒危等级] 无危（LC）

（8）中华石龙子 *Plestiodon chinensis* (Gray, 1838)

Tiliqua chinensis Gray, 1838, Ann. (Mag.) Nat. Hist., London, [Ser. 1], 2: 289. Type locality: China.

Plestiodon pulchrum Duméril and Bibron, 1839, Erpetol. Gen., Paris, 5: 710.

Plestiodon sinense Duméril and Bibron, 1839: 704. Roret/Fain et Thunot, Paris, Vol. 5: 871.

Plestiodon chinensis pulcher Duméril and Bibron, 1839, Tome cinquième. Librairie Encyclopédique de Roret: Paris, VIII: 854.

Mabouia chinensis Günther, 1864, London (Taylor & Francis), XXVII: 452.

Eumeces chinensis Boulenger, 1887, Catalogue of the Lizards in the British Museum (Nat. Hist.) III.

Eumeces chinensis Böttger, 1894, Ber. Senckenb. Ges., 1894: 129-152.

Eumeces chinensxs Stejneger, 1907, Bull. U.S. Natl. Mus., Washington, 58: 208.

Eumeces chinensis formosensis van Denburgh, 1912, Proc. Cal. Ac. Sci. (Series 4), 3 (10): 187-258.

Eumeces chinensis Smith, 1935, Reptiles and Amphibia, Vol. II. Sauria. Taylor & Francis, London: 440.

Eumeces chinensis Taylor, 1936, Univ. Kansas Sci. Bull., 23 (14): 1-643 [1935].

Plestiodon chinensis daishanensis Mao, 1983, Acta Herpetologica Sinica, 2 (3): 59-60.

Eumeces chinensis daishanensis Zhao and Adler 1993, Herpetology of China. SSAR, Oxford/Ohio: 1-522.

Eumeces chinensis Lazell et al., 1999, Postilla, 217: 1-18.
Eumeces chinensis Griffith, Ngo and Murphy, 2000, Russ. J. Herpetol., 7 (1): 1-16.
Plestiodon chinensis Schmitz et al., 2004, Hamadryad, 28 (1-2): 73-89.

［同物异名］ *Tiliqua chinensis*；*Eumeces chinensxs*；*Plestiodon pulchrum*；*Plestiodon sinense*；*Mabouia chinensis*；*Plestiodon chinensis daishanensis*

［俗名］ 山龙子、石龙蜥、猪婆蛇、四脚蛇、山弹

［鉴别特征］ 本种与蓝尾石龙子相似，但本种体较粗壮；有上鼻鳞，无后鼻鳞；第二列下颚鳞楔形；后颏鳞2枚；背面和腹面散布浅色斑点。

［形态描述］ 依据茂兰地区采集的两号标本进行描述 头体长为101.65～115.26mm，尾长94.96～180.89mm。吻钝圆，吻鳞大，背面可见部分小于额鼻鳞；上鼻鳞1对，

成体量度

编号	产地	头长（mm）	头宽（mm）	头高（mm）	头体长（mm）	尾长（mm）	上唇鳞（枚）	下唇鳞（枚）
GZNU20100813 ♀	荔波茂兰	25.87	14.32	13.48	101.65	94.96	7	7
GZNU320995 ♀	荔波茂兰	25.22	17.08	15.16	115.26	180.89	8	8

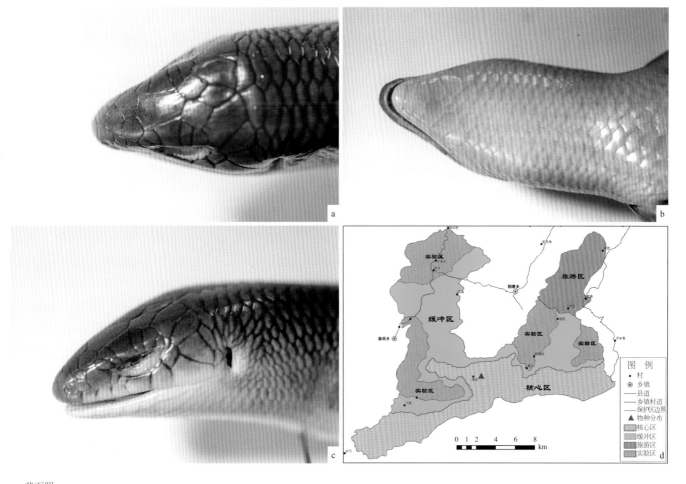

a 背面照
b 腹面照
c 侧面照
d 在茂兰地区的分布

在中线相接；前额鳞显著大于上鼻鳞且相接；间顶鳞1枚，比额顶鳞小，将顶鳞分离，直接与颈鳞相接；顶鳞1对彼此分离；颈鳞3对。鼻鳞小，鼻孔位于鼻鳞中央，将鼻鳞分为前后两半，中间有一条明显的缝线；无后鼻鳞；颊鳞2枚，后颊鳞较大，与第3枚上唇鳞相接；1枚较小的眶前鳞，下面有1列逐渐变小的粒鳞；两枚小的眶后鳞，下1枚大于上1枚；眶上鳞4枚，第3枚较其他3枚大，前2枚与额鳞相接；颞鳞1+2+3，上唇鳞7～8枚，第1枚与前颊鳞相接，后面第7枚唇鳞显著高于前面的唇鳞，并接近耳孔部位；下唇鳞七、八枚；颔片3对。体鳞平滑，圆形，覆瓦状排列。

四肢发达，前后肢贴体相向时，指、趾端不相遇，前肢前伸时指端可达眼，腋区覆盖小鳞，腕部有2枚结节鳞，指长顺序为3＞4＞2＞5＞1，蹠部的一团结节鳞有6枚。趾长顺序为4＞3＞5＞2＞1。生活时成体背面橄榄色，头部棕色，颈侧及体侧红棕色，腹面白色。

[生活习性]　在低海拔的山区、平原耕作区、落叶杂草中、丘陵地区青苔和茅草丛生的路旁易见。住宅附近公路旁边草丛中及树林下的低矮灌木林下和杂草茂密的地方也可见到。时常出没在太阳照射到的碎石堆中。

[地理分布]　茂兰地区分布于翁昂。国内南方各省区广泛分布。

[模式标本]　正模标本（CAS 18605）保存于中国科学院

[模式产地]　中国

[保护级别]　三有保护动物；贵州省重点保护野生动物

[IUCN 濒危等级]　无危（LC）

5. 蜥蜴科　Lacertidae Gray, 1825

Lacertirudae Gray, 1825, Ann. Philos., 25: 200.

Lacertidae Günther,1864, Rest. Brit., India: 68.

体型修长匀称。头部锥形，与蜥体分界明显；吻尖；眼中等大，有发达的活动眼睑，少数种类于下眼睑上具半透明的睑窗，瞳孔圆形。耳孔显露。舌长而薄，前端分叉或具深缺刻，舌面有鳞状乳突或"八"字形横褶。两颌具侧生齿。喉部通常有领围。四肢发达，指、趾5枚。尾长似鞭，有自截能力而易折断，再生力强。头背有对称排列的大鳞；体背被棱鳞或平滑的粒鳞；腹鳞较大，大多呈方形或矩形，纵横有序，排列成行；尾鳞轮生成环。有股孔或鼠蹊孔。

生活于森林、草原、荒漠和平原等各种地带的不同生境，于地面活动穴居地下，也有少数种类具攀爬灌木或树干的能力。昼出夜伏，以昆虫和其他无脊椎动物为主要食源。雄蜥于繁殖季节常出现色彩鲜艳的"婚色"。

本科已知321种，隶属于44属，广泛分布在除马达加斯加岛、塞舌尔群岛之外的非洲、欧洲、亚洲及其邻近岛屿，其中尤以非洲的种类为最多。我国现有4属28种，除西藏外，各地均有分布。茂兰地区有1属2种。

6）草蜥属　*Takydromus* Daudin, 1802

Takydromus Daudin, "X" (1802), Hist. Nat. Rept., Paris, 3: 251. Type species: *Takydromus sexlineatus* Daudin, 1802, of Southeast Asia and the East India, by subsequent designation (Fitaing, 1843, Syst. Rept., Vienna, 1: 21).

本属多数种类体型圆长而略扁平，尾长多数超过头体长的2倍以上，有的达3倍以上。头鳞正常。鼻孔位于鼻鳞、后鼻鳞与第1枚上唇鳞之间，额鼻鳞1枚；前额鳞2枚；额鳞1枚；额顶鳞2枚；顶鳞2枚；顶间鳞1枚；下眼睑被细鳞。多数身体背面被覆起棱大鳞；体侧为颗粒细鳞，腹鳞呈覆瓦状排列，纵横成行，外端游离，全部或仅外侧数行起棱。指、趾末端直伸，最末节近端不侧扁。具鼠蹊孔且多数3对以下。

本属已知超过30种，广泛分布于欧洲、亚洲（西部）和非洲。茂兰地区有2种，分种检索表如下。

茂兰地区种检索表

背部起棱大鳞4行，腹部大鳞12行··南草蜥 *Takydromus sexlineatus*

背部起棱大鳞6行，腹部大鳞8行··北草蜥 *Takydromus septentrionalis*

（9）北草蜥 *Takydromus septentrionalis* Günther, 1864

Tachydromus septentrionalis Günther, 1864, Rept. Brit. India, London: 69. Type locality: Ningbo, Zhejiang, China.

Tachydromus septentrionalis Günther, 1888, The Reptiles of British India., London (Taylor & Francis), XXVII: 452.

Takydromus septentrionalis Boulenger, 1899, Proc. Zool. Soc., London: 159-172.

Takydromus septentrionalis Stejneger, 1907, Bull. U.S. Natl. Mus., 58: xx, 1-577.

Takydromus septentrionalis Schlüter, 2003, Kirschner & Seufer Verlag: 110. [review in Draco, 21: 91, Elaphe, 13 (2): 22.]

Takydromus septentrionalis Wang et al., 2017, Zootaxa, 4338 (3): 441-458.

[同物异名] 无

[俗名] 草蜥

[鉴别特征] 背部起棱大鳞通常6行，腹鳞8行且起棱；尾长为头体长的2～3倍及以上；背面棕绿色。

[形态描述] 依据茂兰地区标本 GZNU320998 描述 头体长51.98mm，断尾，头长13.18mm，头宽7.29mm，头高5.58mm，头长约为头宽的1.8倍。吻部较窄，吻端锐圆，吻鳞较窄，不入鼻孔，与额鼻鳞略相接，将左右上鼻鳞隔开；鼻孔开口于鼻鳞、后鼻鳞与第1枚上唇鳞之间；额鼻鳞较大，长宽几乎相等；前额鳞的前1/3相接，并略小于顶鳞；额鳞长大于宽，小于顶鳞，前端宽，呈三角形，向前突出与前额鳞相接，后端窄且近于平直并和额顶鳞相接；额顶鳞1对，中缝相接，其余部分与额鳞，3、4眶上鳞，顶鳞及顶间鳞相接；顶鳞是头部最大的一对鳞片，外缘有1排较长的棱鳞；顶间鳞很小；顶眼清晰；枕鳞甚小，比顶间鳞短，很少相等，通常为1～2枚小鳞片或顶鳞的中接线把枕鳞与顶间鳞隔开；眶上鳞4枚，第1、第4枚很小，第1枚更小，几乎呈粒状，第2、第3枚较大，大小相近或第2枚稍大；上睫鳞4或5枚，第1、第2枚最长，有的与眶上鳞相接，在上睫鳞与眶上鳞间有一组粒鳞填充；颊鳞2枚，后一枚较前者长；上唇鳞通常6，第5枚最大，位于眼下方；下唇鳞6；颏鳞宽大于长；颏片通常3对（少数为4对或不对称，右4左3、左2右3等），颞鳞小，微棱；耳孔上方边缘有3枚较大鳞，其中一枚尤为窄长；鼓膜上有1枚狭而长的鼓鳞，头后面和成行大鳞之间有许多小粒鳞；领围鳞片由11～12枚较尖且起棱鳞片组成。背鳞起棱大鳞片，其棱首尾相接连续成线；前段鳞多为7行（少数为6或8行）；中段为6行（少数为5或7行）；后段5行（少数为4行），中间一行或两行较小，有的行间有增加的小鳞。腹部起棱大鳞8行，靠外侧2～3行起棱更明显；腹部鳞片近方形，尖端钝，且具短的锐突，排列整齐，纵横成行；从领围到肛前鳞之间，腹鳞为26～31（通常为27～29）横列。肛

成体量度

编号	产地	头长（mm）	头宽（mm）	头高（mm）	头体长（mm）	尾长（mm）	上唇鳞（枚）	下唇鳞（枚）
GZNU320998 ♂	荔波茂兰	13.18	7.29	5.58	51.98	断尾	6	6

前鳞光滑，具有2条隆起线，又在肛前鳞两侧各有两枚鳞片。背腹之间体侧为粒鳞，近背侧1行，近腹侧2～3行具起棱鳞片。四肢背面的鳞片近菱形且起棱，也有粒状鳞。鼠蹊孔1对；趾下瓣单个或部分分开，大多数为23～29个。尾鳞强棱，尖端具短锐突，在尾基背面鳞的棱形成非常硬的脊。

生活时整个背面为棕绿色，腹面灰白色或灰棕色；头侧近口缘处和体侧近腹部色浅；眼至肩部常有一条窄的线状纵纹，边缘齐整。雄性背鳞的外侧从顶鳞后缘起到尾基部有一鳞片宽（实际是相邻两行鳞片各一半形成的一鳞片宽）草绿色齐整的纵纹；体侧有不规则的深色斑。

[生活习性]　生活在山地草丛中，多分布在海拔436～1700m的山坡。4～9月均可见其活动。

[地理分布]　茂兰地区分布于永康。国内除新疆、西藏等地外，广泛分布于黄河以南各省区。

[模式标本]　模式标本（BMNH 1946.8.4.27）保存于英国伦敦自然历史博物馆

[模式产地]　浙江宁波

[保护级别]　三有保护动物；贵州省重点保护野生动物

[IUCN濒危等级]　无危（LC）

a　背面照
b　腹面照
c　侧面照
d　整体照
e　在茂兰地区的分布

（10）南草蜥 *Takydromus sexlineatus* Daudin, 1802

Takydromus sexlineatus Daudin, 1802, Sonnini, Vol. 3. F. Dufart, Paris: 256.

Tachydromus ocellatus Guérin-Méneville, 1829, Avec un text descriptif. Paris & London, [Vol.] 1, Plates. I-XXX, [Vol.] 3, I-XVI: 1-576.

Tachydromus typus Gray, 1838 (fide Schlüter, 2003).

Tachydromus sexlineatus Duméil and Bibron, 1839, Roret/Fain et Thunot, Paris, Vol. 5: 871.

Tachydromus meridionalis Günther, 1864, The Reptiles of British India, London (Taylor & Francis).

Tachydromus sexlineatus de Rooij, 1915, Leiden (E. J. Brill), XIV: 384.

Tachydromus kwangsiensis Ahl, 1930: 327. Sitzungsberichte der Gesellschaft naturforschender Freunde, Berlin, Vol. 1: 310-332.

Takydromus sexlineatus sexlineatus Smith, 1935: 366. Reptiles and Amphibia, Vol. II. Sauria. Taylor & Francis, London: 440.

Takydromus sexlineatus Taylor, 1963, The lizards of Thailand. Vriv. Kansas Sci. Bull., 44: 687-1077.

Takydromus sexlineatus ocellatus Schlüter, 2003: 77. Kirschner & Seufer Verlag: 110.

Takydromus sexlineatus Arnold et al., 2007, Zootaxa, 1430: 1-86.

Takydromus sexlineatus ocellatus Teynié et al., 2010, Zootaxa, 2416: 1-43.

Takydromus sexlineatus Wang et al., 2017, Zootaxa, 4338 (3): 441-458.

[同物异名] *Tachydromus sexlineatus*；*Tachydromus ocellatus*；*Tachydromus typus*；*Tachydromus meridionalis*；*Tachydromus kwangsiensis*

[俗名] 草蜥

[鉴别特征] 本种与北草蜥十分相似，但背部棱起大鳞通常只有4行，腹鳞10~12行且起棱，尾长为体长的3倍以上。

[形态描述] 依据茂兰地区标本 GZNU20170420001 描述 头体长 64.83mm，头长 17.93mm，头宽 8.49mm，头高 7.34mm，头长约为头宽的2倍。吻部较窄，吻端锐圆，吻鳞较窄，不入鼻孔，与额鼻鳞略相接，将左右上鼻鳞隔开；鼻孔开口于鼻鳞、后鼻鳞与第1枚上唇鳞之间；额鼻鳞较大，长宽几乎相等，略小于顶鳞；额鳞长大于宽，前端向前突出与前额鳞相接，后端和额顶鳞相接；额顶鳞1对，中缝相接，其余部分与额鳞，2、3眶上鳞相接，顶鳞是头部最大的1对鳞片，被顶间鳞隔开，顶间鳞很小，顶眼清晰，枕鳞细小，鳞上均有棱粒；眶上鳞3枚。颊鳞2枚，后一枚比前一枚长，上唇鳞通常5~6枚，第5枚最大，位于眼下方；下唇鳞5；颏片3对，颞鳞小，微棱；耳孔上方边缘有3枚较大鳞，其中一枚尤为窄长，头后面和成行大鳞之间有许多小粒鳞；背鳞起棱大鳞片，棱首尾相接连续成线；前段鳞为4行；中段为5行；后段鳞为4行；腹部鳞片具棱，排列整齐，纵横成行；肛前鳞光滑，具有2条隆起线。背腹之间体侧为粒鳞，近背侧1行，近腹侧2行具起棱鳞片。四肢背面的鳞片近菱形且起棱，也有粒状鳞。

生活时头背及体背棕绿色；头侧稍蓝，体侧灰绿色，头、体及背面灰白色，腹部灰棕色。

[生活习性] 栖息于长满长草的低地，常见于干燥空旷的地方。4~9月均可见其活动。

[地理分布] 茂兰地区分布于洞塘、永康。国内主要分布于华中、华南及西南地区。

[模式标本] 全模标本（MNHN-RA 2662）保存于法国国家自然历史博物馆

[模式产地] 中国南方

[保护级别] 三有保护动物

[IUCN 濒危等级] 无危（LC）

成体量度

编号	产地	头长（mm）	头宽（mm）	头高（mm）	头体长（mm）	尾长（mm）	上唇鳞（枚）	下唇鳞（枚）
GZNU20170420001 ♂	荔波茂兰	17.93	8.49	7.34	64.83	227.35	5	5
GZNU320259 ♂	荔波茂兰	17.82	10.05	8.62	66.68	241.03	6	5

a 背面照
b 腹面照
c 侧面照
d 整体照
e 在茂兰地区的分布

6. 蛇蜥科 Anguidae Gray, 1825

Anguidae Gray, 1825, Ann. Phil., xxvi: 201 (in part).

体型圆筒状似蛇，多数种类无四肢，但有肢带的残迹。我国产的种类均无四肢。体两侧各有纵沟1条。头背具有大鳞片，均呈对称排列；躯尾背面被覆瓦状圆鳞。鳞下衬以来源于真皮的骨板。眼小，具活动眼睑。舌很长，舌尖弯，多少可缩入舌鞘内，舌前端有深缺刻或呈分叉状；舌前端覆以鳞状乳突，舌基部较厚，被绒毛状乳突。

头骨具有颞弓及眶后弓。侧生齿，大小形状不一，有圆锥状、结节状或上端尖锐向后微弯。尾较长，长度至少为头体长1倍以上，易断，再生能力强。无肛前孔和股孔。

本科多数种类陆生，少数具有树栖习性。不少种类夜间活动频繁，四处觅食，而白天穴居洞内。以昆虫和其他小型的无脊椎动物为食。卵生，少数为卵胎生。

本科已知 73 种，隶属于 10 属，主要分布在美洲，旧大陆仅有 2 属。我国已知有 1 属 4 种，主要分布于长江以南。茂兰地区仅有 1 种。

7）脆蛇蜥属 *Dopasia* Daudin, 1803

Ophisaurus Daudin, 1803, Bull. Sci. Soc. Philomath., Paris, Ser. 2, 3: 188. Type species: *Anguis zientralis* Linnaeus, 1766, of North America, by monotypy.
Ophisaurus Daudin, 1803, Hist, Nat. Rept., 7: 346 (Type species: *Ophisaurus ventralis* Linnaeus, 1766).
Bipes Oppel, 1811, Ordn. Rept.: 43.
Proctopus Fischer, 1813, Mem. Soc. Imp. Sci. Mosc., Ⅳ: 241.
Pseudopus Merrem, 1820, Tent. Syst. Amphib., 13: 78.
Dopasia Gray, 1853, Ann. Mag. Nat. Hist., 2 (7): 389.
Ophiseps Blyth, 1853, J. Asiat. Soc. Beng., 22: 655.
Hyalosaurus Guenther, 1873, Ann. Mag. Nat. Hist., 4 (11): 351.

本属种类无四肢或仅有退化的后肢（或不显）。体两侧各有纵沟 1 条。体表覆盖方形或菱形的鳞片，纵横排列成行。具翼骨齿。

本属世界已知 13 种，分布于欧洲（东南部）、亚洲（西南部）、非洲（北部）及北美洲。我国目前已知有 4 种。茂兰地区分布 1 种——脆蛇蜥 *Dopasia harti*。

（11）脆蛇蜥 *Dopasia harti* (Boulenger, 1899)

Ophisaurus harti Boulenger, 1899, Pros. Zool. Soc., London: 160. Type locality: Kuatun (Guadun), northwestern Fokien [= Fujian] Prov., China; 3000 to 4000 feet or more.
Ophisaurus formosensis Kishida, 1930, Lansania, Tokyo, 2 (18): 124-128.
Ophisaurus harti Smith, 1935: 394. Reptiles and Amphibia, Vol. Ⅱ. Sauria. Taylor & Francis, London: 440.
Ophisaurus harti Liu, 1970, Biol. Bull., Taiwan Normal Univ., 5: 51-93.
Ophisaurus formosensis Zhao and Adler, 1993, Hevpetology of China: 199.
Ophisaurus harti Lin et al., 2003, Zoological Studies, 42 (3): 411-419.
Ophisaurus harti Pianka and Vitt, 2003, University of California Press, Berkeley: 232.
Ophisaurus harti Bobrov and Semenov, 2008, Moscow: 236.
Dopasia harti Conrad et al., 2011, Cladistics, 27 (3): 230-277.
Dopasia harti Nguyen et al., 2011, Zootaxa, 2894: 58-68.

［同物异名］ *Ophisaurus formosensis*（台湾脆蛇）；*Ophisaurus harti*（脆蛇）

［俗名］ 碎蛇、脆蛇、山黄鳝、小泥鳅（贵州）、金蛇、银蛇、锡蛇

［鉴别特征］ 本种体型较为粗壮；鼻鳞与单枚的前额鳞间有 2 枚小鳞片；体侧纵沟间背鳞 16～18 行；尾长约为头体长的 2 倍。

［形态描述］ 依据茂兰地区标本 GZNU320991 描

述　头体长133.43mm，尾长289.91mm，尾长约为头体长的2倍。无四肢，体侧纵沟自颈后至肛侧。头长约为头宽的1.6倍，头宽大于头高；眼径约为吻长之半；鼻鳞与单枚的前额鳞间仅有2枚小鳞片；上唇鳞9枚，下唇鳞11枚。体侧纵沟间背鳞16纵列，中央9行鳞大而起棱，前后棱相连续成为清晰的纵脊，这些纵脊自颈后一直延伸至尾末端；腹鳞光滑，体侧两纵沟间腹鳞数均为10行。自肛后至尾尖部背、腹全为棱鳞，前后棱相连续成为清晰的纵脊，体背浅褐色及灰褐色。体背前段有20多条不规则蓝黑色或天蓝色的横斑及点斑，自颈部至尾端有色深且较宽的纵线，此纵线延至体后更为清晰，腹面色泽比体背浅，腹部无斑纹。

[生活习性]　生活于茂兰地区的山地中、石块下、玉米地、菜地、草丛中和岩隙间。行动似蛇，但较缓慢，靠身体左右摆动前进，尾极易断，能再生。

[地理分布]　茂兰地区分布于翁昂。国内分布于江苏、浙江、福建、台湾、湖南、广西、四川、贵州、云南。国外分布于越南（北部）。

[模式标本]　选模标本（BMNH 1946.8.3.81）、全模标本（BMNH 1946.8.3.80、BMNH 99.4.24.6-99.4.24.7）保存于英国伦敦自然历史博物馆

[模式产地]　福建

[保护级别]　三有保护动物；贵州省重点保护野生动物

[IUCN 濒危等级]　濒危（EN）

成体量度

编号	产地	头长（mm）	头宽（mm）	头高（mm）	头体长（mm）	尾长（mm）	上唇鳞（枚）	下唇鳞（枚）
GZNU320991	荔波茂兰	14.92	9.36	7.56	133.43	289.91	9	11

a　背面照
b　侧面照
c　腹面照
d　在茂兰地区的分布

7. 鬣蜥科 Agamidae Gray, 1827

Agamidae Gray, 1827, Phil. Mag., 2 (2): 57.

体型中等大小或小型。头背面无对称排列的大鳞，身体表面鳞片多呈覆瓦状排列；鳞常具棱，或有鬣鳞。眼较小，眼睑发达，瞳孔圆形，鼓膜裸露或被鳞。舌厚，短或中等，前端完整或微缺，舌面上具绒毛状乳突。端生齿，异形。头骨具颞弓及眶后弓。前凹型椎体。尾较长，不易断。四肢发达。多数种类无肛前孔或股孔，有的种类雄性具胼胝鳞或喉囊或颞部隆肿。

生活于陆地上，营树栖的种类较多；常在地面生活的种类背、腹略扁平；树栖的体略侧扁。主要以昆虫为食，少数种类兼食植物，极少数专吃植物。

本科已知 445 种（不包括亚种），隶属于 56 属。广泛分布于旧大陆（斐济岛、汤加群岛及马达加斯加岛没有），以东洋界为主。中国有 12 属 62 种。贵州有 3 属（棘蜥属、树蜥属和龙蜥属）4 种。茂兰地区有 2 属 2 种，分属检索表如下。

茂兰地区属检索表

体侧扁，背鳞大小一致，体背常有鬣鳞···拟树蜥属 *Pseudocalotes*
体背、腹扁，背鳞大小不一致，背棘发达···棘蜥属 *Acanthosaura*

8）棘蜥属 *Acanthosaura* Gray, 1831

Gonocephalus Kaup, 1825, Isis: 590.
Lophyrus (not of Poli, 1791) Fitzinger, 1826, Class. Rept.: 49.
Acanthosaura Gray, 1831, in Griffith and Pidgeon (eds.), Animal Kingd. Cuvier, London, 9 (Synops Spec.): 56. Type species: *Agama armata* Hardwicke and Gray, 1827, by monotype.

体型侧扁，背脊部常具棱，雄性背鬣发达；背鳞大小不一，间杂有大鳞，有或无喉囊，肩前褶显著。鼓膜裸露或部分被鳞；尾部略扁，尾鳞起棱，大于背鳞，无股孔，亦无肛前孔。

世界已知 3 种，分布于缅甸、泰国、马来西亚、印度尼西亚等东南亚地区。中国有 2 种，茂兰地区有 1 种，即丽棘蜥 *Acanthosaura lepidogaster*。

（12）丽棘蜥 *Acanthosaura lepidogaster* (Cuvier, 1829)

Calotes lepidogaseter Cuvier, 1829, Refine Animal, ed. 2, 2: 39. Type locality: China, Vietnam.
Acanthosaura lamnidentata Boulenger, 1885: 302. Catalogue of the Lizards in the British Museum (Nat. Hist.) I. Geckonidae, Eublepharidae, Uroplatidae, Pygopodidae, Agamidae. London: 450.
Acanthosaura hairranensis Boulenger, 1900, Proc. Zool. Soc., London: 957. Type locality: Mt. Wuzhi, Hainan, China.
Acantlwsaura braueri Vogt, 1914, Sitz. Ges. Naturf. Fr. Berlin: 97. Type locality: China. Mell, 1922, Archiv. Naturg., Berlin, 88: 112 (east of "Shuichow", Guangdong).

Goniocephalus lepidogaster Smith, 1935: 161. Reptiles and Amphibia, Vol. II. Sauria. Taylor & Francis, London: 440.

Gonocephalus lepidogaster Mell, 1952, Die Aquar. Terrar. Z. (DATZ), Stuttgart, 56: 160-163.

Acanthosaura lepidogaster Taylor, 1963, Univ. Kansas Sci. Bull., 44: 687-1077.

Acanthosaura lepidogaster Brygoo, 1988, Bull. Mus. Nat. Hist. Nat., 10 (ser. 4) A (3), suppl.: 1-56.

Calotes brevipes Zhao and Adler, 1993, Herpetology of China. SSAR, Oxford/Ohio: 1-522.

Acanthosaura fruhstorferi Denzer et al., 1997, Mitt. Zool. Mus. Berlin, 73 (2): 309-332.

Acanthosaura lepidogaster Schlüter and Hallermann, 1997, Stuttgarter Beitr. Naturk. Ser., A (553): 1-15.

Acanthosaura lepidogaster Cox et al., 1998: 93. A photographic guide to snakes and other reptiles of Peninsular Malaysia, Singapore and Thailand. Ralph Curtis Publishing: 144.

Acanthosaura lepidogaster Manthey and Schuster, 1999: 20. Agamen, 2. Aufl. Natur und Tier Verlag (Münster): 120.

Acanthosaura lepidogastra Macey et al., 2000, Systematic Biology, 49 (2): 233-256.

Acanthosaura lepidogaster Barts and Wilms, 2003, Die Agamen der Welt. Draco, 4 (14): 4-23.

Acanthosaura lepidogaster Bobrov and Semenov, 2008, Lizards of Vietnam, Moscow: 236.

［同物异名］ *Calotes lepidogaseter*；*Lophyrus tropidoga-ster*；*Acanthosaura hairranensis*；*Acanthosaura fruhstorferi*；*Acantlwsaura braueri*；*Goniocephalus lepidogaster*；*Calotes fruhstorferi*；*Calotes brevipes*

［俗名］ 丽背蜥、七步跳（福建崇安）

［鉴别特征］ 眼后棘不发达，长度约为眼径的一半；体鳞大小不一，间杂有大棱鳞；颈鬣发达，与背鬣不连续。后肢贴体前伸达吻眼之间。尾长约为头体长的2倍，尾背有棕黑色环纹。

［形态描述］ 依据茂兰地区标本GZNU32005描述 头体长95.52mm，背、腹略扁平，吻钝圆，头长大于头宽，头顶前部较平；头背部鳞片大小几乎一致，仅中央有数枚略大的鳞片。眼大，两眼间的头顶部略有凹陷，吻长大于眼径，吻棱显著。上唇鳞8，下唇鳞11；颊鳞三角形，由颊鳞至口角有3行棱鳞与下唇鳞平行排列。体鳞大小不一，呈不规则覆瓦状排列，间杂有大棱鳞，大鳞后端起强棱，棱尖斜向后上方，肩褶发达；腹鳞比背鳞大，并明显起棱。尾细长而侧扁，约为头体长的1.9倍。基部膨大，雄性尾基腹面突起。

生活时头背部为淡黑灰色，体躯灰棕色，体背具有黑褐色菱形斑，体两侧带有浅绿黄色；四肢背面具黑褐色横纹，并有少数黄色斑；体腹面色浅，有分散不规则的黑斑点。尾背有棕黑色环纹。

［生活习性］ 生活于山区林下，常活动在路旁、溪边、灌丛下及林下落叶处。行动迅速，爬行时常四肢触地，身体略举起。

［地理分布］ 茂兰地区分布于洞塘。国内分布于福建、贵州、云南、海南、广西、江西、广东。国外分布于泰国、越南、柬埔寨。

［模式标本］ 同模标本（ZMB 18180）、选模标本（NMW 11411，成年雌性）、配模标本（SMNS 4155，成年雌性）、正模标本（ZMB 24178）保存于德国柏林洪堡大学自然历史博物馆

［模式产地］ 越南

［保护级别］ 三有保护动物；贵州省重点保护野生动物

［IUCN 濒危等级］ 无危（LC）

成体量度

编号	产地	头长（mm）	头宽（mm）	头高（mm）	头体长（mm）	尾长（mm）	上唇鳞（枚）	下唇鳞（枚）
GZNU32005	荔波茂兰	27.84	17.99	15.74	95.52	230.54	8	11
GZNU32004	荔波茂兰	31.52	18.26	18.26	95.11	189.21	10	12

a 背面照
b 腹面照
c 侧面照
d 整体照
e 在茂兰地区的分布

9）拟树蜥属 *Pseudocalotes* Fitzinger, 1843

Calotes Rafinesque, 1815, Anal. Nat.: 75. Nomen nudum.

Bronchocela Kaup, 1827, Isis: 619.

Pseudocalotes Fitzinger, 1843, Vienna: Braumüller and Seidel.: 46.

Pseudocalotes Hallermann and Böhme, 2000, Amphibia Reptilia, 21 (2): 193-210.

本属物种原来一直被放在树蜥属 *Calotes* 中，后基于形态学和分析系统学研究，从原树蜥属 *Calotes* 新拆分出拟树蜥属 *Pseudocalotes*（Hallermann and Böhme, 2000; Mahony, 2010; Pyron et al., 2013）。

本属物种体侧扁。背鳞排列规则，大小一致（短肢拟树蜥与蚌西拟树蜥除外）；背鬣较发达，与颈鬣相连续；通常具有喉囊；肩前褶或有或无；鼓膜裸露；尾细长，不易断，长度常为头体长的 2～3 倍。雄蜥成体喉囊及鬣鳞较发达，尾基部常膨大，该处鳞片增大变厚。不具肛前孔或股孔。

本属已知27种，分布于印度、中国、越南、马来西亚、印度尼西亚及巴布亚新几内亚等地。中国有4种。茂兰地区有1种，即细鳞拟树蜥 *Pseudocalotes microlepis*。

（13）细鳞拟树蜥 *Pseudocalotes microlepis* **(Boulenger, 1887)**

Calotes microlepis Boulenger, 1887, Ann. Mus. Civ. Stor. Nat. Genova, Ser. 2, 5: 476. Type locality: Pla-poo, Tenasserim, Burma.

Pseudocalotes microlepis Cox et al., 1998: 99. A Photographic Guide to Snakes and Other Reptiles of Peninsular Malaysia, Singapore and Thailand. Ralph Curtis Publishing: 144.

Pseudocalotes microlepis Hallermann and Böhme, 2000, Herpetologica, 57 (3): 255-265.

Pseudocalotes microlepis Nguyen et al., 2009, Chimaira, Frankfurt: 768.

[同物异名] *Calotes microlepis*（细鳞树蜥）

[俗名] 无

[鉴别特征] 头高与头宽约相等，无肩褶，背鳞棱尖斜向后下方，环体中段鳞61～73行，后肢贴体前伸最长趾端达腋部或肩部。

[形态描述] 依据茂兰地区标本 GZNU320260 描述 头体长为65.41mm，尾长为125.97mm。头高与头宽约相等，无肩褶。背面浅金褐色，具黑色斑纹，眼四周有黑色辐射纹。前额较平，头背鳞片大小不一，具棱，前额大鳞排成倒"Λ"形，吻棱和上睑脊明显；上唇鳞8。鼓膜明显，直径约为眼眶直径的1/2。体明显侧扁，背鳞大小一致，排列规则整齐成行，鳞片后半部起棱，上部3行鳞棱尖直向后，其余棱尖向后下方，腹鳞起强棱，环体中段鳞65行。咽喉部鳞起棱，小于腹鳞，喉囊较小。颈鬣由7枚鳞片组成，背鬣不发达。四肢细弱，后肢贴体前伸最长趾端达腋部或肩部。尾侧扁，被棱鳞；尾背正中一行鳞扩大形成锯齿状棱。

[生活习性] 生活于山间草坡、杂草灌丛下或乱石间。

[地理分布] 茂兰地区调查时见于翁昂，荔波还见于小七孔景区。国内分布于海南、云南、贵州。国外分布于缅甸、越南。

[模式标本] 全模标本（BMNH 1946.8.11.21）保存于英国伦敦自然历史博物馆

[模式产地] 缅甸，1200m

成体量度

编号	产地	头长（mm）	头宽（mm）	头高（mm）	头体长（mm）	尾长（mm）	上唇鳞（枚）	下唇鳞（枚）
GZNU320260	荔波茂兰	22.69	11.11	11.25	65.41	125.97	8	6
GZNU320098	荔波茂兰	15.84	8.76	8.35	46.45	81.94	9	10

[保护级别] 三有保护动物；贵州省重点保护野生动物

[IUCN 濒危等级] 无危（LC）

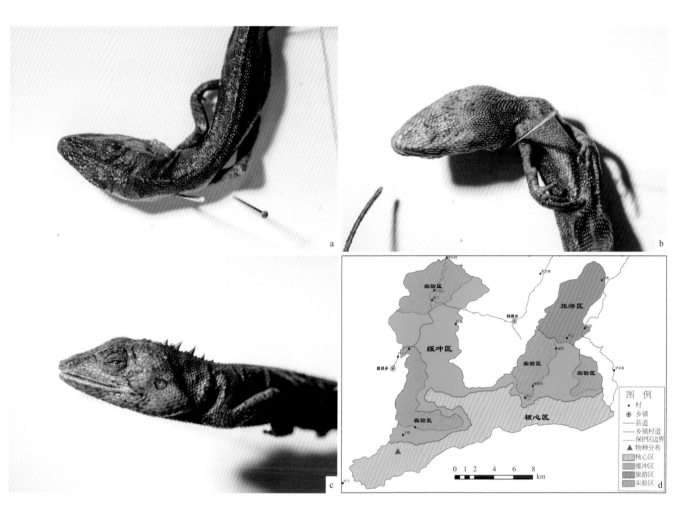

a 背面照
b 腹面照
c 侧面照
d 在茂兰地区的分布

II. 蛇亚目 Serpentes

Serpentes Linnaeus, 1758, Syst. Nat., 10th ed. I: 214.

蛇亚目体表被鳞；无四肢及肢带，无胸骨；左右下颌骨在前端以韧带相连；上下颌骨表面着生有齿；尾长明显短于体长；无活动眼睑，视力较差；鼓膜和外耳孔退化消失；舌末端分叉，有很强的伸缩性。雄性有成对的交接器。体内受精，卵生或卵胎生。树栖、穴居、陆栖或水栖。茂兰地区有5科，分科检索表如下。

茂兰地区科检索表

1 { 背鳞较小，体中段 30 行以上，腹鳞窄，仅比相邻的背鳞稍宽，泄殖肛孔两侧有爪状后肢残余 ················ 蚺科 Pythonidae
   { 背鳞较大，体中段不超过 30 行，腹鳞较宽，无爪状后肢残余 ·· 2
2 { 上颌骨前端没有毒牙 ·· 3
   { 上颌骨前端有较大的毒牙 ·· 4
3 { 有颊沟 ··· 游蛇科 Colubridae
   { 无颊沟 ··· 钝头蛇科 Pareatidae
4 { 上颌骨前端有沟牙，上颌骨不能竖立，瞳孔圆形 ·· 眼镜蛇科 Elapidae
   { 上颌骨前端有管牙，上颌骨能竖立，瞳孔直立椭圆形 ··· 蝰科 Viperidae

8. 蚺科 Pythonidae Fitzinger, 1826

Pythonidae Fitzinger, 1826, Neue Classification der Reptilien nach ihren natürlichen Verwandtschaften. Vienna: J. G. Hübner: vii + 66 (in German and Latin).

蚺科蛇类体型较大，躯体一般很长，有后肢的残余，雄性蛇身上的这一特征更为突出。头部一般鳞片较小。卵生或卵胎生。

本科世界共有 19 属 75 种，中国有 1 属 1 种。

10）蚺属 *Python* Daudin, 1803

Python Daudin, 1803 Hist. Nat. Rept., 5: 226. Type species: *Coluber molurus* Linnaeus, 1758.

体型较大的无毒蛇类，头、颈分明，头背有对称的大鳞。吻鳞及前两枚上唇鳞具唇窝。眼大，瞳孔纵置椭圆形；外鼻孔仰开或向侧上开孔于半裂的大鼻鳞上，鼻鳞被一对鼻间鳞分隔；前颌骨有齿，上下颌前段齿裂较长，向后逐渐变短。体稍侧扁，背鳞小且平滑无棱，尾下鳞大部或全部双行。雄蛇泄殖肛孔两侧有较明显的爪状后肢残余。卵生。树栖，具缠绕性。以夜间活动为主，善捕食温血动物。

本属世界约 10 种，分布于非洲、亚洲（南部）及澳大利亚。中国只有 1 种。

（14）蟒蛇 *Python bivittatus* Kuhl, 1820

Python bivittatus Kuhl, 1820, Beitr. Zool. Vergkeich. Anat Frankfert am Main, (2): 94. Type localtiy: none given.

Python molurus bivittatus Mertens, 1930, Abh. Senckenb. Narurforsch. Gesellsch., Frankfurt am Main, 44 (2): 287.

Python molurus bivittatus Manthey and Grossmann, 1997: 429. Natur und Tier Verlag (Münster): 512.

Python molurus bivittatus Cox et al., 1998: 15. A photographic guide to snakes and other reptiles of Peninsular Malaysia, Singapore and Thailand. Ralph Curtis Publishing: 144.

Python molurus bivittatus Zhao, 2006. The snakes of China.

Python bivittatus Jacobs et al., 2009, Sauria, 31 (3): 5-16.

Python bivittatus progschai Jacobs, Aulyia and Böhme, 2009, Sauria, 31 (3): 5-16.
Python bivittatus progschai Koch, 2011, Edition Chimaira: 1-374.
Python bivittatus Wallach et al., 2014: 618. Snakes of the World: A catalogue of living and extinct species. Taylor and Francis, CRC Press: 1237.

[同物异名] *Python molurus bivittatus*（蚺双带亚种、蟒蛇）；*Python bivittatus progschai*

[俗名] 黑尾蟒、金花蟒、南蛇（福建）、琴蛇（广东）

[形态描述] 据《中国蛇类（上）》（赵尔宓，2006）描述 雄性次成体体全长/尾长为2390mm/360mm；雌性次成体体全长/尾长为1275mm/153mm，活重900g。成体体型巨大，一般全长3～4m，最长可达6～7m。头较小，吻端较窄而略扁，头、颈区分明显。鼻孔开于鼻鳞上部；眼大小适中，瞳孔直立椭圆形；部分上唇鳞和下唇鳞有唇窝（热测位器官）；泄殖肛孔两侧有呈爪状的后肢残迹，雄性者较为明显。通体棕褐色，体背及两侧有镶黑边的云豹斑纹；腹面黄白色。头颈背面有暗褐色矛形斑，头侧眼前后有1黑色线纹向后斜达口角，眼下另有1黑色纹斜达口缘；头腹黄白色。

眶前鳞2，眶后鳞4，围眼一周有鳞片8；唇鳞9～12；下唇鳞16～20；头背除鼻间鳞、前额鳞、额鳞及头侧的前述各鳞外，头部其余都是小鳞片。背鳞较小，颈部一周55～57，躯干中部一周61～70，肛前一周41～46；腹鳞窄小不发达，255～270；肛鳞完整；尾下鳞59～69对。上颌齿每侧17～18。

[生活习性] 栖息环境为林木茂盛的低山或中山地区，喜攀援于树上或浸泡于水中。多于夜晚活动。

[地理分布] 茂兰地区分布于翁昂。国内分布于福建、海南、广西、四川、贵州、云南。国外分布于缅甸、泰国、越南、马来西亚、印度尼西亚。

[模式标本] 正模标本（ZFMK 87481）保存地未知

[模式产地] 印度尼西亚爪哇

[保护级别] 国家I级重点保护野生动物

[IUCN濒危等级] 极危（CR）

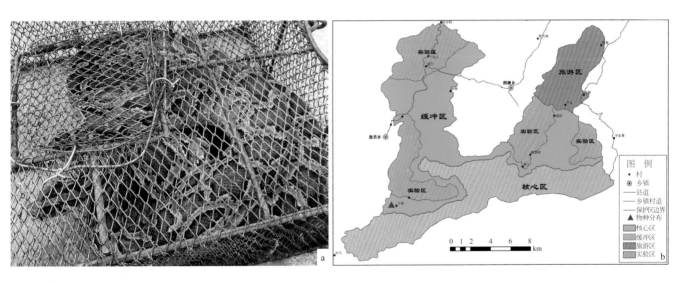

a 侧面照
b 在茂兰地区的分布

9. 钝头蛇科 Pareatidae Romer, 1956

Pareinae Romer, 1956, Osteology of the Reptiles: 583.
Pareatidae Vidal, Delmas, David, et al., 2007, The phylogeny and classification of caenophidian snake inferred from seven nuclear protein-coding genes. C. R. Biologies, 330: 182-187.

本科物种无毒，头较大，吻钝而圆，体略侧扁，眼大。颔片交错，无颔沟。陆栖或树栖。卵生。

近年来的形态学和分子系统学研究，都认为蛇亚目是单系群（Townsend et al., 2004; Conrad, 2008; Pyron et al., 2013）。科级分类阶元中，不少研究认为原游蛇科 Colubridae 不是单系，而应该被拆分为游蛇科 Colubridae、闪皮蛇科 Xenodermatidae、钝头蛇科、水蛇科 Homalopsidae 和鳗形蛇科 Lamprophiidae（Vidal et al., 2007; Wiens et al., 2008; Zaher et al., 2009; Pyron et al., 2011, 2013）。本书采用上述的分类建议，将钝头蛇科从游蛇科中划分出来独立为科。

11）钝头蛇属 *Pareas* Wagler, 1830

Pareas Wagler, 1830, Syst. Amphib: 181. Type species: *Amblycephalus carinatus* Boie, 1828.

头较大，吻宽钝，头、颈区分明显；眼大，瞳孔直立椭圆形；有1枚长的眶下鳞，常常由眼前下角弯向眼后方；上唇鳞多不入眶。颔片左右交错排列，不形成颔沟。躯干略侧扁。背鳞通体15行，平滑或起棱，肛鳞完整，尾下鳞双行。

本属世界已知约14种，分布于东南亚地区。我国已知9种。茂兰地区有2种，即平鳞钝头蛇 *Pareas boulengeri* 和福建钝头蛇 *Pareas stanleyi*，分种检索表如下。

茂兰地区种检索表

头部无黑色巨斑，背部鳞片平滑无棱，腹鳞176～189，尾下鳞62～77···平鳞钝头蛇 *Pareas boulengeri*
头部有1黑色巨斑，背部鳞片中央均起棱，腹鳞146～175，尾下鳞44～60···福建钝头蛇 *Pareas stanleyi*

（15）平鳞钝头蛇 *Pareas boulengeri* (Angel, 1920)

Amblycephalus boulengeri Angel, 1920, Bull. Mus. Hist. Nat., Paris, 26: 113. Type locality: Koeï Tchéou [= Guizhou], China.
Amblycephalus monticola boulengeri Mell, 1931, Lingnan Sci. Jour., Canton, 8 [1929]: 199-219.
Pareas boulengeri Hu, Zhao and Liu, 1973, Acta Zool. Sinica, Beijing, 19 (2): 158.
Pareas boulengeri Wallach et al., 2014: 535. Taylor & Francis, CRC Press: 1237.

[同物异名] *Amblycephalus boulengeri*; *Amblycephalus monticola boulengeri*

[俗名] 黄狗蛇（贵州）、钝头蛇（甘肃）

[鉴别特征] 前额鳞入眶，没有眶前鳞，颊鳞入眶甚多，背鳞平滑无棱。

[形态描述] 头较大，吻端宽钝，头、颈区分明显，躯干略侧扁。雄蛇体全长（380+110）mm，雌蛇体全长（488+125）mm，背面黄褐色，其上有由黑点缀连成的横纹，腹面颜色浅淡，头背面自眶上鳞有1黑线纹，延伸至头后与由顶鳞起始向后延伸的黑线相会合成粗黑线，而后断断续续；头侧有1黑色细线纹，从眼后到口角。吻鳞宽与高略相等，从背面能见到它的上缘；鼻间鳞宽胜于长；前额鳞宽度超过长度，外侧与颊鳞相接，向后入眶甚多，额鳞长度大于宽度，鼻鳞大，鼻孔位于单枚鼻鳞中央；颊鳞1枚，长大于宽，后端入眶，没有眶前鳞，或个别一侧有眶前鳞1枚。眼大，瞳孔纵置，椭圆形，眼径远大于从它到口缘的距离。眶下鳞与眶后鳞愈合。颞鳞2+3，个别1+2，上唇鳞7或8，不入眶，由前向后依次增大，最后一枚最长；下唇鳞8，少数一侧为7或9，前4枚切前颔片，个别一侧为5或3；颔片3对，交错排列，第1对长大于宽，第2对宽大于长，第3对长宽约相等；背鳞平滑，通体15行，脊鳞不扩大，肛鳞完整。上颌齿通常为每侧4～5。

a 背面照
b 整体照
c 腹面照
d 在茂兰地区的分布

雄蛇半阴茎伸达第 14 枚尾下鳞，在第 4 枚尾下鳞处分叉。据伍律等（1985）记载：一幼蛇上唇鳞右侧为 39，一雌下唇鳞为 10，有的一侧具 1 眶前鳞，有的前颞鳞一侧为 3，后颞鳞一侧为 2 或 4，上颌齿一侧为 3。

［生活习性］ 生活于林间山地，在考察中发现其静止立在林间的石头上。

［地理分布］ 茂兰地区分布于洞塘。国内分布于江苏、浙江、安徽、福建、江西、广东、广西、四川、贵州、云南、陕西、甘肃等地。国外无分布。

［模式标本］ 全模标本（MNHN 1912.349）保存于法国国家自然历史博物馆

［模式产地］ 中国（可能在东南部或贵州）

［保护级别］ 三有保护动物；贵州省重点保护野生动物

［IUCN 濒危等级］ 无危（LC）

（16）福建钝头蛇 *Pareas stanleyi* (Boulenger, 1914)

Amblycephalus stanleyi Boulenger, 1914, Ann. Mag. Nat. Hist., London, (Ser. 8), 14: 484. Type locality: Northwestern Fokien [= Fujian], China.

Amblycephalus sinensis Stanleyi, "1916" (1917), Jour. North-China Banch Royal Asiatic Soc., Shanghai, New Ser., 47: 83. Type locality: Kuatun [= Guadun] Hills, northwest Fokien [= Fujian], China.

Amblycephalus (*Formosensis*) *stanleyi* Mell, 1931, Lingnan Sci. Jour., Canton, 8 [1929]: 199-219.

Pareas stanleyi Hu, Huang, Xie et al., 1980, Atlas Chin. Snakes, Shanghai: 113.

Pareas stanleyi Guo et al., 2004, Asiatic Herpetological Research, 10: 280-281.

Pareas stanleyi Wallach et al., 2014: 537. Taylor & Francis, CRC Press: 1237.

[同物异名] *Amblycephalus stanleyi*；*Amblycephalus sinensis*；*Amblycephalus*（*Formosensis*）*stanleyi*

[俗名] 棱脊钝头蛇、崇安钝头蛇

[鉴别特征] 前额鳞入眶，没有眶前鳞，颊鳞入眶甚多，背鳞中央 5～7 行起微棱。

[形态描述] 依据茂兰地区标本 GZNU320275 描述 体全长 491.5mm。生活时背面淡黄色，具有略呈横行的黑点，或形成断续的横带；腹面带黄白色，散有稀疏的

成体量度

编号	产地	体全长（mm）	头体长（mm）	尾长（mm）	背鳞（枚）	腹鳞（枚）	肛鳞（枚）	上唇鳞（枚）	下唇鳞（枚）	颊鳞（枚）	眶前鳞（枚）	眶后鳞+框下鳞（枚）	前颞鳞（枚）	后颞鳞（枚）
GZNU320275	荔波茂兰	491.5	400.3	91.2	14-15-15	175	1	8	7	1	0	2	2	3

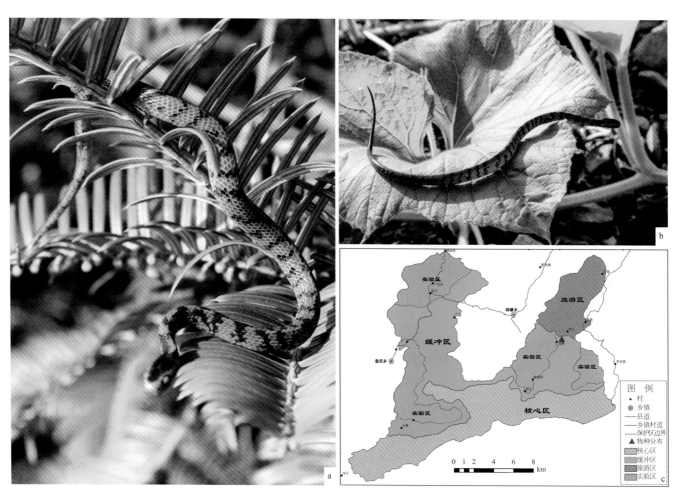

a 侧面照（祝非摄）
b 腹面照（祝非摄）
c 在茂兰地区的分布

棕黑斑点，头背面从鼻鳞之后到颈部有一大黑斑块，自颈向后分裂为二纵线，有一黑线纹从眼眶向后达颈部。吻鳞宽度超过高度，从背面隐约可见；鼻间鳞短于前额鳞，前额鳞后外侧入眶，额鳞六边形，长超过宽，长于从吻端到它的距离，短于顶鳞。没有眶前鳞，颊鳞1，长大于宽，向后入眶甚多，眶后鳞1，眶下鳞1。颞鳞2+3。背鳞14-15-15，上唇鳞8，最后一枚最长，均不入眶；下唇鳞7，前4枚切前颏片，颏片3对，第1对长大于宽，除两侧最外一行其余均具棱。腹鳞175，肛鳞完整，尾下鳞48对。

雄蛇半阴茎伸达第17枚尾下鳞，在第6枚尾下鳞处分叉，具萼片状物而无刺。上颌齿每侧4～8枚。

[生活习性] 生活于海拔700～1000m的山区。以蜗牛、蛞蝓为食。卵生。

[地理分布] 茂兰地区分布于洞塘。国内分布于四川、浙江、福建、江西、贵州。

[模式标本] 正模标本（BMNH 1946.1.20.4）保存于英国伦敦自然历史博物馆

[模式产地] 福建

[保护级别] 三有保护动物；贵州省重点保护野生动物

[IUCN 濒危等级] 无危（LC）

10. 蝰科 Viperidae Bonaparte, 1840

Viperidae Bonaparte, 1840, Mem. Acc. Torin., 2 (2): 393.

头较大，略呈三角形，头、颈区分明显，眼与鼻孔之间无颊窝（蝮亚科 Crotalinae 蛇类为有颊窝类），躯干部较粗而短，尾短小。上颌骨短而高，能竖立，前端有一对长而弯曲的管牙及若干预备牙。一般行动较迟缓。陆栖、树栖或穴居，少数树栖种类尾能缠绕。吃各种脊椎动物，主要以鸟类和兽类为食。有些种类的幼蛇能吃昆虫等节肢动物。卵生或卵胎生。

茂兰地区有蝰科蝮亚科 Crotalinae 5属6种，分属检索表如下。

茂兰地区属检索表

1 { 头部背面覆盖有大型的对称鳞片···亚洲蝮属 Gloydius
  { 头部背面覆盖的全为细小鳞片··2
2 { 体背以绿色为主··3
  { 体背不是绿色··4
3 { 鼻间鳞较小，彼此不相连，鼻鳞不与第一上唇鳞相接··绿蝮属 Viridovipera
  { 鼻间鳞较大，彼此相连，或偶间一枚小鳞，鼻鳞与第一上唇鳞相接··竹叶青属 Trimeresurus
4 { 头部尖端呈突出状··尖吻蝮属 Deinagkistrodon
  { 头部尖端不呈突出状···原矛头蝮属 Protobothrops

12）原矛头蝮属 Protobothrops Hoge and Romano-Hoge, 1983

Protobothrops Hoge and Romano-Hoge, 1983, Mem. Instituto Butantan, Sao Paulo, 44/45: 87.

头背全被小鳞，光滑无棱。鼻小孔位于鼻腔后壁。尾下鳞双行。鼻骨大小一般，呈"刀"字形。额骨近长方形，顶骨不明显下凹，与前额骨关节面较大。顶骨骨脊一般，呈三角形。鳞骨一般长条形或后端较向背方突出，超过枕骨大孔。后额骨大小一般，与额骨接触。上颌骨颊窝前缘突起不明显。腭骨无齿或少齿，与翼骨关节突退化而不呈马鞍形。翼骨齿列仅达外翼关节处的前方或下方。外翼骨前端不特别扩大。

本属现有15种，分布于印度、孟加拉国、缅甸、中国（南部）、越南等地。中国有9种。茂兰地区有2种，分种检索表如下。

茂兰地区种检索表

体型较小，体色灰褐色；体背有55～60行相互错开的不呈波浪状紫褐色纵纹；背鳞外行平滑，其余均起棱……………………………………………………………………………………茂兰原矛头蝮 Protobothrops maolanensis

体型较大，体黄褐色或红褐色；体背有1行暗紫色逗点状斑组成的波浪形纵纹；背鳞均起棱………………原矛头蝮 Protobothrops mucrosquamatus

（17）原矛头蝮 Protobothrops mucrosquamatus (Cantor, 1839)

Trigonocephalus mucrosquamatus Cantor, 1839, Proc. Zool. Soc., London: 31-34.

Trimeresurus mucrosquamatus Günther, 1864, The Reptiles of British India. London: 390.

Trimeresurus mucrosquamatus Fischer, 1886, Abh. Geb. Naturw. Hamb., 9: 51-67.

Trimeresurus mucrosquamatus Boulenger, 1890, Reptilia and Batrachia. Taylor & Francis, London: 428.

Lachesis mucrosquamatus Boulenger, 1896, Reptilia and Batrachia. Taylor & Francis, London: 552.

Trimeresurus mucrosquamatus Stejneger, 1907: 467. Bull. U.S. Natl. Mus., 58: xx, 1-577.

Trimeresurus mucrosquamatus Smith, 1943, Reptilia and Amphibia. 3 (Serpentes). Taylor & Francis, London: 507.

Protobothrops mucrosquamatus Ziegler, 2002, Natur und Tier Verlag (Münster): 282.

Protobothrops mucrosquamatus Wallach et al., 2014: 573. Taylor & Francis, CRC Press: 1237.

[同物异名] *Trigonocephalus mucrosquamatus*; *Lachesis mucrosquamatus*

[俗名] 烙铁头

[鉴别特征] 吻部有颊窝；头背都是小鳞片；体色棕黄或红褐，背脊有1行暗紫色波状纹。本种头步左右眼眶上鳞间1横排有小鳞11～18枚，左右鼻间鳞相隔2～6枚小鳞片。

[形态描述] 依据茂兰地区标本GZNU20160805013描述 头较窄长，三角形，吻棱明显；尾末端较细，有缠绕性。体全长737.5mm，尾长147.4mm。背面棕褐色，正背有1行镶浅黄色边的大点状暗紫色斑，斑周黄色较深，中心色略浅，大多地方前后连续，形成波状脊纹。体侧尚各有1行暗紫色斑块。腹面浅褐色，每一腹鳞有由深棕色细点组成的若干斑块，整体上交织成深浅错综的网纹。头背棕褐色，眼后到颈侧有1棕褐色纵线纹，唇缘色稍浅；头腹浅褐色。鼻间鳞大于头背其他鳞片，彼此相隔5枚小鳞；眶上鳞为头背面最大鳞片，宽度约为左右眶上鳞间距的1/3，其间1横排有小鳞11～18；头背其余鳞片粒状。鼻鳞较大，不分为前后两半，鼻孔直立椭圆形，开口朝向后外方；鼻鳞与窝前鳞间隔2枚小鳞；瞳孔直立呈椭圆形；眶前鳞2，眶下鳞为若干枚小鳞。上唇鳞9，第1枚较小，略呈三角形，不与鼻鳞相接；第2枚高，构成颊窝前鳞；第3枚最大，第4枚位于眼正下方，与眼间相隔数排小鳞；第4枚后的上唇鳞均较低小。下唇鳞14，第1对在颏鳞之后相切，前2对切颔片。背鳞较窄长，末端尖出，颈部24行，中段为21行，肛前段17行。上颌骨着生中空的管牙，有颊窝。

[生活习性] 生活于丘陵及山区，栖于竹林、灌丛或溪边；在住宅周围（如草丛、垃圾堆、柴草、石缝间）活动，有时进入室内，保护色极强，不易被发现。为毒蛇。

[地理分布] 茂兰地区分布于翁昂。国内分布于云南、甘肃、湖南、江西、浙江、福建、安徽、贵州、四川、海南、台湾等地。国外分布于孟加拉国、印度、缅甸、越南、老挝等地。

[模式标本] 未知（可能已丢失） 动物

[模式产地] 印度 [IUCN 濒危等级] 无危（LC）

[保护级别] 三有保护动物；贵州省重点保护野生

成体量度

编号	产地	体全长（mm）	体长（mm）	尾长（mm）	背鳞（枚）	腹鳞（枚）	肛鳞（枚）	上唇鳞（枚）	下唇鳞（枚）	颊鳞（枚）	眶前鳞（枚）
GZNU20160805013	荔波茂兰	737.5	590.1	147.4	24-21-17	189	1	9	14	2	2

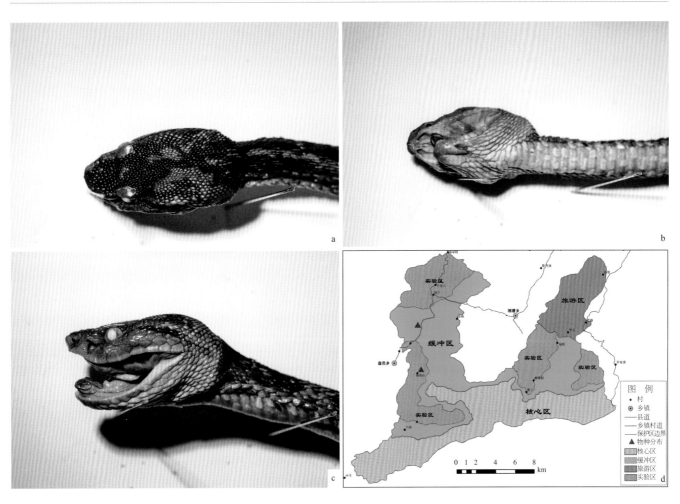

a 背面照
b 腹面照
c 侧面照
d 在茂兰地区的分布

（18）茂兰原矛头蝮 *Protobothrops maolanensis* Yang, Orlov and Wang, 2011

Protobothrops maolanensis Yang, Orlov and Wang, 2011, Zootaxa, 2936: 59-68. Holotype: SYS r000211, collected by Sheng Zheng and Jian-Huan Yang on 18 May, 2010.

Protobothrops maolanensis Wallach et al., 2014, Snakes of the World: A catalogue of living and extinct species. Taylor & Francis, CRC Press: 573.

[同物异名] 无

[俗名] 无

[鉴别特征] 身体细长，略侧扁；头略呈长三角形，头部与颈区分极其明显，瞳孔纵置，鼻间鳞与吻鳞不相接，头背、体背和尾背棕褐色，横斑有时在体背和尾背中央断开。头侧 1 条棕黑色细纹从眼眶后达颞部。头腹、体腹和尾腹灰白色；腹鳞和尾下鳞均具方斑，两侧为棕黑色，中央为棕灰色。

[形态描述] 依据茂兰地区标本 GZNU20160728002 描述 中等体型，体全长 811.39mm，尾长 137.53mm；身体细长，略侧扁；头略呈长三角形，头被覆以不规则细鳞，头部与颈区分极明显；眼较大并凸起，瞳孔纵置。吻鳞略

成体量度

编号	产地	体全长（mm）	头体长（mm）	尾长（mm）	背鳞（枚）	腹鳞（枚）	肛鳞（枚）	上唇鳞（枚）	下唇鳞（枚）	颊鳞（枚）	眶前鳞（枚）	眶后鳞（枚）
GZNU20160728002	荔波茂兰	811.39	673.86	137.53	21-21-17	198	1	8	11	2	2	2

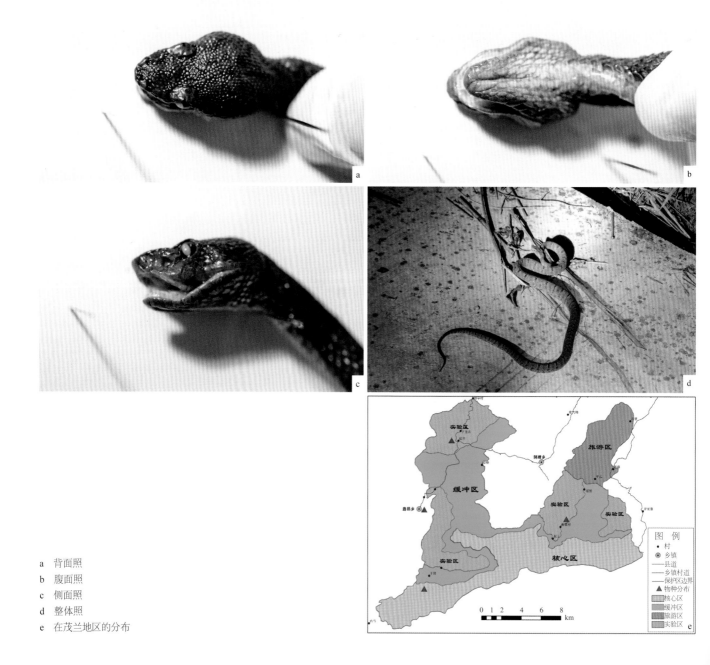

a 背面照
b 腹面照
c 侧面照
d 整体照
e 在茂兰地区的分布

呈三角形，从头背面可见；鼻间鳞与吻鳞不相接，鼻间鳞之间相隔 3 枚小鳞；鼻鳞呈梯形，不二分，鼻孔在中间，圆形；鼻间鳞和眶上鳞间隔 5 枚小鳞；颊鳞 2 枚；眶前鳞 2，上一枚形成颊窝上缘，下面一枚形成颊窝后缘；眶后鳞 2；上唇鳞 8，第 2 枚形成颊窝前缘，第 3 枚最大；下唇鳞 11，第 1 对相接，前 3 对切前颏片；背鳞 21-21-17，背鳞狭长，鳞尖向后，起强棱，但最外一行略弱；腹鳞 198，肛鳞完整。头背、体背和尾背棕褐色，体背和尾背镶黄边的棕褐色横斑，宽 1～2 枚背鳞，大部分横斑在体背和尾背中央断开。头侧 1 条棕黑色细纹从眼眶后达颞部。头腹、体腹和尾腹灰白色；腹鳞和尾下鳞均具方斑，两侧为棕黑色，中央为棕灰色。

［生活习性］ 生活于丘陵及山区，栖于竹林、灌丛中，夜晚喜出没于林缘、马路边、水沟旁，捕食蛙类和小型啮齿动物。白天隐于林间，不易被发现。

［地理分布］ 茂兰地区分布于翁昂、永康、洞塘等地。国内分布于贵州、广西。

［模式标本］ 正模标本（SYS r000211）、配模标本（SYS r000210、SYS r000276、SYS r000277，采集于 2010 年）保存于中山大学生物博物馆

［模式产地］ 茂兰保护区

［保护级别］ 贵州省重点保护野生动物

［IUCN 濒危等级］ 数据缺乏（DD）

13）尖吻蝮属 *Deinagkistrodon* Gloyd, 1979

Deinagkistrodon Gloyd, 1979, Proc. Biol, Soc., Washington, 91: 963. Type species: *Halys acutus* Cünther, 1888, by original designation.

蛇亚目蝰科蝮亚科下的一个有毒单型蛇属，属下有尖吻蝮 *D. acutus* 1 种。头大，三角形，颈较细，两者区分明显；吻端尖出上翘，是此属的主要特征。

（19）尖吻蝮 *Deinagkistrodon acutus* (Günther, 1888)

Halys acutus Günther, 1888: 171, Ann. Mag. Nat. Hist., (1) 6: 165-172. Type species: *Deinagkistrodon actulls* Gloyd, 1979. Type locality: Wusueh, Hupeh [= Hubei], China.

Ancistrodon acutus Boulenger, 1896, Catalogue of the snakes in the British Museum, Vol. 3. London (Taylor & Francis): 727.

Ancistrodon acutus Smith, 1943, Reptilia and Amphibia. Taylor & Francis, London: 501.

Calloselasma acutus Burger, 1971, Genera of Pitvipers. Ph.D. Thesis, University of Kansas, Lawrence, USA.

Deinagkistrodon acutus Gloyd, 1979, Proc. Biol. Soc., Washington, 91: 961.

Agkistrodon acutus Harding and Welch, 1980, Venomous snakes of the word: a clecklist, pergamon press: 188.

Deinagkistrodon acutus Mcdlarmid, Campbell and Touré, 1999, Snake species of the world, Vol. 1: 301.

Deinagkistrodon acutus Wallach et al., 2014, Snakes of the World: A catalogue of living and extinct species. Taylor & Francis, CRC Press: 213.

［同物异名］ *Halys acutus*；*Ancistrodon acutus*；*Agkistrodon acutus*；*Calloselasma acutus*

［俗名］ 百花蛇、百步蛇、五步蛇、七步蛇、蕲蛇、中华蝮

［鉴别特征］ 吻端尖而略翘向前上方；头呈三角形。头背具对称大鳞片，有颊窝，体型粗短，背面正中 1 行有多个方形大斑块（又称方胜纹）。

［形态描述］ 头大，明显呈三角形，吻端尖出；颈较细；体型粗短；尾较短而细，背面深棕色或棕褐色，头背具对称大鳞片，正背有 16+21+26 个方形大斑块，每前后 2

个方斑以尖角彼此相接；有的方斑不完整，形成"乙"字形纹，方斑边缘浅褐色，中央略深。腹面白色，有交错排列的黑褐色斑块，略呈纵列；每一斑块跨 1～3 枚腹鳞，有的斑块淡而不显，有的若干斑块互相连续而界限不清。头背黑褐色；自吻棱经眼斜至口角以下为黄白色，偶有少许黑褐色点；头腹及喉部为白色，散有稀疏黑褐色点斑。尾背后段纯黑褐色，看不出方形斑；尾腹面白色散有疏密不等的黑褐色。

吻鳞甚高，上部窄长，构成尖吻的腹面；鼻间鳞 1 对，亦窄长，构成尖吻的背面。前额鳞长，眶上鳞甚宽。鼻鳞较大，分为较大的前半和较小的后半；鼻孔直立椭圆形，位于前后二半鼻鳞之间，开口朝后略偏外。上颌骨具管牙，有颊窝。

[生活习性]　生活于山区或丘陵林木茂盛的阴湿地方，曾发现于山溪旁阴湿岩石上或落叶间、瀑布下的大岩缝中、路边岩下、路边草丛中。为毒蛇。

[地理分布]　茂兰地区分布于洞塘。国内分布于台湾、贵州、四川、湖北、湖南、浙江、福建、广东等地。国外分布于越南、老挝。

[模式标本]　模式标本 [BMNH 1946.1.19.56-58（3）] 保存于英国伦敦自然历史博物馆

[模式产地]　湖北武穴

[保护级别]　国家 II 级重点保护野生动物；贵州省重点保护野生动物

[IUCN 濒危等级]　濒危（EN）

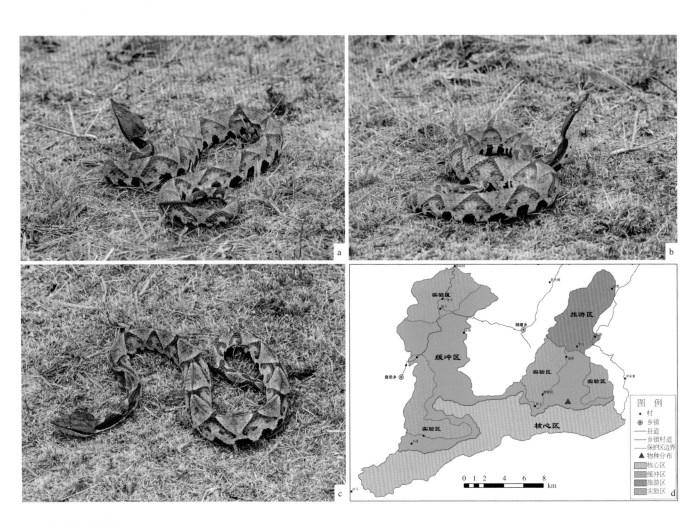

a～c　整体照（祝非摄）
d　在茂兰地区的分布

14）绿蝮属 *Viridovipera* Malhotra and Thorpe, 2004

Viridovipera Malhotra and Thorpe, 2004, Biological Journal of the Linnean Society, 82 (2): 219.

上腭具有管牙，有颊窝。体型中等。头三角形，吻棱明显，躯干细长。尾较长，具缠绕性。背面通体绿色，尾背及尾尖焦红色，体两侧各有1条纵线纹；腹面色稍浅淡。

本属世界共4种，均在中国有分布。茂兰地区有1种，即福建绿蝮 *Viridovipera stejnegeri*。

（20）福建绿蝮（竹叶青） *Viridovipera stejnegeri* (Schmidt, 1925)

Trimeresurus gramineus Stejneger, 1907: 480. Scientific books: Herpetology of Japan and adjacent territory.

Trimeresurus stejnegeri Schmidt, 1925, American Museum Novitates, (157): 1-5. Distribution: S China [from Yunnan north, SE Gansu and east to Jiangsu; Jilin, Kwangsi [= Guangxi], Kwantung [= Guangdong], Hainan, Fukien [= Fujian], Chekiang [= Zhejiang], Taiwan (incl. Lanyu).], Nepal, India (Sikkim, Assam), Myanmar, Vietnam.

Trimeresurus gramineus kodairai Maki, 1931, Zool. Inst. Coll. of Sci. Kyoto Imp. Univ., Dai-ichi Shobo, Tokyo.

Trimeresurus graminaeus formosensis Maki, 1931 (non *T. monticola formosensis* Mell, 1929).

Trimeresurus stejnegeri stejnegeri Pope, 1935, The reptiles of China.

Trimeresurus stejnegeri makii Klemmer, 1963, Liste der rezenten Giftschlangen. Elapidae, Hydropheidae [sic], Viperidae und Crotalidae.

Trimeresurus stejnegeri stejnegeri Welch, 1994: 117. Current psychology of cognition, 13 (1): 117-123.

Trimeresurus stejnegeri chenbihuii Zhao, 1995, Infraspecific classification of some Chinese snakes. Sichuan Journal of Zoology, 14 (3): 107-112.

Trimeresurus stejnegeri Cox et al., 1998, Ralph Curtis Publishing: 21.

Trimeresurus stejnegeri Mcdlarmid, Campbell and Touré, 1999, Snake species of the world. Vol. 1: 344.

Trimeresurus stejnegeri Tu et al., 2000, Zoological Science, 17: 1147-1157.

Viridovipera stejnegeri Malhotra and Thorpe, 2004, Biological Journal of the Linnean Society, 82 (2): 219.

Trimeresurus stejnegeri stejnegeri Gumprecht et al., 2004, Asian pitvipers. Geitje Books, Berlin: 368.

Viridovipera stejnegeri Dawson et al., 2008, Molecular Phylogenetics and Evolution, 49: 356-361.

Trimeresurus stejnegeri stejnegeri Sang et al., 2009, Herpetofauna of Vietnam. Chimaira, Frankfurt: 768.

Trimeresurus (*Viridovipera*) *stejnegeri* David et al., 2011, Zootaxa, 2992: 1-51.

Viridovipera stejnegeri Wallach et al., 2014, Snakes of the World: A catalogue of living and extinct species. Taylor & Francis, CRC Press: 790.

Trimeresurus stejnegeri chenbihuii David and Vogel, 2015, An updated list of Asian pitvipers and a selection of recent publications, in Visser: Asian Pitvipers. Breeding Experience & Wildlife. Frankfurt am Main, Edition Chimaira: 571.

[同物异名] *Trimeresurus stejnegeri*；*Trimeresurus graminaeus*；*Trimeresurus graminaeus formosensis*；*Trimeresurus stejnegeri stejnegeri*；*Trimeresurus stejnegeri chenbihuii*；*Trimeresurus gramineus kodairai*

[俗名] 竹叶青蛇

[鉴别特征] 通体绿色，头呈三角形，尾背及尾尖焦红色，头侧有颊窝，管牙式毒牙。

[形态描述] 依据茂兰地区标本 GZNU20100808021

描述 体全长 644.5mm。通体绿色，头呈三角形，尾背及尾尖焦红色，头侧有颊窝，管牙式毒牙。头、颈区分明显，头背密布细小鳞片，仅眶上鳞与鼻间鳞稍大，左右眶上鳞间有 1 横排小鳞，眼小，瞳孔椭圆形，纵置。鼻孔与眼间有 1 深凹的颊窝，有窝上鳞 2，窝下鳞 1 枚狭长；颊鳞无；眶前鳞 3；眶后鳞 2。上唇鳞 10，第 1 枚较小，略呈三角形，不与鼻鳞相接；第 2 枚高，构成颊窝前鳞，且与鼻鳞有一凸点相接；第 3 枚最大，呈梯形，上边缘与眼眶相隔 1 枚

成体量度

编号	产地	体全长（mm）	头体长（mm）	尾长（mm）	背鳞（枚）	腹鳞（枚）	肛鳞（枚）	上唇鳞（枚）	下唇鳞（枚）	颊鳞（枚）	眶前鳞（枚）	眶后鳞+眶下鳞（枚）
GZNU20100808021	荔波茂兰	644.5	515.2	129.3	21-20-13	159	1	10	9	0	3	2

a 背面照
b 腹面照
c 侧面照
d 整体照
e 在茂兰地区的分布

小鳞，第 4～第 6 枚位于眼正下方，与眼间相隔 2 排小鳞；第 4 枚后的上唇鳞均较低小。下唇鳞 9，第 1 对在颔鳞之后相切，前 2 对切颔片，第 4 对最大。颔片 2 对，前对约为后对 7 倍大小，后对颔片尾端断开为第 3 对小粒鳞，颔沟明显。背鳞 21-20-13，除最外 1 行平滑外，其余都有棱；腹鳞 159；肛鳞完整。

[生活习性] 生活于林间溪流旁，喜攀附于灌丛、竹林、岩壁等离开地面的物体上，多在水域边捕食蛙类、蜥蜴、鸟类、小型兽类等。

[地理分布] 茂兰地区分布于洞塘。国内分布于安徽、重庆、福建、浙江、云南、四川、贵州、广西、广东、台湾、江苏、江西、海南、河南、湖北、湖南、吉林、甘肃。国外分布于印度（大吉岭、阿萨姆邦）、缅甸、泰国（东南部）、越南。

[模式标本] 正模标本（AMNH 21054）保存于美国自然历史博物馆；正模标本（CIB 64III5599 [*chenbihuii*]）保存于中国科学院成都生物研究所；（NSMT H02839 [*kodairai*]）保存于日本国家自然科学博物馆（National Science Museum, Tokyo）

[模式产地] 福建邵武

[保护级别] 三有保护动物；贵州省重点保护野生动物

[IUCN 濒危等级] 无危（LC）

15）竹叶青属 *Trimeresurus* Lacepède, 1804

Trimeresurus Lacepède, 1804, Ann. Mus. Natl. Hist. Nat., Paris, 4: 209. Type species: *Vipera tiridis* Daudin, 1803 (= *Colube gramineus* Shawi, 1802), of India by subsequent designation (Stejneger, 1907, Bull. U.S. Natl. Mus., Washington, 58: 465).

头背全被小鳞，顶区前部鳞片光滑无棱，喉区鳞片亦光滑。鼻小孔位于鼻腔后壁。尾下鳞单行或双行。鼻骨大小一般或偏小。额骨近方形或长方形，顶部一般不明显下凹。顶骨骨脊一般，大多呈三角形，也有近 "T" 形者。鳞骨一般过枕骨大孔。上颌骨颊窝前缘一般有一小突起。腭骨一般近三角形，一般有齿 3～5 枚，也有无齿者，有齿者与翼骨关节一般较明显，为马鞍形。翼骨齿列一般达外翼骨关节处下方。外翼骨有的前端不明显宽大、有的前端特别宽大。

本属现有 56 种，分布于南亚和东南亚地区及我国（南部）。我国已知有 2 种。茂兰地区有 1 种，即白唇竹叶青蛇 *Trimeresurus albolabris*。

（21）白唇竹叶青蛇 *Trimeresurus albolabris* (Gray, 1842)

Coluber gramineus Raffles, 1822 (non *Coluber gramineus* Shaw, 1802).

Trigonocephalus viridis Müller and Schlegel, 1842 (non *Vipera viridis* Daudin, 1803).

Trimeresurus [sic] *albolabris* Gray, 1842, Zool. Misc., London, 2: 47-51. Type locality: China.

Trigonocephalus gramineus Cantor, 1847 (non *Coluber gramineus* Shaw, 1802).

Trimeresurus gramineus Boulenger, 1890 (non *Coluber gramineus* Shaw, 1802).

Trimeresurus purpureomaculatus var. *bicolor* Boulenger, 1890 (non *Trimeresurus bicolor* Gray, 1853, a synonym of
    Trigonocephalus erythrurus Cantor, 1839).

Lachesis gramineus Boulenger, 1896 (non *Coluber gramineus* Shaw, 1802).

Lachesis grammineus [sic] Brongersma, 1929.

Trimeresurus gramineus albolabris Mell, "1929" (1931), Lingnan Sei. Jour., Canton, 8: 218.

Lachesis gramineus albolabris Mell, 1931, Lingnan Sci. Journ VIII: 218.

Trimeresurus albolabris Pope and Pope, 1933, American Museum novitates: 620.

Trimeresurus albolabris albolabris Kramer, 1977.

Trimeresurus albolabris Manthey and Grossmann, 1997, Natur und Tier Verlag (Münster): 407.
Trimeresurus albolabris albolabris Cox et al., 1998: 20. Ralph Curtis Publishing: 144.
Trimeresurus albolabris Mcdlarmid, Campbell and Touré, 1999: 329. Snake species of the world. Vol. 1. Herpetologists' League: 511.
Trimeresurus (*Trimeresurus*) *albolabris* David and Vogel, 2000, Senckenbergiana Biologica, 80 (1/2): 225-232.
Cryptelytrops albolabris Malhotra and Thorpe, 2004, Molecular Phylogenetics and Evolution, 32: 83-100.
Cryptelytrops albolabris Chan-ard et al., 2015: 284. A field guide to the reptiles of Thailand. Oxford University Press, N.Y.: 352.

[同物异名] *Trimeresurus albolabris*；*Lachesis gramineus albolabris*；*Trimeresurus albolabris albolabris*；*Coluber gramineus*；*Trigonocephalus viridis*；*Trigonocephalus gramineus*；*Trimeresurus gramineus*；*Lachesis gramineus*；*Lachesis grammineus* [sic]；*Bothrops erythrurus*；*Trimeresurus purpureomaculatus* var. *bicolor*；*Trimeresurus gramineus albolabris*；*Trimeresurus albolabris albolabris*；*Trimeresurus albolabris*；*Cryptelytrops albolabris*

[俗名] 竹叶青、小青蛇、青竹蛇、小绿蛇

[鉴别特征] 有颊窝，头背都是小鳞片；通体绿色，体侧有的具白色纵线纹；眼睛红色，尾背及尾尖焦红色。本种与竹叶青相近，区别在于本种鼻鳞与第1枚上唇鳞完全愈合或仅有极短的鳞沟；鼻间鳞较大，显著区别于头背其他鳞片，彼此相切或偶有隔1枚小鳞。

[形态描述] 头大，呈三角形，颈细，尾较短。最大体全长雄性（600+142）mm，雌性（585+105）mm。通体背面绿色，体侧多有1条白色纵线，一般位于最外行 D_1（指背鳞外侧第1行）中央，少数标本位于 D_1 与 D_2（指背鳞外侧第2行）之间；侧线前达颈侧、颌角、眼后或第2枚上唇鳞，向后止于肛前、尾基部、尾中段或尾末；腹面黄白色。头侧全为绿色或眼以下部分黄白色，在后一种情况下，有的标本上唇鳞仍呈绿色；头腹带白色，但下唇部分有不同程度的绿色。眼睛红色，尾后段背面及尾尖焦红色。液浸标本背面多显示灰黑色粗横斑。

鼻间鳞较大，在吻鳞之后彼此相切，少数标本有1枚小鳞相隔；眶上鳞较窄小，宽度仅为左右眶上鳞间距的 1/6～1/4，其间一横排有小鳞8～12；头背其余鳞片较小，位于额前区的较平，位于枕区及颞区的中央隆起；鼻鳞较大，不分裂，中部亦不隘窄；鼻孔圆形，位于鼻鳞中央或中部的下半，开口斜向外后方；鼻鳞与窝前鳞常相切，仅少数标本相隔1枚小鳞，个别标本相隔2或3枚极小鳞片。眼较小，瞳孔直立椭圆形；眶前鳞2，上枚与鼻鳞间相隔1枚鳞片；眶后鳞2，极少为1或3枚；眶下鳞为1枚新月形较大鳞片，前接窝下鳞。上唇鳞10～11枚者较多，少数为9或10，第1枚较小，略呈三角形，与鼻鳞完全愈合或两端甚至一端仅有极短鳞沟，仅个别标本的一侧鳞沟较完整；第2枚高，构成颊窝前鳞；第3枚最大；第4枚位于眼的正下方，与眶下鳞相隔1枚鳞；第4枚以后的上唇

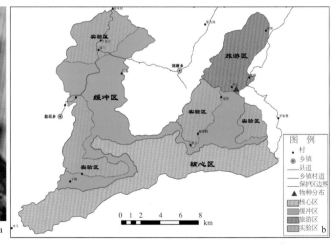

a 整体照（祝非摄）
b 在茂兰地区的分布

鳞均较低小。下唇鳞 11～14，第 1 对在颔鳞之后相切；前 3 对切颏片。背鳞 21 行，中段起棱鳞雄性 17 或 19 行，雌性 13～17 行，棱弱；腹鳞雄性平均 158.7，雌性平均 159.9；肛鳞完整；尾下鳞雄性 56～74 对，雌性 44～73 对。

上颌骨具管牙，有颊窝。

[生活习性] 生活于平原、丘陵或低山区等生境，常栖于溪沟、水塘、田埂边杂草中或低矮的灌木上、田野杂草或树枝上。白天、晚上都可发现，但主要于晚上活动、捕食，尾不具缠绕性。为毒蛇。

[地理分布] 茂兰地区分布于洞塘。国内分布于云南、广东、海南、安徽、福建、广西、贵州、江西等地。国外分布于印度、尼泊尔、缅甸、泰国、柬埔寨、老挝、越南、印度尼西亚等地。

[模式标本] 正模标本（BMNH 1946.1.19.85）保存于英国伦敦自然历史博物馆

[模式产地] 中国

[保护级别] 三有保护动物；贵州省重点保护野生动物

[IUCN 濒危等级] 无危（LC）

16）亚洲蝮属 *Gloydius* Hoge and Romano-Hoge, 1981

Gloydius Hoge and Homano-Hoge, 1981, Mem. Butantan, Sao Paulo, 42/43: 194. Type species: *Trigonocephalus halys* Boie, 1827 (= *Coluber halys* Pallas, 1776).

有颊窝蛇类，吻端不尖出；头背具 9 枚对称大鳞的小型到中等大小的管牙类毒蛇。体全长 500～800mm。头部略呈三角形，上唇鳞不伸入颊窝，不构成窝前鳞；躯体不特别粗壮，背鳞具棱；尾较短，尾下鳞成对。头骨略窄长，上颚骨后端未超出脑匣，外翼骨无钩，腭骨前端不分叉。通体背面棕褐色或灰褐色，体背有成对排列的深褐色大圆斑，圆斑中央颜色较浅。

本属已知 14 种，主要分布于亚洲。我国已知 10 种，但存在分歧（见短尾蝮讨论）。

（22）短尾蝮 *Gloydius brevicaudus* (Stejneger, 1907)

Agkistrodon blomhoffii brevicaudus Stejneger, 1907, Bull. U.S. Natl. Mus., Washington, 58: 463. Type locality: Fusan [= Pusan], Korea.

Ancistrodon halys brevicaudus Nikolsky, 1916, Faune de la Russie, Reptiles Vol. 2. Ophidia, Petrograd: 396.

Agkistrodon blomhoffii siniticus Gloyd, 1977, Proc. Biol. Soc., Washington, 90: 1004. Type locality: Ningkwo [= Ningguo Co.l] Anhweila Anhui, China.

Agkistrodon blomhoffuii dubitatus Gloyd, 1977, Proc. Biol. Soc., Washington, 90: 1007. Type locality: Hsinglungshan [= Mt. Xinglong] Eastern Tombs, Hopei [= Hebei], now part of Beijing Manicipality, China

Gloydius blomhoffii brevicaudus Hoge and Romano-Hoge, 1981, Mem. Inst. Butantan, 42/43: 195.

Agkistrodon blomhoffii brevicaudus Zhao and Adler, 1993, Herpetol. China, Oxford, Ohio: 272.

Agkistrodon blomhoffii brevicaudus Welch, 1994, Snakes of the world. A checklist. 1: 11. Venomous snakes. KCM Books, Somerset, England.

Gloydius brevicaudus Zhao, 1998, in Zhao et al. (eds.), Fauna Sinica, Reptilia, Vol. 3. Serpentes, Beiing: 394..

Gloydius brevicaudus Wallach et al., 2014, Snakes of the World: A catalogue of living and extinct species. Taylor & Francis, CRC Press: 310.

Gloydius blomhoffii siniticus Orlov et al., 2014, Distribution of Pitvipers of *Gloydius blomhoffii* Complex in Russia with the First Records of *Gloydius blomhoffii blomhoffii* at Kunashir Island (Kuril Archipelago, Russian Far East). Russ. J. Herpetol., 21 (3): 169-178.

Gloydius brevicaudus brevicaudus David and Vogel, 2015, An updated list of Asian pitvipers and a selection of recent publications, in Visser: Asian Pitvipers. Breeding Experience & Wildlife. Frankfurt am Main, Edition Chimaira.

Gloydius brevicaudus siniticus David and Vogel, 2015, An updated list of Asian pitvipers and a selection of recent publications, in Visser: Asian Pitvipers. Breeding Experience & Wildlife. Frankfurt am Main, Edition Chimaira.

[同物异名] *Agkistrodon blomhoffii brevicaudus*；*Ancistrodonhalys brevicaudus*；*Gloydius blomhoffii brevicaudus*；*Agkistrodon blomhoffii brevicaudus*；*Gloydius brevicaudus*；*Gloydius brevicaudus siniticus*；*Agkistrodon blomhoffii siniticus*；*Gloydius blomhoffii siniticus*

[俗名] 五步蛇、草上飞、地扁蛇

[鉴别特征] 头侧有颊窝的管牙类毒蛇。头略呈三角形，与颈区分明显，头背有9枚对称排列的大鳞。体略粗，尾较短。躯尾背面浅褐色，有两行粗大、周围暗棕色、中心色浅而外侧开放的圆斑，圆斑左右交错或并列；尾后端略呈白色，但尾尖常黑。

[形态描述] 依据茂兰地区调查资料记载 最大体全长/尾长：雄性625mm/81mm，雌性623mm/73mm。躯尾背面灰褐色、黄褐色或肉红色，左右两侧各有1行大圆斑，各斑边缘深棕色，中央色较浅，外侧开放（无深色边缘，因而看似马蹄形斑），左右圆斑交错、并列，或内缘几乎相切，个别甚至连接，有的地区的个体背脊尚有1条棕红色细纵线，体侧D_1及腹鳞外侧位置有1行不规则黑色粗大点斑，略呈星状；腹面灰白色，密布灰褐色或黑褐色细点。头背深棕色，枕背有1浅褐色桃形斑，眼后到颈侧有1黑褐色纵纹，纵纹上缘镶以白色细纹，故俗称白眉蝮；上唇缘及头腹灰褐色。尾后段略呈白色，但尾尖常黑。

头背左右鼻间鳞相接的鳞沟较长，但外缘尖细而略后弯，形似逗点。头侧鼻孔与眼之间有颊窝，窝上鳞2，上下并列，窝下鳞1，均窄长且位于眼前，相当于眶前鳞位置；鼻鳞与窝前鳞相接，其间无小鳞相隔；上颊鳞1，介于鼻鳞与上枚窝上鳞之间；眼大小适中，瞳孔直立椭圆形；眶后鳞2（3）；下枚眶后鳞新月形弯至眼后下角；颞鳞2+4，颞区鳞片平滑。上唇鳞7（1-2-4式），个别一侧6（1-1-4式）或8（3-1-4式）；第2枚最小，不伸入颊窝，其上缘与窝前鳞相接；第3枚最大且入眶；第4枚位于眼正下

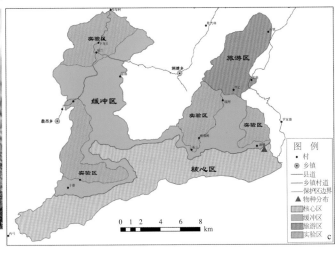

a 侧面照（祝非摄）
b 整体照（钟光辉摄）
c 在茂兰地区的分布

方，与眼以眶后下鳞相隔。下唇鳞9～12，以10～11为多，第1对在颊鳞之后相接，前4（个别为3）枚接颔片；颔片1对，前宽后窄；颔部其他小鳞排成数排，正中者往往对称排列，形成颔沟。背鳞颈部平滑或全部具棱，中段21行，肛前17（个别15）行，中段最外行平滑或全部具棱。腹鳞134～152；肛鳞完整；尾下鳞29～46对，少数个体有数枚成单。

［生活习性］ 栖息环境为平原、丘陵、低山。平时栖息于坟堆、灌丛、草丛、石堆或任何有洞穴的地方。春秋多于白天活动，炎热夏季则在晚上活动。食物来源依据环境而异。

［地理分布］ 茂兰地区分布于洞塘。国内分布于安徽、北京、重庆、福建、甘肃、贵州、河北、河南、湖北、湖南、江苏、江西、辽宁、山西、陕西、上海、四川、台湾、天津、云南、浙江等地。国外分布于朝鲜、韩国。

［模式标本］ 正模标本（AMNH 25554）保存于美国自然历史博物馆

［模式产地］ 韩国

［保护级别］ 三有保护动物；贵州省重点保护野生动物

［IUCN 濒危等级］ 近危（NT）

［讨论］ 关于蝮蛇最早的报道是由 Pallas（1776）于俄罗斯叶尼塞河流域上游采集并命名为"*Coluber halys*"，后被归入美洲的蝮属 *Agkistrodon*。Hoge 和 Sarwl（1978/1979）基于骨骼等形态学特征将亚洲的蝮蛇种类从美洲蝮属中划分出来，独立成一新属，即亚洲蝮属。近年分子生物学的研究结果亦支持这一划分。赵尔宓（2006）认为我国有8种（2亚种），蔡波等（2015）对我国爬行动物分类进行了整合，认为我国分布亚洲蝮属10种。史静耸等（2016）在对西伯利亚蝮 - 中介蝮复合种在中国的分布及其种下分类（蛇亚目：蝮亚科）研究中，经过广泛采样并结合形态和线粒体 *ND4* 和 *Cyt b* 基因，对西伯利亚蝮 - 中介蝮复合种进行了厘定，认为亚洲蝮属有18种，我国分布14种（3亚种）。

11. 眼镜蛇科 Elapidae Boie, 1827

Elapidae Boie, 1827, Isis: 510.

头椭圆形，头背具有对称大鳞片，眼大，瞳孔圆形，尾下鳞单行或双行；上颌骨较短，具有1对较大的前沟牙（不包括后备沟牙）；全部为毒蛇。

茂兰地区有3属3种，分属检索表如下。

茂兰地区属检索表

1 { 脊鳞较其余背鳞大，呈六角形，尾下鳞单行 ······ 环蛇属 *Bungarus*
  { 脊鳞不比其余背鳞大，尾下鳞全部或大部分成对 ······ 2

2 { 颈部不能膨扁，体背以红褐色、白色为主，形成横条纹，鼻间鳞不切鼻孔 ······ 中华珊瑚蛇属 *Sinomicrurus*
  { 颈部能膨扁，体背以黑色或黑褐色为主，鼻间鳞切鼻孔 ······ 眼镜蛇属 *Naja*

17）中华珊瑚蛇属 *Sinomicrurus* Slowinski, Boundy and Lawson, 2001

Sinomicrurus Slowinski, Bouncy and Lawson, 2001, Herrpetolngica, 57 (2): 233-245. Type species: *Eldys macclellandi* Reinhardt, 1844.

本属蛇类体型中等偏小，属前沟牙类毒蛇，全长600mm左右。上颌骨较短，前端有沟牙，一齿间隙后有0～5颗小牙。头较小，与颈区分不明显；头侧没有颊鳞，眶前鳞前面与鼻鳞相接；躯尾圆柱形，通体粗细相似；背鳞平滑，通体13或15行；脊鳞不扩大；肛鳞二分或完整；尾下鳞双行。背面红褐色，有细窄的黑色横纹或纵纹。本属原隶属于丽纹蛇属，Slowinski等（2001）从形态学和分子生物学两方面论证，并与美洲珊瑚蛇类 Miruruides 和 Micrrurus 比较，认为亚洲珊瑚蛇类可分为3支，其中分布于东亚的一支（Calliophis hatoei、Calliophis japonicus、Calliophis macclellandi、Calliophis sautrer）与美洲珊瑚蛇类关系最接近，不宜将它们与分布于南亚和菲律宾的另两支共置于同一属内，有必要给它们重新命名。同时推测美洲珊瑚蛇类的祖先可能是经白令陆桥扩散到美洲的亚洲珊瑚蛇类。

本属世界共有5种，中国有4种。茂兰地区有1种，即中华珊瑚蛇。

（23）中华珊瑚蛇 *Sinomicrurus macclellandi* (Reinhardt, 1844)

Calliophis macclellandi Reinhardt, 1844, Description of a new species of venomous snake, *Elaps macclellandi*. Calcutta J. Nat. Hist., 4: 532-534.

Elaps personatus Biyth, 1855, Notices and descriptions of various reptiles, new or little known [part 2]. J. Asiatic Soc. Bengal, Calcutta, 23 (3): 287-302.

Callophis [sic] *annularis* Günther, 1864, The Reptiles of British India. London (Taylor & Francis): 349.

Callophis macclellandii Anderson, 1871, On some Indian reptiles. Proc. Zool. Soc., London, 1871: 149-211.

Callophis maclellandii [sic] Wall, 1908, Remarks on some recently acquired snakes. J. Bombay Nat. Hist. Soc., 18: 778-784.

Calliophis macclellandii Stejneger, 1910, U.S. Natl. Mus., 38: 91-114.

Callophis macclellandi var. *typica* Venning, 1910, Further notes on snakes from the Chin Hills. J. Bombay Nat. Hist. Soc., 20: 770-775.

Calliophis swinhoei van Denburgh, 1912, Proc. Cal. Ac. Sci. (Series 4), 3 (10): 187-258.

Calliophis iwasakii Maki, 1935. A new poisonous snake (*Calliophis iwasakii*) from Loo-Choo. Transactions of the Natural History Society of Formosa, Taihoku, 25: 216-219.

Callophis macclellandi Smith, 1943: 423. The Fauna of British India, Ceylon and Burma, Including the Whole of the Indo-Chinese Sub-Region. Reptilia and Amphibia. 3 (Serpentes). Taylor & Francis, London.

Calliophis macclellandi Cox et al., 1998: 32. A Photographic Guide to Snakes and Other Reptiles of Peninsular Malaysia, Singapore and Thailand. Ralph Curtis Publishing: 144.

Hemibungarus macclellandi macclellandi Chan-ard et al., 1999: 209. A field guide to the reptiles of Thailand. Oxford University Press, N. Y.: 352.

Sinomicrurus macclellandi Slowinski, Boundy and Lawson, 2001, Herpetologica, 57: 233-245.

Hemibungarus macclellandii Orlov et al., 2003, Venomous snakes in Southern China. Reptilia, (31): 22-29.

Sinomicrurus macclellandi Leviton et al., 2003, The Dangerously Venomous Snakes of Myanmar Illustrated Checklist with Keys. Proc. Cal. Acad. Sci., 54 (24): 407-462.

Calliophis macclellandi iwasakii Staniszewski, 2003, Field trip to Japan. Part II: Western Honshu and subtropical Islands. Reptilia, (26): 48-51.

Sinomicrurus macclellandi Ziegler et al., 2007, The diversity of a snake community in a karst forest ecosystem in the central Truong Son, Vietnam, with an identification key. Zootaxa, 1493: 1-40.

Sinomicrurus macclellandi iwasakii Sang et al., 2009, Herpetofauna of Vietnam. Chimaira, Frankfurt: 768.

Calliophis macclellandi macclellandi Murthy, 2010, The reptile fauna of India. B.R. Publishing, New Delhi: 332.

[同物异名] *Calliophis macclellandi*；*Elaps personatus*；*Callophis* [sic] *annularis*；*Calliophis swinhoei*；*Callophis* [sic] *formosensis*；*Micrurus macclellandi*；*Hemibungarus macclellandii*；*Calliophis iwasakii*；*Calliophis macclellandi iwasakii*；*Sinomicrurus macclellandi iwasakii*

[俗名] 丽纹蛇

[鉴别特征] 小型前沟牙类毒蛇。头较小，与颈区分不明显；躯干圆柱形。头背黑色，有1黄白色倒"Λ"形斑。体背红褐色，间有黑色横斑，背鳞通体15行。

[形态描述] 依据茂兰地区标本 GZNU20170513003

成体量度

编号	产地	头体长（mm）	体全长（mm）	尾长（mm）	背鳞（枚）	腹鳞（枚）	肛鳞（枚）	上唇鳞（枚）	下唇鳞（枚）	颊鳞（枚）	眶前鳞（枚）	眶后鳞（枚）	前颞鳞（枚）	后颞鳞（枚）
GZNU20170513003	荔波茂兰	485	513	46	13-13-13	228	2	7	6	0	1	2	1	1

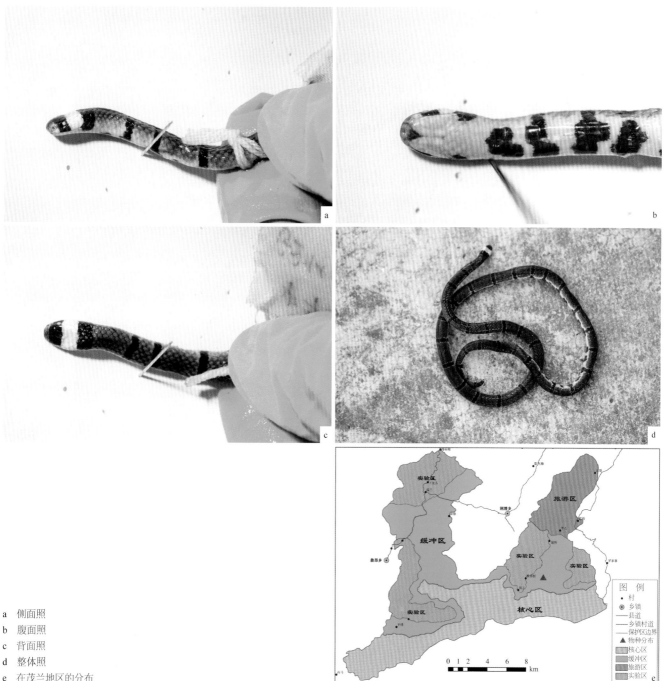

a 侧面照
b 腹面照
c 背面照
d 整体照
e 在茂兰地区的分布

描述　体全长/尾长：雄性513mm/46mm。头较小，与颈区分不明显；躯干圆柱形。头背黑色，有两条黄白色横纹，前条细，横跨两眼；后条较粗，呈倒"Λ"形。体背红褐色，间有1枚鳞宽的黑色横斑（19+3）条，背鳞通体15行；腹面白色，腹鳞有长短不等的黑横斑。鳞被没有颊鳞；眶前鳞1，眶后鳞2；颞鳞1+1；上唇鳞7（2-2-3式）；下唇鳞6，前3枚接前颏片；颏片2对。背鳞平滑，13-13-13；腹鳞228；肛鳞二分。

[生活习性]　栖息于山区森林地区。夜晚活动。于2016年5月在翁根坡采集到一条，性情温顺，不攻击人。

[地理分布]　茂兰地区分布于洞塘。国内分布于重庆、福建、广东、广西、贵州、海南、湖南、江西、浙江等地。国外分布于越南、老挝（北部）。

[模式标本]　正模标本（CAS 14978 [*swinhoei*]、CAS 18864 [*formosensis*]）保存于中国科学院

[模式产地]　中国台湾及印度阿萨姆邦

[保护级别]　贵州省重点保护野生动物

[IUCN濒危等级]　易危（VU）

18）眼镜蛇属 *Naja* Laurenti, 1768

Naja Laurenti, 1768, Synops. Rept. Viena: 90. Type species: *Coluber naja* Linnaeus, 1758.

体型偏大的前沟牙类毒蛇，体全长约2m。上颌骨前端超出腭骨，较短，前端有沟牙，齿间隙后有1～3枚小牙。头椭圆形，与颈区分不明显；无颊鳞；第3上唇鳞高，其前切鼻鳞后入眶；第4与第5下唇鳞之间的口缘嵌有1枚极小鳞片。背鳞平滑，体中段13～25行，排成斜行；腹鳞圆形；肛鳞完整或二分。尾下鳞一般成对，个别成单。

本属世界约28种，分布于亚洲（南部）及非洲。中国有2种，即舟山眼镜蛇 *Naja atra* 和孟加拉眼镜蛇 *N. kaouthia*。茂兰地区仅发现一种，即孟加拉眼镜蛇。

（24）孟加拉眼镜蛇 *Naja kaouthia* Lesson, 1831

Naja kaouthia Lesson, 1831, Catalogue des reptiles qui font partie d'une collection zoologique recueille dans l'Inde continentale ou en Afrique, et apportée en France par M. Lamare-Piquot. Bulletin des Sciences Naturelles et de Géologie, Paris, 25 (2): 119-123. Type locality: Begal [India].

Naja kaouthia Manthey and Grossmann, 1997: 423. Amphibien & Reptilien Südostasiens. Natur und Tier Verlag (Münster): 512.

Naja kaouthia Cox et al., 1998: 26. A Photographic Guide to Snakes and Other Reptiles of Peninsular Malaysia, Singapore and Thailand. Ralph Curtis Publishing: 144.

Naja naja kaouthia Sharma, 2004. Handbook Indian Snakes. Akhil Books, New Delhi: 292.

Naja (*Naja*) *kaouthia* Wallach et al., 2009. In praise of subgenera: taxonomic status of cobras of the genus *Naja* Laurenti (Serpentes: Elapidae). Zootaxa, 2236: 26-36.

Naja kaouthia Wallach et al., 2014: 460. Snakes of the World: A Catalogue of Living and Extinct Species. Taylor & Francis, CRC Press: 1237.

[同物异名]　*Naja tripudians* var. *fasciata*；*Naja naja sputatrix*；*Naja naja kaouthia*；*Naja kaouthia kaouthia*；*Naja kaouthia suphanensis*；*Naja*（*Naja*）*kaouthia*；*Naja kaouthia*

[俗名]　眼镜蛇

[鉴别特征]　大型前沟牙类毒蛇。受惊扰时，常竖立前半身，颈部平扁扩大，作攻击姿态，同时颈背露出呈单圈的眼镜状斑纹。与其他眼镜蛇的区别是：眼镜王蛇颈背没有眼睛状斑纹，而头背顶鳞正后有1对较大的枕鳞；相

近种舟山眼镜蛇颈背的眼镜状斑纹是双圈，而孟加拉眼镜蛇为单圈。

[形态描述]　依据茂兰地区标本 GZNU20100726004 描述　体全长/尾长为493mm/83mm。体色暗褐色，背面有白色细横纹，受惊扰时，常竖立前半身，颈部平扁扩大，作攻击姿态，同时颈背露出呈单圈的眼镜状斑纹。颊鳞1；眶前鳞1，眶后鳞+眶下鳞3；颞鳞2+3；上唇鳞7（2-2-3式），第3枚最高大；下唇鳞8，前4枚接前颏片；颏片2对。脊鳞两侧数行较窄长，斜列；腹鳞168；肛鳞完整。

[生活习性]　栖息环境为平原、丘陵和低山。见于耕作区、路边、池塘附近及住宅院内。

[地理分布]　茂兰地区分布于永康。国内分布于湖南、安徽、香港、澳门、重庆、福建、广东、广西、贵州、海南、湖北、浙江。国外分布于印度。

[模式标本]　未知

[模式产地]　孟加拉国

[保护级别]　三有保护动物；贵州省重点保护野生动物

[IUCN 濒危等级]　无危（LC）

成体量度

编号	产地	体长（mm）	体全长（mm）	尾长（mm）	背鳞（枚）	腹鳞（枚）	肛鳞（枚）	上唇鳞（枚）	下唇鳞（枚）	颊鳞（枚）	眶前鳞（枚）	眶后鳞+眶下鳞（枚）	前颞鳞（枚）	后颞鳞（枚）
GZNU20100726004	荔波茂兰	410	493	83	21-19-15	168	1	7	8	1	1	3	2	3

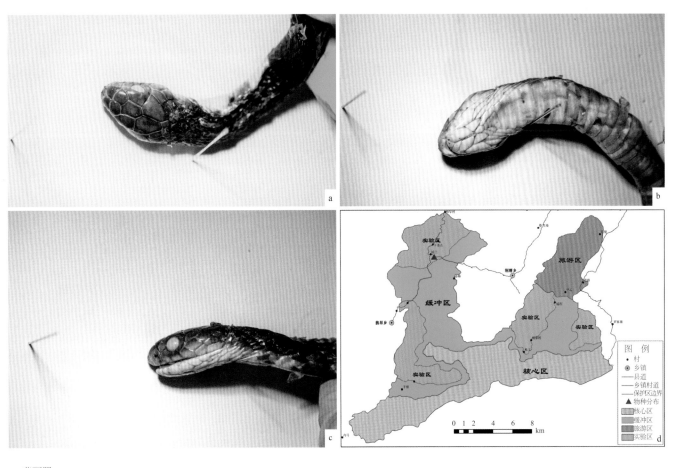

a　背面照
b　腹面照
c　侧面照
d　在茂兰地区的分布

19）环蛇属 *Bungarus* Daudin, 1803

Bungarus Daudin, 1803, Hist. Nat. Rept., Ⅴ: 263. Type species: *Pseudohoa fasciata* Schneider, 1801.

体型中等到较大的前沟牙类毒蛇，体全长 1.5m 左右。上颌骨较短（与游蛇科比较），前端有沟牙，齿间隙后有 1~4 颗小牙。头椭圆形，与颈略可区分，眼中等或较小，瞳孔圆形，头侧没有颊鳞，眶前鳞前伸与鼻鳞相切；躯尾修长适度，背鳞平滑，通体 13~19 行，除个别种外，背面正中一行脊鳞扩大呈六角形；腹鳞圆形；肛鳞完整；尾下鳞单行，少数种成对。

本属世界现有 15 种，分布于亚洲（南部）。我国有 3 种。茂兰地区有 1 种，即银环蛇 *Bungarus multicinctus*。

（25）银环蛇 *Bungarus multicinctus* Blyth, 1861

Bungarus semifasciatus Günther, 1858: 221. Catalogue of Colubrine snakes of the British Museum. London.

Bungarus multicinctus Blyth, 1861, Proceedings of the Society. Report of the Curator. J. Asiatic Soc. Bengal [1860]: 87-115.

Bungarus semifasciatus Günther, 1864: 344. The Reptiles of British India. London, Taylor & Francis.

Bungarus semifasciatus Günther, 1888, On a collection of reptiles from China. Ann. Mag. Nat. Hist. (1) 6: 165-172.

Bungarus candidus var. *multicinctus* Boulenger, 1896: 369. Catalogue of the snakes in the British Museum, Vol. 3. Taylor & Francis, London.

Bungarus multicinctus Stejneger, 1907, Herpetology of Japan and adjacent territory. Bull. U.S. Natl. Mus.: 397.

Bungarus multicinctus wanghaotingi Pope, 1928. Four new snakes and a new lizard from South China. American Museum Novitates, 325: 1-4.

Bungarus multicinctus Smith, 1943: 416. The Fauna of British India, Ceylon and Burma, Including the Whole of the Indo-Chinese Sub-Region. Reptilia and Amphibia. 3 (Serpentes). Taylor & Francis, London: 583.

Bungarus cf. *multicinctus* Chan-ard et al., 1999: 209. Amphibians and reptiles of peninsular Malaysia and Thailand-an illustrated checklist [bilingual English and German]. Bushmaster Publications, Würselen, Germany: 240.

Bungarus wanghaotingi Leviton et al., 2003. The Dangerously Venomous Snakes of Myanmar Illustrated Checklist with Keys. Proc. Cal. Acad. Sci., 54 (24): 407-462.

Bungarus multicinctus Sang et al., 2009. Herpetofauna of Vietnam. Chimaira, Frankfurt: 768.

Bungarus wanghaotingi Wallach et al., 2014: 130. Snakes of the World: A Catalogue of Living and Extinct Species. Taylor & Francis, CRC Press：1237.

[同物异名] *Bungarus semifasciatus*；*Bungarus caeruleus*；*Bungarus candidus* var. *multicinctus*；*Bungarus candidus multicinctus*；*Bungarus* cf. *multicinctus*；*Bungarus multicinctus multicinctus*；*Bungarus multicinctus wanghaotingi*；*Bungarus wanghaotingi*

[俗名] 百步蛇、金钱白花蛇

[鉴别特征] 体型中等略偏大的前沟牙类毒蛇。头椭圆而略扁，吻端圆钝，与颈略可区分；鼻孔较大；眼小，瞳孔圆形；躯干圆柱形；尾短，末端略尖细。背面黑色或黑褐色，通体背面有黑白相间的横纹。

[形态描述] 依据茂兰地区标本 GZNU20160805009 描述 体型中等略偏大。体全长/尾长为 1240mm/70mm。头椭圆而略扁，吻端圆钝，与颈略可区分；鼻孔较大；眼小，瞳孔圆形；躯干圆柱形；尾短，末端略尖细。背面黑色，通体背面有黑白相间的横纹；腹面白色。头背黑色，枕及颈背有白色的"Λ"形斑。颊鳞 1；眶前鳞 1，眶后鳞 2；

颞鳞 2+3；上唇鳞 7（2-2-3 式）；下唇鳞 7，前 3 枚或前 4 枚接前颏片；颏片 2 对。背鳞平滑，20-19-15；脊鳞扩大呈六角形，背脊无明显棱粒；腹鳞 211；肛鳞完整。

[生活习性]　平原、丘陵、山地有分布，几乎无处不在。晚上在田埂上多见，不怕人，爬行缓慢。

[地理分布]　茂兰地区分布于翁昂、洞塘。国内分布于安徽、澳门、重庆、福建、广东、广西、贵州、海南、湖北、湖南、江西、台湾、香港、云南、浙江。国外分布于缅甸、越南（北部）及老挝。

[模式标本]　正模标本（AMNH 35230）保存于美国自然历史博物馆

[模式产地]　厦门

成体量度

编号	产地	头体长（mm）	体全长（mm）	尾长（mm）	背鳞（枚）	腹鳞（枚）	肛鳞（枚）	上唇鳞（枚）	下唇鳞（枚）	颊鳞（枚）	眶前鳞（枚）	眶后鳞（枚）	前颞鳞（枚）	后颞鳞（枚）
GZNU20160805009	荔波茂兰	1170	1240	70	20-19-15	211	1	7	7	1	1	2	2	3

a　背面照
b　腹面照
c　侧面照
d　整体照
e　在茂兰地区的分布

[**保护级别**]　三有保护动物；贵州省重点保护野生动物

[**IUCN 濒危等级**]　无危（LC）

12. 游蛇科 Colubridae Cope, 1893

Colubridae Cope, 1893, Amer. Nat.: 480.

　　本科是蛇亚目中最大的一科，约占蛇类物种数目的 2/3。本科蛇类头背被覆对称的大鳞，覆瓦状排列；腹鳞宽大；前颌骨无齿，上颌骨具齿，有沟牙；无腰带。陆栖、树栖或水栖。

　　茂兰地区有 20 属 37 种，分属检索表如下。

茂兰地区属检索表

1 ┌ 无鼻间鳞及颊鳞 ·· 两头蛇属 *Calamaria*
　└ 有鼻间鳞及颊鳞 ·· 2

2 ┌ 具有 2~4 枚的颊鳞 ·· 鼠蛇属 *Ptyas*
　└ 颊鳞只有 1 枚或没有 ·· 3

3 ┌ 前额鳞 1 枚，颊鳞切前颏片 ·· 后棱蛇属 *Opisthotropis*
　└ 前额鳞 2 枚，颊鳞不切前颏片 ·· 4

4 ┌ 吻鳞显著，在背面能见部分较多 ·· 小头蛇属 *Oligodon*
　└ 吻鳞不显著，在背面看不到或仅能见到其上端 ·· 5

5 ┌ 中段背鳞 15 行 ·· 6
　└ 中段背鳞 15 行以上 ·· 7

6 ┌ 颈背正中有 1 明显的颈槽 ·· 颈槽蛇属 *Rhabdophis*
　└ 颈背正中无颈槽 ·· 翠青蛇属 *Ophcodrys*

7 ┌ 中段背鳞 17 行 ·· 8
　└ 中段背鳞 17 行以上 ·· 9

8 ┌ 背鳞通体 17 行 ·· 剑蛇属 *Sibynophis*
　└ 背鳞不是通体 17 行 ·· 链蛇属 *Lycodon*

9 ┌ 中段背鳞 19 行，眶后鳞 3 枚或以上 ·· 10
　└ 中段背鳞 19 行或更多，眶后鳞 2 枚或以上 ·· 13

10 ┌ 半阴茎分叉，精沟分叉或不分叉 ·· 11
　 └ 半阴茎和精沟均不分叉（草腹链蛇半阴茎顶端分叉）·· 12

11 ┌ 精沟分叉 ·· 异色蛇属 *Xenochrophis*
　 └ 精沟不分叉 ·· 华游蛇属 *Sinonatrix*

12 ┌ 成体头呈黄褐色，颌下黄色 ·· 腹链蛇属 *Amphiesma*
　 └ 成体头不呈黄褐色，颌下不呈黄色 ·· 东亚腹链蛇属 *Hebius*

13 ┌ 脊鳞扩大或稍扩大 ·· 林蛇属 *Boiga*
　 └ 脊鳞不扩大 ·· 14

14 ┌ 体前部背鳞成明显斜行 ·· 斜鳞蛇属 *Pseudoxenodon*
　 └ 体前部背鳞不成斜行 ·· 15

15 ┌ 颏鳞起棱，背鳞明显起棱 ·· 颈棱蛇属 *Macropisthodon*
　 └ 颏鳞不起棱，背鳞平滑或微棱 ·· 16

16 ┌ 全身体色翠绿色 ·· 绿蛇属 *Rhadinophis*
　 └ 全身体色不为翠绿色 ·· 17

17	头侧有 3 条放射状黑线纹	三索蛇属 *Coelognathus*
	头侧无 3 条放射状黑线纹	18

18	体背部有连续纵纹，无明显的斑纹	紫灰蛇属 *Oreocryptophis*
	体背部无连续纵纹，有明显的斑纹	19

19	体背中央有 1 行菱形大块斑，块斑和中心颜色不同	玉斑蛇属 *Euprepiophis*
	体背部非菱形块斑，块斑和中心颜色一般相同	晨蛇属 *Orthriophis*

20）两头蛇属 *Calamaria* Boie, 1826

Calamaria Boie, 1826, Isis: 981. Type species: *Calamaria linnaei* Boie, 1827.

小型穴居无毒蛇类，体全长 300mm 左右。体呈圆柱形。头小，与颈不分；眼小，瞳孔圆形；没有颊鳞、鼻间鳞和颞鳞。躯干圆柱形，前后粗细一致，背鳞通体 13 行，平滑。尾极短，尾下鳞双行。颈侧和尾基部两侧各有 1 浅黄色斑。由于尾短而末端钝，尾基部黄斑又与颈侧黄斑相似，被误认为蛇两端各有一头，故称为两头蛇。

本属现有 63 种，分布于南亚及东亚地区。我国目前已知 3 种，即云南两头蛇 *Calamaria yunnanensis*、钝尾两头蛇 *Calamaria septentrionalis* 和尖尾两头蛇 *Calamaria pavimentata*。茂兰地区仅 1 种——钝尾两头蛇。

（26）钝尾两头蛇 *Calamaria septentrionalis* Boulenger, 1890

Calamaria septentrionalis Boulenger, 1890. List of the reptiles, batrachians, and freshwater fishes collected by Professor Moesch and Mr. Iversein in the district of Deli, Sumatra. Proc. Zool. Soc., London: 31-40. Type locality: Kiukiang [= Jiujiang], Hong Kong.

Calamaria septentrionalis Mell, 1931, Lingnan Sci. Jour., Canton, 8 [1929]: 199-219.

Calamaria septentrionalis Smith, 1943: 239. The Fauna of British India, Ceylon and Burma, Including the Whole of the Indo-Chinese Sub-Region. Reptilia and Amphibia. 3 (Serpentes). Taylor & Francis, London: 583.

Calamaria septentrionalis Sang et al., 2009. Herpetofauna of Vietnam. Chimaira, Frankfurt: 768.

Calamaria septentrionalis Wallach et al., 2014: 141. Snakes of the World: A catalogue of living and extinct species. Taylor & Francis, CRC Press: 1237.

［同物异名］ 无

［俗名］ 无

［鉴别特征］ 小型穴居无毒蛇。头小，与颈不分。通体圆柱形，尾极短而末端略尖锐。

［形态描述］ 体全长/尾长：雄性 439mm/19mm，雌性 190mm/11mm。通体圆柱形，尾极短而末端略尖锐。体尾背面略带红褐色，有深色纵线纹；腹面色较浅淡。头小，与颈不分。颈侧各有 1 黄色斑，尾基两侧也各有 1 黄色斑，尾腹面正中有 1 条短黑色纵线。鳞被吻鳞宽，从背面可见；没有鼻间鳞、颊鳞和颞鳞；前额鳞大，前伸与吻鳞相接，下延与上唇鳞相切；额鳞长大于宽；眶前鳞 1，眶后鳞 1；上唇鳞 4（1-2-1 式）；下唇鳞 5，前 3 枚接前颏片，个别为 7，前 4 枚接前颏片。背鳞平滑，通体 13 行；腹鳞 167～192；肛鳞完整；尾下鳞 14～23 对。上颌齿每侧 8～11，几乎等大。

［生活习性］ 生活于低山丘陵或隐匿于地表之下，晚上或雨天到地面活动。

［地理分布］ 茂兰地区分布于洞塘。国内分布于福建、广东、广西、贵州、海南、四川、台湾、云南、浙江。国外分布于印度（阿萨姆邦）、越南、印度尼西亚等地。

［模式标本］ 正模标本［BMNH 1947.3.6.60-61（91.1.28.1-2）］

保存于英国伦敦自然历史博物馆

[模式产地] 香港九江

[保护级别] 三有保护动物；贵州省重点保护野生动物

[IUCN 濒危等级] 无危（LC）

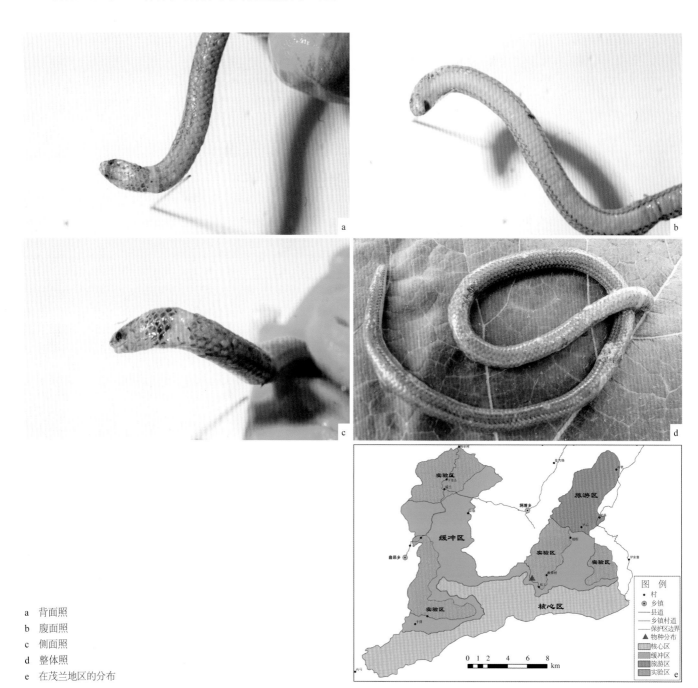

a 背面照
b 腹面照
c 侧面照
d 整体照
e 在茂兰地区的分布

21）斜鳞蛇属 *Pseudoxenodon* Boulenger, 1890

Pseudoxenodon Boulenger, 1890, Fauna Brit. India: 340. Type species: *Pseudoxenodon macrops* Blyth, 1855. Type locality: India.

头、颈区分明显，眼大，瞳孔圆形。背鳞明显起棱。体前部背鳞成明显斜行，体中部背鳞绝大多数 17～19 行，肛前 15 行。尾下鳞成对。所有椎骨均有椎体下突。上颌齿 19～27，最后两个颌齿明显增大。

本属中国分布 4 种。茂兰地区有 2 种，即大眼斜鳞蛇和横纹斜鳞蛇。

茂兰地区种检索表

体及尾部具有明显的黑色横纹，第一横纹的两端分别沿颈侧前方伸成较狭窄的黑纹，在顶鳞处相接汇成一黑色环··横纹斜鳞蛇 *Pseudoxenodon bambusicola*

体及尾部不呈黑色横纹，体背斑纹带有黄色、橘黄色、红色或棕色，颈部具有白色细线纹和黑色箭形斑············ 大眼斜鳞蛇 *Pseudoxenodon macrops*

（27）大眼斜鳞蛇 *Pseudoxenodon macrops* (Blyth, 1855)

Tropidonotus macrops Blyth, 1855. Notices and descriptions of various reptiles, new or little known. J. Asiatic Soc. Bengal, Calcutta, 23 (3): 287-302.

Tropidonotus sikkimensis Anderson, 1871, J. Asiat. Soc. Bengal, Calcutta, 40, part 11 (1): 12-39.

Pseudoxenodon macrops Boulenger, 1890. Reptilia and Batrachia. Taylor & Francis, London.

Pseudoxenodon macrops Boulenger, 1893: 270. Catalogue of the snakes in the British Museum (Nat. Hist.) I. London (Taylor & Francis).

Tropidonotus macrops Wall, 1908, Bombay Nat. Hist. Soc., 18: 312-337.

Tropidonotus handeli Werner, 1922. Neue Reptilien aus Süd-China, gesammelt von Dr. H. Handel-Mazzetti. Anz. Akad. Wissensch. Wien, ser. Mathemat.-Naturwissensch., Vienna, 59 (24-25): 220-222.

Pseudoxenodon macrops fukiensis Pope, 1928, American Museum Novitates, 320: 1-6.

Natrix handeli Mell, 1931, Lingnan Sci. Jour., Canton, 8 [1929]: 199-219.

Pseudoxenodon macrops Smedley, 1932, Bull. Raffles Mus., No. 7: 9-17.

Pseudoxenodon macrops Smith, 1943: 311. Reptilia and Amphibia. 3 (Serpentes). Taylor & Francis, London.

Pseudoxenodon fukienensis [sic] Marx, 1958, Fieldiana Zool., 36: 407-496.

Pseudoxenodon macrops fukienensis Zhao and Adler, 1993, Herpetology of China. SSAR, Oxford/Ohio: 1-522.

Pseudoxenodon sinensis Boulenger, 1904, Ann. Mag. Nat., Hist., 13 (74): 130-134.

Pseudoxenodon macrops Manthey and Grossmann, 1997: 383. Amphibien and Reptilien Südostasiens. Natur und Tier Verlag (Münster).

Pseudoxenodon macrops Cox et al., 1998: 49. A photographic guide to snakes and other reptiles of Peninsular Malaysia, Singapore and Thailand. Ralph Curtis Publishing.

Pseudoxenodon macrops macrops Chan-ard et al., 1999: 182. Amphibians and reptiles of peninsular Malaysia and Thailand-an illustrated checklist. Bushmaster Publications, Würselen, Germany: 240.

Pseudoxenodon macrops macrops Zhao, 2006. The snakes of China. Hefei, China, Anhui Science & Technology Publ. House, Vol. 1:

372., Vol. II (color plates): 280.

Pseudoxenodon macrops Sang et al., 2009. Herpetofauna of Vietnam. Chimaira, Frankfurt.

Pseudoxenodon macrops Wallach et al., 2014: 609. Snakes of the world: A catalogue of living and extinct species. Taylor & Francis, CRC Press.

[同物异名] *Tropidonotus macrops*；*Xenodon macrophthalmus*；*Tropidonotus sikkimensis*；*Tropidonotus macrophthalmus*；*Tropidonotus tigrinus* var. *niger*；*Tropidonotus handeli*；*Natrix handeli*；*Pseudoxenodon macrops macrops*；*Pseudoxenodon macrops fukiensis*；*Pseudoxenodon fukiensis*；*Pseudoxenodon fukienensis* [sic]；*Pseudoxenodon sinensis*

[俗名] 大斜鳞蛇、中华斜鳞蛇、臭蛇（云南）、草上飞（云南独龙江）、气扁蛇（四川）

[鉴别特征] 脊鳞两侧的背鳞窄长，排列成斜行，故名斜鳞蛇。颈背有1尖端向前的粗大黑色箭形斑，该斑两前缘未镶白边。此蛇遇惊扰常扁平其身体，有时呼呼作响，故有"气扁蛇"之名。如握手中，有特殊难闻的气味。

[形态描述] 依据茂兰地区标本 GZNU20170423097 描述 体全长/尾长为395.2mm/74.1mm。体背鳞片棕褐色，

成体量度

编号	产地	头体长（mm）	体全长（mm）	尾长（mm）	背鳞（枚）	腹鳞（枚）	肛鳞（枚）	上唇鳞（枚）	下唇鳞（枚）	颊鳞（枚）	眶前鳞（枚）	眶后鳞（枚）	前颞鳞（枚）	后颞鳞（枚）
GZNU20170423097	荔波茂兰	321.1	395.2	74.1	19-17-17	140	2	9	9	1	1	3	2	3

a 背面照
b 腹面照
c 侧面照
d 在茂兰地区的分布

各鳞边缘呈黑色；腹面玉白色，每一腹鳞不规则散布着或长或短的黄褐色斑，相互组合成大黄褐色斑。头腹白色无斑。脊鳞两侧的背鳞窄长，排列成斜行；颈背有1尖端向前的粗大黑色箭形斑，该斑两前缘未镶白边；上下唇黄白色。颊鳞1；眶前鳞1；眶后鳞3；颞鳞2+3；上唇鳞9（3-3-3式）；下唇鳞9，第1对在颏鳞后相接，前5枚接前颏；颏片2对。背鳞19-17-17，除最外行均具棱，脊鳞两侧各数行鳞窄长，斜向排列；腹鳞140；肛鳞二分。

[生活习性] 本种分布范围广，数量多，平原、丘陵、山区都可见到。其栖息环境也极其多样，灌丛、草地、农田，特别是水域附近多见，可能与其食性有关。

[地理分布] 茂兰地区分布于翁昂、洞塘。国内分布于重庆、福建、甘肃、广东、广西、贵州、河南、湖北、湖南、陕西、四川、西藏、云南等地。国外分布于缅甸、越南、尼泊尔、泰国、印度（阿萨姆邦、大吉岭）。

[模式标本] 正模标本（AMNH R34650）保存于美国国家自然历史博物馆；共模标本（ZSI 7506-07、ZSI 7556-57）保存于爱尔兰动物学协会

[模式产地] 印度大吉岭

[保护级别] 贵州省重点保护野生动物

[IUCN 濒危等级] 无危（LC）

（28）横纹斜鳞蛇 *Pseudoxenodon bambusicola* Vogt, 1922

Pseudoxenodon bambusicola Vogt, 1922, Zur Reptilien-und Amphibienfauna Südchinas. Archiv für Naturgeschichte 88A (10): 135-146. Type locality: Mountains of North Kwangtung [= Guangdong], China.

Pseudoxenodon melli Vogt, 1922, Archiv für Naturgeschichte, 88A (10): 135-146.

Pseudoxenodon dorsalis melli Mell, 1931, Lingnan Sci. Jour., Canton, 8 [1929]: 199-219.

Pseudoxenodon dorsalis bambusicola Mell, 1931, Lingnan Sci. Jour., Canton, 8 [1929]: 199-219.

Pseudoxenodon bambusicola Smith, 1943: 313. Reptilia and Amphibia. 3 (Serpentes). Taylor & Francis, London.

Pseudoxenodon bambusicola Ziegler et al., 2006, Sauria, 28 (2): 29-40.

Pseudoxenodon bambusicola Wallach et al., 2014: 609. Snakes of the World: A catalogue of living and extinct species. Taylor & Francis, CRC Press.

[同物异名] *Pseudoxenodon melli*; *Pseudoxenodon dorsalis melli*; *Pseudoxenodon dorsalis bambusicola*

[俗名] 无

[鉴别特征] 脊鳞两侧的背鳞窄长，排列成斜行；头背有1尖端向前的黑色箭形斑，其后分叉成两纵线沿颈侧向后延伸约一个半头长，再弯至体背成一环（是本种的典型特征），但个别标本无此环。

[形态描述] 依据茂兰地区标本GZNU20160805011描述 体全长/尾长为350.2mm/63.1mm。背面黄褐色，有黑色粗大横纹16+4条，横跨整个背面；脊鳞两侧的背鳞窄长，排列成斜行；腹面黄白色，前部往往有深褐色横纹。头背有1尖端向前的黑色箭形斑，始于额鳞后缘，向后分叉成两纵线沿颈侧延伸约一个半头长，再弯向体背连成一环，头侧另有1粗黑纹起自鼻间鳞经眼达口角；头腹灰白色。

颊鳞1；眶前鳞2，眶后鳞3；颞鳞2+2；上唇鳞8（3-2-3式）；下唇鳞9，前4接前颏片；颏片2对。背鳞19-17-15，最外行平滑；腹鳞138；肛鳞二分。

[生活习性] 栖息于山区森林、竹林、草丛、路边或溪流和稻田附近。昼夜均活动。

[地理分布] 茂兰地区见于干排。国内分布于福建、广东、广西、贵州、海南、湖南、江西、浙江。国外分布于越南（北部）。

[模式标本] 正模标本（ZMB38070）保存于德国柏林洪堡大学自然历史博物馆

[模式产地] 广东

[保护级别] 三有保护动物；贵州省重点保护野生动物

[IUCN 濒危等级] 无危（LC）

成体量度

编号	产地	头体长（mm）	体全长（mm）	尾长（mm）	背鳞（枚）	腹鳞（枚）	肛鳞（枚）	上唇鳞（枚）	下唇鳞（枚）	颊鳞（枚）	眶前鳞（枚）	眶后鳞（枚）	前颞鳞（枚）	后颞鳞（枚）
GZNU20160805011	荔波茂兰	287.1	350.2	63.1	19-17-15	138	2	8	9	1	2	3	2	2
GZNU20170508033	荔波茂兰	285.1	341.2	55.1	19-17-15	135	2	8	9	1	2	3	2	2

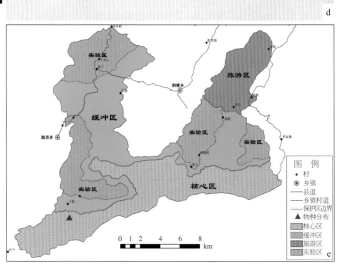

a 背面照
b 腹面照
c 侧面照
d 整体照
e 在茂兰地区的分布

22）剑蛇属 *Sibynophis* Fitzinger, 1843

Sibynophis Fitzinger, 1843, Syst. Rept., Vienna, 1: 26. Type species: *Coluber geminatus* Oppel and Boie, 1826, of Southeast Asia and Indonesia, by original designation.

体型细长的中等偏小的无毒蛇类。上颌牙齿细小均匀且数目多，每侧25～56枚，各齿较扁而上端略平，整体形成锐利的切缘；全部脊椎都有发达的椎体下突。背面棕褐色，腹面白色略带黄绿色，头背有1黑色大斑，后缘镶白，黑斑与躯背正中的黑色脊线相连，上唇鳞黑白相间。

本属已知10种，分布于南亚及东南亚等地。我国分布3种。茂兰地区有1种，即黑头剑蛇。

（29）黑头剑蛇 *Sibynophis chinensis* (Günther, 1889)

Ablabes chinensis Günther, 1889. Third contribution to our knowledge of reptiles and fishes from the Upper Yangtsze-Kiang. Ann. Mag. Nat. Hist., (6) 4 (21): 218-229. Type locality: Miyi, Sichuan, 880m elevation.

Sibynophis chinensis grahami Boulenger, 1904, Ann. Mag. Nat. Hist., (7) 13 (74): 130-134.

Sibynophis collaris chinensis Mell, 1931, Lingnan Sci. Jour., Canton, 8 [1929]: 199-219.

Sibynophis chinensis Pope, 1935: 88. Amer. Mus. Nat. Hist., New York, Nat. Hist. Central Asia, 10: lii, 1-604.

Sibynophis grahami Pope, 1935: 88. Amer. Mus. Nat. Hist., New York, Nat. Hist. Central Asia, 10: 1-604.

Scaphiodontophis sumichrasti Taylor et al., 1943, Univ. Kansas Sci. Bull., 29 (6): 301-337.

Sibynophis chinensis miyiensis Zhao, 1987, Chinese Herpetological Research, 1: 1-6.

Sibynophis chinensis miyiensis Guo et al., 1999. Asiatic Herpetological Research, 8: 43-47.

Sibynophis chinensis grahami Sang et al., 2009. Herpetofauna of Vietnam. Chimaira, Frankfurt: 768.

Sibynophis chinensis Wallach et al., 2014: 672. Snakes of the World: A catalogue of living and extinct species. Taylor & Francis, CRC Press: 1237.

[同物异名] *Ablabes chinensis*；*Sibynophis chinensis chinensis*；*Henicognathus sumichrasti*；*Sibynophis hainanensis*；*Sibynophis collaris formosensis*；*Sibynophis collaris chinensis*；*Scaphiodontophis sumichrasti*；*Polyodontophis grahami*；*Sibynophis chinensis graham*；*Sibynophis graham*；*Sibynophis chinensis miyiensis*

[俗名] 黑头蛇

[鉴别特征] 小型无毒蛇。头背部黑色，与体背中央黑褐色脊线相连，体背棕褐色。

[形态描述] 依据茂兰地区标本GZNU20160806001描述 体型细长。体全长/尾长为642.2mm/210.1mm。尾具缠绕性。体背棕褐色；头背部黑色，与体背中央黑褐色脊线相连；头背有1黑斑，上唇及腹面白色，鳞沟色黑，两色对比非常鲜明。吻鳞宽，鼻孔较大，位于鼻鳞中央；眼适中，瞳孔圆形。颊鳞1；眶前鳞1，眶后鳞2；上唇鳞9（3-3-3式）；下唇鳞9，前4枚切前颌片；颌片2对，大小几乎相等。背

成体量度

编号	产地	头体长（mm）	体全长（mm）	尾长（mm）	背鳞（枚）	腹鳞（枚）	肛鳞（枚）	上唇鳞（枚）	下唇鳞（枚）	颊鳞（枚）	眶前鳞（枚）	眶后鳞（枚）	前颞鳞（枚）	后颞鳞（枚）
GZNU20160806001	荔波茂兰	432.3	642.2	210.1	17-17-17	213	2	9	9	1	1	2	2	2

鳞通体17行，平滑无棱；腹鳞213；肛鳞二分。

[**生活习性**] 生活于平原、丘陵和山区，见于路边、河边或茶山草丛中，也见于林下或山林中石板路上。白天活动。

[**地理分布**] 茂兰地区分布于翁昂。国内分布于山西、安徽、重庆、福建、甘肃、广东、广西、贵州、海南、河南、湖北、湖南、江苏、江西、陕西、上海、四川、台湾、香港、云南、浙江等地。国外分布于老挝、越南。

[**模式标本**] 正模标本（CIB 105027）保存于中国科学院成都生物研究所

[**模式产地**] 四川

[**保护级别**] 贵州省重点保护野生动物

[**IUCN 濒危等级**] 无危（LC）

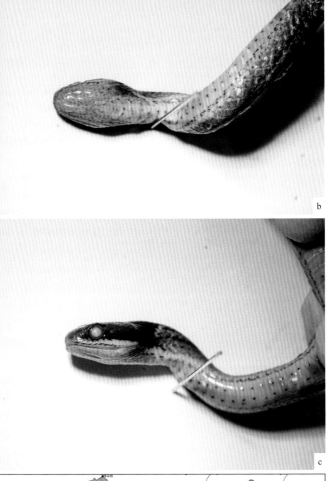

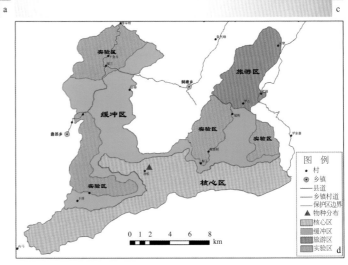

a 背面照
b 腹面照
c 侧面照
d 在茂兰地区的分布

23）林蛇属 *Boiga* Fitzinger, 1826

Boiga Fitzinger, 1826, Neue Classif. Rept., Vienna: 60. Type species: *Coaluer irregu* Lari, B. Merrem in Bechsstein, 1802, of Australia, by subsequent designation (Stejneger, 1907, Bull. U.S. Natl. Mus., Washington, 58: 381).

后沟牙类毒蛇。体型细长，全长 1m 左右。头大，与颈区分十分明显；眼大，瞳孔直立椭圆形。躯干略侧扁，背鳞平滑，斜列，脊鳞扩大；腹面被圆鳞，具不明显的侧棱，亦略侧扁，具缠绕性。

本属世界已知 32 种，分布于澳大利亚、亚洲（南部）及热带非洲。我国分布 4 种。茂兰地区有繁花林蛇 *Boiga multomaculata* 和绞花林蛇 *Boiga kraepelini* 2 种。

茂兰地区种检索表

中段背鳞 19 行；颞部鳞片正常，脊鳞明显较背鳞扩大··繁花林蛇 *Boiga multomaculata*
中段背鳞 21 行；颞部鳞片小，脊鳞不扩大或略大于背鳞··绞花林蛇 *Boiga kraepelini*

（30）繁花林蛇 *Boiga multomaculata* (Boie, 1827)

Dipsas multomaculata Boie, 1827, Isis van Oken, 20: 508-566. Type locality: Java.

Dipsas multimaculata [sic] Duméril et al., 1854: 1139. Paris, Librairie Encyclopédique de Roret.

Dipsadomorphus multimaculatus Boulenger, 1896. Catalogue of the snakes in the British Museum, Vol. 3. London (Taylor & Francis), XIV: 727.

Boiga multimaculata Smith, 1923. Taylor & Francis, London: 583.

Boiga multomaculata hainanensis Mell, 1931, Lingnan Sci. Jour., Canton, 8 [1929]: 199-219.

Boiga multomaculata sikiangensis Mell, 1931, Lingnan Sci. Jour., Canton, 8 [1929]: 199-219.

Boiga multomaculata Manthey and Grossmann, 1997: 323. Amphibien & Reptilien Südostasiens. Natur und Tier Verlag (Münster): 512.

Boiga multomaculata Cox et al., 1998: 76. Ralph Curtis Publishing: 144.

Boiga multomaculata Groen, 2008, Het *Boiga*-genus. Lacerta, 66 (1-3): 64-79.

Boiga multomaculata Wallach et al., 2014: 105. Taylor & Francis, CRC Press: 1237.

[曾用名]　繁花蛇

[同物异名]　*Dipsas multomaculata*; *Dipsas multimaculata* [sic]; *Dipsadomorphus multimaculatus Boiga multimaculata*; *Boiga multimaculata hainanensis*; *Boiga multimaculata sikiangensis*

[鉴别特征]　本种与绞花林蛇相似，区别在于本种颞部鳞片正常，前颞鳞 1～3；脊鳞较相邻的背鳞显著扩大；肛鳞多完整。

[形态描述]　依据茂兰地区标本描述　体全长/尾长：雄性 728mm/159mm，雌性 852mm/163mm。体型中等偏大，头大，与颈区分十分明显，躯尾细长，适于缠绕。颞部鳞片正常，脊鳞显著大于相邻背鳞。头背有 1 深棕色尖端向前的 "V" 形斑，始自吻端，分支达枕部；另有两深棕色纵纹自吻端分别沿头侧经眼斜达颌角。通体背面浅褐色，正背有深棕色粗大点斑 2 行，彼此交错排列，体侧各有 1 行较小的深棕色点斑；每一腹鳞上有数个略呈三角形的浅褐色斑。颊鳞 1；眶前鳞 1，眶后鳞 2；颞鳞 2+2；上唇鳞 9（2-3-4 式），下唇鳞 11，前 4 枚接前颌片；颔片 2 对，

前对大于后对。背鳞平滑，19-19-15，排列成斜行，脊鳞显著大于相邻背鳞；腹鳞 196～230；肛鳞完整。

[生活习性]　生活于山区或有树林的丘陵地区。有攀援习性，树栖，常在夜间出来活动。主要吃鸟、蜥蜴及鼠类。卵生。有轻微毒性，为后沟牙类毒蛇。

[地理分布]　茂兰地区分布于洞塘。国内分布于澳门、福建、广东、广西、贵州、海南、湖南、江西、香港、云南、浙江等地。国外分布于印度（东北部）、缅甸、泰国、老挝、越南、柬埔寨、马来西亚（西部）、新加坡和印度尼西亚。

[模式标本]　未知（Sang et al.，2009）
[模式产地]　印度尼西亚爪哇
[保护级别]　三有保护动物；贵州省重点保护野生动物
[IUCN 濒危等级]　无危（LC）
[讨论]　在以前的历次调查中没有在茂兰地区发现该种。此次发现为该地区新纪录，为繁花林蛇的分布增加了新的地点。

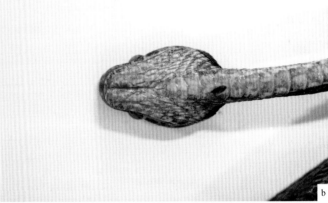

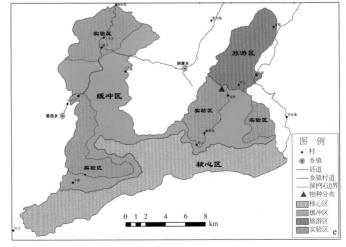

a　背面照
b　腹面照
c　侧面照
d　整体照
e　在茂兰地区的分布

（31）绞花林蛇 *Boiga kraepelini* Stejneger, 1902

Boiga kraepelini Stejneger, 1902, Proc. Biol. Soc., Washington, 15: 15-17. Type locality: Kelung, Formosa [= Taiwan], China.
Dipsadomorphus kraepelini Wall, 1903, Proc. Zool. Soc., London, 1903: 84-102.
Dinodon multitemporalis Oshima, 1910, Annot. Zool. Japon., Tokyo, 7 (3): 185-207.
Boiga kraepelini sinensis Mell, 1931, Lingnan Sci. Jour., Canton, 8 [1929]: 199-219.
Boiga multitemporalis Bourret, 1935, Bull. Gen. Instr. Pub. Hanoi, 8: 1-13.
Boiga kraepelini Ziegler, 2002: 221. Natur und Tier Verlag (Münster): 342.
Boiga kraepelini Wallach et al., 2014: 105. Taylor & Francis, CRC Press: 1237.

[曾用名] 绞花蛇

[同物异名] *Dipsadomorphus kraepelini*；*Dinodon multitemporalis*；*Boiga kraepelini sinensis*；*Boiga multitemporalis*

[鉴别特征] 本种与繁花林蛇相似，区别在于本种颞区鳞片较小，脊鳞不扩大或略大于相邻背鳞，肛鳞多二分。与原矛头蝮易混淆，区别在于本种没有颊窝，头背具对称大鳞片；原矛头蝮在眼与鼻孔之间有颊窝，头背都是小鳞片。

[形态描述] 依据茂兰地区标本 GZNU20100726010 描述 最大体全长/尾长为 1208mm/338mm。头大，与颈区分明显，躯干甚长而略侧扁，尾细长，适于缠绕。头背具对称大鳞片，颞部鳞片较小，不成列；脊鳞不扩大或略大于相邻背鳞。通体背面灰褐色或浅紫褐色，躯尾正背有1行粗大而不规则、镶黄边的深棕色斑。颊鳞1；眶前鳞1，眶后鳞2；颞鳞为多数小鳞而不成列；上唇鳞9（2-3-4式）；下唇鳞13，前3枚接前颏片；颏片2对。肛鳞二分。

[生活习性] 生活于岩石、山区、丘陵，有攀援习性。常栖息于沟旁灌木上或茶山矮树上，也见于溪流旁岩石及住宅附近。多于夜晚活动。属大型林栖后沟牙类毒蛇。

[地理分布] 茂兰地区分布于洞塘。国内分布于安徽、重庆、福建、甘肃、广东、广西、贵州、海南、湖南、江西、四川、台湾、浙江等地。国外无记录。

[模式标本] 模式标本（ZMH R04377，以前编号为 no. 1565 ZNH）保存地未知；标本（NWHN-RA）保存于法

成体量度

编号	产地	头体长（mm）	体全长（mm）	尾长（mm）	背鳞（枚）	腹鳞（枚）	肛鳞（枚）	上唇鳞（枚）	下唇鳞（枚）	颊鳞（枚）	眶前鳞（枚）	眶后鳞（枚）	前颞鳞（枚）	后颞鳞（枚）
GZNU20100726010	荔波茂兰	870	1208	338	21-21-14	205	2	9	13	1	1	2	0	0

a 背面照
b 腹面照

国国家自然历史博物馆（Sang et al., 2009）

[模式产地] 台湾基隆

[保护级别] 三有保护动物；贵州省重点保护野生动物

[IUCN 濒危等级] 无危（LC）

c 侧面照
d 整体照
e 在茂兰地区的分布

24）小头蛇属 *Oligodon* Boie, 1827

Oligodon Boie, 1827, Isision Uken, Jena, 20: col. 519. Type species: *Coluber bitorquatus* Reinwardt in Boie, 1827, of Indonesia, by original designation.

体型较小或中等大小的无毒蛇类。头部较为小，与颈区分不明显；眼中等大。瞳孔圆形；吻鳞大，显露于头背面，吻鳞高大于宽，可见部分的长度大于额鳞到吻鳞的长度；颊鳞有或无。鼻孔在长形的鼻鳞中央；鼻间鳞1对或缺失；额鳞六角形，长略大于宽；颊鳞1或缺失；眶前鳞1（2）；眶后鳞2（1）；下唇鳞7～9，前3（4）对与前颏片接触。上颌齿6～16，在后方的大而窄扁；腭齿发达或成遗迹。躯干圆柱形，背鳞平滑，从头背到体中段13～21行；腹鳞或两侧具侧棱；肛鳞完整或二分。尾下鳞双行。卵生。

本属现存76种，分布于南亚、中国（南部）、东南亚。我国分布12种，茂兰地区有3种。

茂兰地区种检索表

1. { 体前背鳞行数 19；头背略似"灭"字形斑 ··· 台湾小头蛇 Oligodon formosanus
   { 体前背鳞行数 17；头背无"灭"字形斑 ·· 2
2. { 头后和颈背有 1 黑褐色箭形斑；腹面有两行黑褐色方形斑 ··························· 中国小头蛇 Oligodon chinensis
   { 头后和颈背无黑褐色箭形斑；腹面也无黑褐色方形斑 ····································· 紫棕小头蛇 Oligodon cinereus

（32）台湾小头蛇 *Oligodon formosanus* (Günther, 1872)

Simotes formosanus Günther, 1872: 20. Ann. Mag. Nat. Hist., (4) 9: 13-37. Type locality: Takou [= Kaohsiung], Formosa [= Taiwan].

Simotes bicatenatus Müller, 1878, Verh. Naturf. Ges. Basel, 6: 390-411.

Simotes formosanus Fischer, 1886: 60. Herpetologische Notizen. Abh. Geb. Naturw. Hamb., 9: 51-67 (1-19).

Simotes hainanensis Boettger, 1894: 133. Ber. Senckenberg. Naturf. Ges.: 129-152.

Simotes formosanus Boulenger, 1894: 222. British Mus. (Nat. Hist.), London, XI: 382.

Holarchus nesiotis Barbour, 1908: 318. Bull. Mus. Comp. Zool. Harvard, 51 (12): 315-325.

Holarchus formosanus hainanensis Barbour, 1909, Proc. New England Zool. Club, 4: 53-78.

Holarchus formosanus brunneus Mell, 1931, Lingnan Sci. Jour., Canton, 8 [1929]: 199-219.

Oligodon formosanus Smith, 1943. Reptilia and Amphibia. 3 (Serpentes). Taylor & Francis, London: 583.

Oligodon formosanus Lazell et al., 1999, Postilla, 217: 1-18.

Oligodon formosanus Green et al., 2010, Asian Herpetological Research, 1 (1): 1-21.

Oligodon formosanus Wallach et al., 2014: 497. Taylor & Francis, CRC Press: 1237.

[同物异名] *Simotes formosanus*；*Simotes bicatenatus*；*Simotes hainanensis*；*Holarchus formosanus*；*Holarchus nesiotis*；*Holarchus formosanus hainanensis*；*Holarchus formosanus violaceoides*；*Holarchus formosanus brunneus*

[俗名] 台湾秤杆蛇

[鉴别特征] 体背有距离相等的黑褐色波浪状横纹，约 1 片鳞宽；背鳞 19-19-17（15）。与紫棕小头蛇的区别是本种头背有略似"灭"字形斑。

[形态描述] 依据茂兰地区标本 GZNU320527 描述 体全长 / 尾长为 504.6mm/87.4mm。头较小，与颈区分不明显；吻鳞从头背可见甚多；有鼻间鳞和颊鳞。体前中段背鳞 19 行。头背有略似"灭"字形斑。体尾背面紫褐色，由于部分背鳞沟色黑而形成多数约等间距排列的黑褐色横纹，有的个体背面有 2 条红褐色纵线；腹面黄白色。颊鳞 1；眶前鳞 2，眶后鳞 2；颞鳞 1+2；上唇鳞 8（3-2-3 式），下唇鳞 8，前 3 枚接前颏片；颏片 2 对。背鳞平滑，19-19-17；腹鳞 186；肛鳞完整；尾下鳞 40～60 对。上颌齿每侧 10～11。

[生活习性] 生活于平原及高山，嗜食爬行动物的卵。

[地理分布] 茂兰地区分布于洞塘。国内分布于澳门、福建、广东、广西、贵州、海南、湖南、江苏、江西、台湾、香港、云南、浙江等地。国外分布于越南。

[模式标本] 正模标本（MCZR-7107）保存于英国伦敦自然历史博物馆

[模式产地] 台湾高雄

[保护级别] 三有保护动物；贵州省重点保护野生动物

[IUCN 濒危等级] 无危（LC）

成体量度

编号	产地	头体长（mm）	体全长（mm）	尾长（mm）	背鳞（枚）	腹鳞（枚）	肛鳞（枚）	上唇鳞（枚）	下唇鳞（枚）	颊鳞（枚）	眶前鳞（枚）	眶后鳞（枚）	前颞鳞（枚）	后颞鳞（枚）
GZNU320527	荔波茂兰	417.2	504.6	87.4	19-19-17	186	1	8	8	1	2	2	1	2

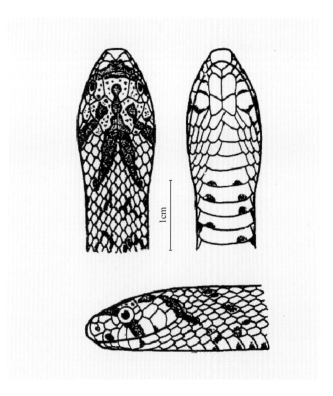

头部特征（引自《中国动物志 爬行纲 第三卷 有鳞目 蛇亚目》）

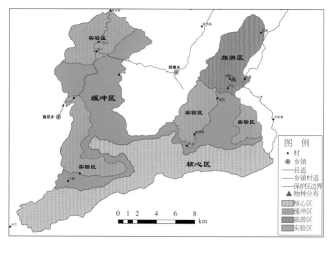

在茂兰地区的分布

（33）中国小头蛇 *Oligodon chinensis* (Günther, 1888)

Simotes chinensis Günther, 1888: 169. Ann. Mag. Nat. Hist., (6) 1: 165-172. Type locality: Mountain, north of Kiu Kiang [= Jiujiang], Changjiang River, Jiangxi, China.

Simotes longicauda Boulenger, 1903: 351. Ann. Mag. Nat. Hist., (7) 12: 350-354.

Holarchus chinensis Schmidt, 1927, Bull. Amer. Mus. Nat. Hist.: 537.

Oligodon chinensis Smith, 1943: 206. Taylor & Francis, London: 583.

Oligodon chinensis Green et al., 2010, Asian Herpetological Research, 1 (1): 1-21.

Oligodon chinensis Wallach et al., 2014: 495. Taylor & Francis, CRC Press: 1237.

[同物异名] *Simotes chinensis*；*Holarchus chinensis*；*Simotes longicauda*；*Holarchus violaceus longicauda*

[俗名] 秤杆蛇

[鉴别特征] 中段背鳞17行；头后和颈背有1箭形黑褐色斑，背面有约等距的黑褐色宽横纹14～20条。与紫棕小头蛇的区别为本种背有略似"人"字形斑。

[形态描述] 依据茂兰地区标本GZNU20160808002描述 体全长/尾长为555.3mm/98.2mm。头较小，与颈区分不明显；吻鳞从头背可见甚多；肛鳞完整。头背有略似"人"字形斑。体尾背面有粗大横斑10余个。

体尾背面褐色或灰褐色，有约等距排列的黑褐色粗横纹14～20条；每两条横纹之间有黑色细纹，犹如秤杆上的秤花，各横纹之间具有多数波状纤细的黑纹，有的个体背脊正中还有1条红黄色纵脊纹；腹面淡黄色，腹鳞有侧棱，棱处色白，整体呈白色纵纹。吻背有1略呈箭形的黑褐色斑，吻两外侧黑色斑经眼斜达第5、第6上唇鳞；头后及颈背有1黑褐色箭斑。颊鳞1；眶前鳞1，眶后鳞2；前颞鳞2，后颞鳞2；上唇鳞8（3-2-3式）；下唇鳞8，前4枚接前颏片；颏片2对，前对大于后对2倍以上。背鳞平滑，17-17-15；腹鳞188；肛鳞完整。

成体量度

编号	产地	头体长（mm）	体全长（mm）	尾长（mm）	背鳞（枚）	腹鳞（枚）	肛鳞（枚）	上唇鳞（枚）	下唇鳞（枚）	颊鳞（枚）	眶前鳞（枚）	眶后鳞（枚）	前颞鳞（枚）	后颞鳞（枚）
GZNU20160808002	荔波茂兰	457.1	555.3	98.2	17-17-15	188	1	8	8	1	1	2	2	2

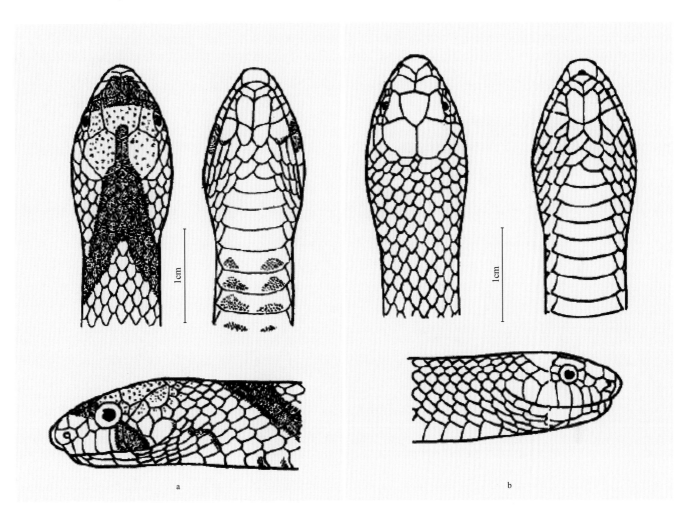

中国小头蛇（a）与紫棕小头蛇（b）头部特征比较（引自《中国动物志 爬行纲 第三卷 有鳞目 蛇亚目》）

a 背面照
b 腹面照

[**生活习性**] 栖息于平原或山区。嗜食爬行动物的卵。

[**地理分布**] 茂兰地区分布于永康、洞塘。国内分布于安徽、福建、广东、广西、贵州、海南、河南、湖南、江苏、江西、四川、云南、浙江等地。国外分布于越南。

[**模式标本**] 正模标本（BMNH 1946.1.3.28）保存于英国伦敦自然历史博物馆

[**模式产地**] 江西九江

[**保护级别**] 三有保护动物；贵州省重点保护野生动物

[**IUCN 濒危等级**] 无危（LC）

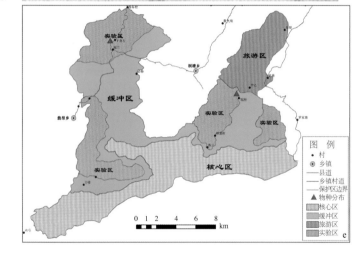

c 侧面照
d 整体照
e 在茂兰地区的分布

（34）紫棕小头蛇 *Oligodon cinereus* (Günther, 1864)

Simotes cinereus Günther, 1864, Taylor & Francis, London, XXVII: 452. Type locality: Gamboja, now Cambodia.

Simotes swinhonis Günther, 1864, Taylor & Francis, London, 17: 1-452.

Simotes multifasciatus Jan, 1865. Iconographie générale des ophidiens. 12. Livraison. Bailière et Fils, Paris.

Simotes semifasciatus Anderson, 1871, J. Asiat. Soc. Bengal, Calcutta, 40, part 11 (1): 12-39.

Simotes violaceus Boulenger, 1890. Reptilia and Batrachia. Taylor & Francis, London, 18: 1-541.

Holarchus violaceus swinhonis Mell, 1931, Lingnan Sci. Jour., Canton, 8 [1929]: 199-219.

Oligodon swinhonis tamdaoensis Hu et al., 1973, Acta Zool. Sinica., 19 (2): 149-181.

Oligodon cinereus Stuart et al., 2006, Raffles Bull. Zool., 54 (1): 129-155.

Oligodon cinereus Das, 2012. A Naturalist's Guide to the Snakes of South-East Asia: Malaysia, Singapore, Thailand, Myanmar, Borneo, Sumatra, Java and Bali. Oxford, John Beaufoy Publishing.

Oligodon cinereus Wallach et al., 2014: 495. Taylor & Francis, CRC Press: 1237.

[同物异名] *Simotes cinereus*；*Simotes swinhonis*；*Simotes multifasciatus*；*Simotes semifasciatus*；*Simotes violaceus*；*Holarchus dolleyanus*；*Holarchus violaceus swinhonis*；*Holarchus violaceus tamdaoensis*；*Oligodon swinhonis tamdaoensis*

[俗名] 棕色小头蛇、棕秤杆蛇

[鉴别特征] 体紫棕色，头后无斑纹，体背有黑褐色波浪状横纹；背鳞17-17-15。体色和花纹都与台湾小头蛇相似，但显著的区别是本种的头背、眼间及头后都无斑纹。

[形态描述] 依据茂兰地区标本GZNU320075描述 体全长/尾长为484.4mm/57.2mm，头较小，与颈区分不明显；吻鳞从头背可见甚多；体腹面紫棕色，背鳞有许多鳞片边缘黑色，形成多数约等距排列的黑褐色横纹（18+3）条，横纹之间夹有小的细横纹；头背面和颈背部无斑纹，腹面黄白色，并杂有黑褐色斑点，头部腹面及尾部斑点小。颊鳞1；眶前鳞2，眶后鳞2；颞鳞2+2；上唇鳞8(3-2-3式)；下唇鳞8，前3枚接前颏片；颏片2对。背鳞平滑，17-17-15；腹鳞161；肛鳞完整。

[生活习性] 多生活于山区。

[地理分布] 茂兰地区分布于洞塘。国内分布于福建、广东、广西、贵州、海南、香港、云南等地。国外分布于柬埔寨、老挝、缅甸、泰国、越南、印度（阿萨姆邦）。

[模式标本] 正模标本（BMNH RR 1946.1.1.25）保存于英国伦敦自然历史博物馆

[模式产地] 柬埔寨

[保护级别] 三有保护动物；贵州省重点保护野生动物

[IUCN濒危等级] 无危（LC）

成体量度

编号	产地	头体长(mm)	体全长(mm)	尾长(mm)	背鳞(枚)	腹鳞(枚)	肛鳞(枚)	上唇鳞(枚)	下唇鳞(枚)	颊鳞(枚)	眶前鳞(枚)	眶后鳞(枚)	前颞鳞(枚)	后颞鳞(枚)
GZNU320075	荔波茂兰	427.2	484.4	57.2	17-17-15	161	1	8	8	1	2	2	2	2

a 整体照
b 在茂兰地区的分布

25）三索蛇属 *Coelognathus* Fitzinger, 1843

Coelognathus Fitzinger, 1843. Vienna: Braumüller & Seidel.: 26. Type species: *Coluber radiatus* Boie, 1827. Type locality: Java.

本属物种有鼻间鳞及颊鳞，体背鳞中段 19 行，枕部有 1 黑横纹。其因眼后及眼下方有 3 条放射状黑线纹而得名。

本属现有 7 种，我国分布 1 种。茂兰地区有 1 种，即三索蛇 *Coelognathus radiatus*。

（35）三索蛇 *Coelognathus radiatus* (Boie, 1827)

Coluber radiatus Boie, 1827. Lieferung: Ophidier. Isis van Oken, 20: 508-566. Type locality: Java.

Coluber radiatus Schelgel, 1837, Phys. Serp. II. Type locality: Java.

Coluber quadrifasciatus Cantor, 1839, Proc. Zool. Soc., London: 31-34.

Tropidonotus quinque Cantor, 1839, Proc. Zool. Soc., London: 31-34.

Elaphis radiatus Duméril, 1853, Mém. Acad. Sci., Paris, 23: 399-536.

Plagiodon radiata Duméril, 1853, Mém. Acad. Sci., Paris, 23: 399-536.

Elaphis (*Compsosoma*) *radiatum* Bleeker, 1857. Natuurkundig Tijdschrift voor Nederlandsch Indie, Batavia, (3) 14 (4): 235-244.

Compsosoma radiatum Stoliczka, 1870, J. Asiat. Soc. Bengal, 39: 134-228.

Coluber radiatus Boulenger, 1894: 61. Catalogue of the Snakes in the British Museum (Natural History). Volume II, Containing the Conclusion of the Colubridæ Aglyphæ. British Mus. (Nat. Hist.), London, XI: 382.

Coluber (*Compsosoma*) *radiatus* Müller, 1895, Verh. Naturf. Ges. Basel., 10: 195-215 [1892].

Coluber radiatus Wall, 1908, J. Bombay Nat. Hist. Soc., 18: 312-337.

Elaphe radiata Barbour, 1912, Memoirs of the Museum of Comparative Zoölogy, 40 (4): 125-136.

Elaphe radiata Smith, 1943. Reptilia and Amphibia. 3 (Serpentes). Taylor & Francis, London: 583.

Elaphe radiata Schulz, 1986, Sauria, 8 (4): 3-6.

Elaphe radiata Manthey and Grossmann, 1997: 344. Natur und Tier Verlag (Münster): 512.

Elaphe radiata Cox et al., 1998: 51. Ralph Curtis Publishing: 144.

Elaphe radiata Lazell et al., 1999, Postilla, 217: 1-18.

Elaphe radiata Chan-ard et al., 1999: 166. Amphibians and reptiles of peninsular Malaysia and Thailand-an illustrated checklist. Bushmaster Publications, Würselen, Germany: 240.

Coelognathus radiatus Gumprecht, 2000, Sauria (Suppl.), 23 (3): 532-534.

Coelognathus radiatus Utiger et al., 2002, Russ. J. Herpetol., 9 (2): 105-124.

Elaphe radiata Ziegler, 2002: 231. Natur und Tier Verlag (Münster): 342.

Coelognathus radiatus Winchell, 2003, Reptilia (Münster), 8 (44): 20-29.

Coelognathus radiatus Gumprecht, 2003, Reptilia (Münster), 8 (44): 37-41.

Elaphe radiata Pauwels et al., 2003, Natural History Journal of Chulalongkorn University, 3 (1): 23-53.

Elaphe radiata Zhao, 2006. Hefei, China, Anhui Science & Technology Publ. House, Vol. I: 372, Vol. II (color plates): 280.

Coelognathus radiata Ziegler et al., 2007, Zootaxa, 1493: 1-40.

Murthy, 2010. The reptile fauna of India. B. R. Publishing, New Delhi: 332.

Coelognathus radiatus Wallach et al., 2014: 172. Taylor & Francis, CRC Press: 1237.

[同物异名] *Elaphe radiatus*; *Coluber radiates*; *Coluber quadrifasciatus*; *Tropidonotus quinque*; *Elaphis radiatus*; *Plagiodon radiata*; *Elaphis* (*Compsosoma*) *radiatum*; *Spilotes radiatus*; *Coluber* (*Compsosoma*) *radiatus*; *Compsosoma radiatum*; *Coluber radiatus*; *Elaphe radiatus*

[俗名] 白花锦蛇、白花蛇、三索线、三索线蛇、三索锦蛇、三索颌腔蛇

[鉴别特征] 大型无毒蛇类。头体背棕黄色，枕部有1黑横纹，眼后及眼下方有3条放射状黑线纹，故名三索。体背前段两侧各有3条黑色纵纹；背鳞中段19行。中央数行起棱；腹鳞超过200。

[形态描述] 体全长在1000mm以上。最大体全长/尾长：雄性1502mm/308mm，雌性1610mm/323mm。头略大，与颈明显区分，头、颈之间有1较细的黑色横线。眼大小适中，瞳孔圆形；躯尾修长适度。头体背棕黄色，头部两侧各有始于眼向后及眼下方的3条放射状黑线纹，顶鳞后有黑横纹；体背前段两侧各有3条黑色纵纹，靠近背中央的一条较粗，腹面淡棕色，具有淡灰色的斑。颊鳞1；眶前鳞1（或一侧为2），眶后鳞2（3），颞鳞2（3）+2（3），或1+2，2+1；上唇鳞9（4-2-3、3-3-3式）或8（3-2-3式），少数10（4-3-3、5-2-3式）；下唇鳞8～11，前4～6枚切前颏片；颏片2对。背鳞21-19-17，中央5～9行弱棱，其余平滑；腹鳞200～245；肛鳞完整；尾下鳞72～100对。上颌齿每侧22。

[生活习性] 栖息于平原、丘陵、山区河谷地带，也见于山坡草丛、甘蔗地、公路边。卵生。7月产卵5～12枚，卵呈长椭圆形。

[地理分布] 茂兰地区分布于永康。国内分布于福建、广东、广西、贵州、香港、云南等地。国外分布于印度、缅甸、马来西亚、印度尼西亚。

[模式标本] 正模标本（RMNH）保存于荷兰莱顿大学

[模式产地] 印度尼西亚爪哇（Boie，1827）

[保护级别] 三有保护动物；贵州省重点保护野生动物

[IUCN濒危等级] 无危（LC）

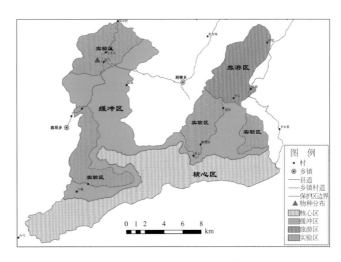

在茂兰地区的分布

26）翠青蛇属 *Cyclophiops* Boulenger, 1888

Cyclophiops Boulenger, 1888, Ann. Mus. Cis Stor. Nat. Genova, Ser. 2, 6: 7. Type species: *Cyclophiops aestivus* Boulenger, 1888, by monotypy.

中等大小无毒蛇类，体全长1m左右。体型修长适度，头、颈略可区分，尾长适中。通体绿色，或体后部有黄色短横斑；腹面黄白色。背鳞通体15行，平滑或有微棱；腹鳞宽大而无侧棱；尾下鳞双行。

本属已知6种，分布于南亚及东亚地区。我国有3种。茂兰地区有1种，即翠青蛇 *Cyclophiops major*。

（36）翠青蛇 *Cyclophiops major* (Günther, 1858)

Cyclophis major Günther, 1858: 120. Catalogue of Colubrine snakes of the British Museum. London, I-XVI: 281. Type locality: Ningpo, China.

Herpetodryas chloris Hallowell, 1861, John Rogers, U. S. N. Proc. Acad. Nat. Sci. Philadelphia, 12 [1860]: 480-510.

Cyclophis major Günther, 1864: 230. London (Taylor & Francis), 27: 452.

Cyclophis major Günther, 1888, Ann. Mag. Nat. Hist., (1) 6: 165-172.

Ablabes major Boettger, 1894: 140. Ber. Senckenberg. Naturf. Ges.: 129-152.

Ablabes major Boulenger, 1894: 279. British Mus. (Nat. Hist.), London, XI: 382.

Liopeltis major Stejneger, 1907: 338. Bull. U.S. Natl. Mus., 58 (20): 577.

Coluber delacouri Smith, 1930, Ann. Mag. Nat. Hist., (6) 10: 681-683.

Liopeltis major bicarinata Maki, 1931, Daiichi Shobo, Tokyo, (1) 7: 240.

Opheodrys major Smith, 1943: 178. Reptilia and Amphibia. 3 (Serpentes). Taylor & Francis, London: 583.

Cyclophiops major Ziegler et al., 2006, Herpetologica Bonnensis II: 247-262.

Cyclophiops major Wallach et al., 2014: 202. Taylor & Francis, CRC Press: 1237.

[同物异名] *Herpetodryas chloris*；*Cyclophis major*；*Ablabes major*；*Entechinus major*；*Liopeltis major*；*Coluber delacouri*；*Liopeltis major bicarinata*；*Eurypholis major*；*Opheodrys major*

[俗名] 青竹标、青蛇、小青

[鉴别特征] 体背面草绿色，腹面黄绿色；头背有对称大鳞，背鳞通体15行，肛鳞2。

[形态描述] 依据茂兰地区标本描述 体全长/尾长为820.3mm/232.1mm，头略大，与颈区分明显；头背大鳞对称分布，眼大，瞳孔圆形；躯尾修长适度。前额鳞大于鼻间鳞，额鳞长大于宽，顶鳞大。体背面草绿色，上唇下半、下唇、颔部及躯尾腹面黄绿色，下颌边缘及颔沟有绿色斑

成体量度

编号	产地	头体长（mm）	体全长（mm）	尾长（mm）	背鳞（枚）	腹鳞（枚）	上唇鳞（枚）	下唇鳞（枚）	颊鳞（枚）	眶前鳞（枚）	眶后鳞（枚）	前颞鳞（枚）	后颞鳞（枚）
GZNU20170501002	荔波茂兰	588.2	820.3	232.1	15-15-15	171	8	6	1	1	2	1	2

a 背面照
b 腹面照

点。颊鳞1；眶前鳞1；眶后鳞2；颞鳞1+2；上唇鳞8（3-2-3），下唇鳞6，前4枚切前颔片，第6枚最长；颔片2对，前大于后。背鳞平滑，通体15行，腹鳞171；肛鳞二分。

[生活习性] 栖息环境多样，包括平原、丘陵、山区。多在农耕区的地面，或攀附在作物、竹木上，或藏于石下。路边、溪畔、河岸、草丛、住宅附近都有发现。茂兰地区发现于白鹇坡，缠绕于一灌木上。

[地理分布] 茂兰地区分布于翁昂、洞塘。国内分布于陕西、安徽、重庆、福建、四川、台湾、甘肃、广东、广西、贵州、海南、河南、湖北、湖南、江苏、江西、香港、浙江等地。国外分布于越南（北部）。

[模式标本] 正模标本（NSM H02569）保存于英国伦敦自然历史博物馆

[模式产地] 宁波

[保护级别] 三有保护动物；贵州省重点保护野生动物

[IUCN 濒危等级] 无危（LC）

c 侧面照
d 整体照
e 在茂兰地区的分布

27）鼠蛇属 *Ptyas* Fitzinger, 1843

Ptyas Fitzinger, 1843, Syst. Rept., Vienna, 1: 26. Type species: *Caluber blumenbachii* Merrem, 1820 (= *Coluber mucosus* Linnaeus, 1758).

本属物种为大型陆栖无毒蛇类。体长而呈圆柱形，体尾修长，头长，头、颈可以区分。主要识别特征是：体长1～2m；眼圆而较大，瞳孔圆形；颊部略内陷，有1眼前下鳞，颊鳞每侧2～4（颊鳞1枚以上是鼠蛇属蛇类的典型特征）；腹鳞与尾下鳞均具侧棱，上颌齿10～28，连续排列，向后逐渐增大；具有缠绕性。

本属共 8 种，广泛分布于中国、西亚及东南亚地区。我国分布 5 种。茂兰地区有滑鼠蛇 Ptyas mucosa、灰鼠蛇 Ptyas korros 和黑线乌梢蛇 Ptyas nigromarginata 3 种。

茂兰地区种检索表

1. 背鳞通常为偶数（16 或 14）；颊鳞 1 枚 ·· 黑线乌梢蛇 Ptyas nigromarginata
   背鳞行数通常为奇数；颊鳞 1 枚以上 ·· 2
2. 背鳞 15-15（13，14）-11，腹鳞 185 以下；背有深浅相间的纵纹；腹部浅黄色 ·············· 灰鼠蛇 Ptyas korros
   背鳞 19（18～21）-17（16，18）-14（11，13），腹鳞 185 以上；背有黑色网状斑纹；腹部黄白色 ·············· 滑鼠蛇 Ptyas mucosa

（37）滑鼠蛇 *Ptyas mucosa* (Linnaeus, 1758)

Coluber mucosus Linnaeus, 1758: 226. Tomus I. Editio decima, reformata. Laurentii Salvii, Holmiæ. 10th Edition: 824. Type locality: India.

Coluber blumenbachii Merrem, 1820: 119. Versuch eines Systems der Amphibien I (Tentamen Systematis Amphibiorum). Kriegeri, Marburg: 191.

Coluber dhumna Cantor, 1839, Proc. Zool. Soc., London: 31-34.

Leptophis trifrenatus Hallowell, 1861: 503. John Rogers, U. S. N. Proc. Acad. Nat. Sci. Philadelphia, 12 [1860]: 480-510.

Ptyas mucosus Cope, 1861, Proc. Acad. Nat. Sci. Philadelphia, 12 [1860]: 553-566.

Ptyas mucosus Günther, 1864: 249. The Reptiles of British India. London (Taylor & Francis).

Zamenis mucosus Boulenger, 1893: 385. London (Taylor & Francis): 448.

Ptyas mucosus Stejneger, 1907: 345. Bull. U.S. Natl. Mus., 58: XX: 577.

Zaocys mucosus Wall, 1921: 172. Colombo Mus. (Cottle, govt. printer), Colombo, XXII: 581.

Ptyas mucosus Smith, 1943: 159. Reptilia and Amphibia. 3 (Serpentes). Taylor & Francis, London: 583.

Ptyas mucosus Manthey and Grossmann, 1997: 386. Natur und Tier Verlag (Münster): 512.

Coluber mucosus Lazell, 1998, Herpetological Review, 29 (3): 134.

Ptyas mucosus Cox et al., 1998: 54. Ralph Curtis Publishing: 144.

Ptyas mucosus Pinou and Dowling, 2000, Herpetological Review, 31 (3): 136-138.

Ptyas mucosa David and Das, 2004, Hamadryad, 28 (1-2): 113-116.

Ptyas mucosus Sharma, 2004. Akhil Books, New Delhi: 292.

Ptyas mucosus Zhao, 2006. The snakes of China. Hefei, China, Anhui Science & Technology Publ. House, Vol. I: 372.

Ptyas mucosa Wallach et al., 2014: 617. Taylor & Francis, CRC Press: 1237.

[同物异名] *Coluber mucosus*；*Natrix mucosa*；*Coluber blumenbachii*；*Coluber dhumna*；*Ptyas blumenbachii*；*Coryphodon blumenbachii*；*Leptophis trifrenatus*；*Zamenis mucosus*；*Zaocys mucosus*；*Ptyas mucosus*

[俗名] 乌肉蛇、草锦蛇、长标蛇、山蛇

[鉴别特征] 大型无毒蛇。头背黑褐色，体背棕灰色，体后有不规则的黑色横斑，横斑至尾部形成网纹。腹面黄白色，腹鳞后缘黑色。颊鳞一般 3 枚，与本属相近种灰鼠蛇的区别是：本种背鳞一般 19（18～21）-17（16，18）-14（11，13），腹鳞一般 185 枚以上。

[形态描述] 体长而粗大，一般在 1500mm 以上，甚至到 2000mm 左右，最大体全长 / 尾长：雄性 2051mm/541mm，雌性 1894mm/447mm。头较长，鼻孔大，开口于长大鼻鳞的后上方，上切鼻间鳞，下切第一上唇鳞；眼大而圆，瞳孔圆形；前额鳞弯向头侧。头背黑褐色，上唇鳞浅灰色，后缘黑色有粗大黑斑，前 5 枚的黑斑贯穿上下唇鳞。体背棕灰色，

体后部由于鳞片的边缘或半枚鳞片为黑色而形成不规则的黑色横斑，横斑至尾部呈网纹。腹面黄白色，腹鳞后缘黑色，身体前段、后段及尾部的腹鳞黑色，后缘更为明显；颊部略内凹。颊鳞3，少数2；眶前鳞1，眶前下鳞1，眶后鳞2；颞鳞2+2（3）；上唇鳞8（3-2-3式）为主，个别9（4-2-3式）；下唇鳞9或10，第1对在颏鳞后相接，前5（偶为4或6）枚接前颔片，个别8（4），第5及第6枚最大；颔片2对，前对略小于后对，后对后半隔以并列的2枚小鳞。背鳞平滑，19（18～21）-17（16，18）-14（11，13）；腹鳞170～198，略具侧棱；肛鳞二分；尾下鳞104～118对。上颌颌齿每侧21～22枚。

[生活习性] 平原、丘陵和山区都有发现。白天多在水域附近活动。

[地理分布] 茂兰地区分布于洞塘。国内分布于安徽、澳门、香港、云南、浙江、江西、福建、广东、广西、贵州、海南、湖北、湖南、台湾、四川、西藏等地。国外分布于印度、斯里兰卡、中南半岛及印度尼西亚等地。

[模式标本] 正模标本 [NRM（NHRM）] 保存于瑞典国家自然历史博物馆（斯德哥尔摩）；正模标本（USNM 7510）保存于美国自然历史博物馆

[模式产地] 印度

[保护级别] 三有保护动物；贵州省重点保护野生动物

[IUCN 濒危等级] 濒危（EN）

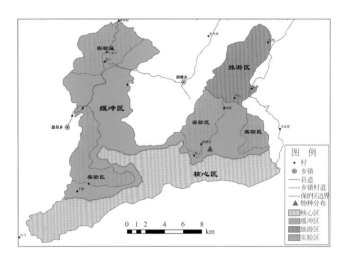

在茂兰地区的分布

（38）灰鼠蛇 *Ptyas korros* (Schlegel, 1837)

Coluber korros Schlegel, 1837: 139. Essai sur la physionomie des serpens. Partie Descriptive. La Haye (Kips and van Stockum), 606 S.
    Type locality: Java.

Ptyas korros Cope, 1861, Proc. Acad. Nat. Sci. Philadelphia, 12 [1860]: 553-566.

Zamenis korros Boulenger, 1890: 324. Reptilia and Batrachia. Taylor & Francis, London, 18: 541.

Zamenis korros Boulenger, 1893: 384. London (Taylor & Francis): 448.

Ptyas korros Stejneger, 1907: 348. Bull. U.S. Natl. Mus., 58 (20): 1-577.

Zamenis korros Wall, 1908: 326. J. Bombay Nat. Hist. Soc., 18: 312-337.

Liopeltis libertatis Barbour, 1910, Proc. Biol. Soc., Washington, 23: 169-170.

Ptyas korros Smith, 1943: 162. Reptilia and Amphibia. 3 (Serpentes). Taylor & Francis, London: 583.

Ptyas korros Manthey and Grossmann, 1997: 385. Amphibien & Reptilien Südostasiens. Natur und Tier Verlag (Münster): 512.

Coluber korros Lazell, 1998, Herpetological Review, 29 (3): 134.

Ptyas korros Cox et al., 1998: 54. Ralph Curtis Publishing: 144.

Ptyas korros Ziegler, 2002: 251. Natur und Tier Verlag (Münster): 342.

Ptyas korros Wallach et al., 2014: 616. Taylor & Francis, CRC Press: 1237.

[同物异名] *Coluber korros*；*Coryphodon korros*；*Liopeltis libertatis*；*Ptyas korros chinensis*；*Zamenis korros*

[俗名] 黄梢蛇、索蛇、过树龙

[鉴别特征] 大型无毒蛇。头及体背灰黑色，有深浅相间的纵纹，唇缘及腹部浅黄色；背鳞一般15-15（13）-11，腹鳞185枚以下，与本属相近种滑鼠蛇的区别是本种背鳞15-15（13，14）-11。

[形态描述] 体长而粗大，一般在1m以上。头较长，

吻鳞高，从吻背可以看到；鼻孔大，位于鼻鳞中央，其上下缘几乎都近鼻鳞边缘。前额鳞弯向头侧。眼大，瞳孔圆形。躯尾修长。色斑背面由于每一背鳞的中间色深，游离缘略黑，而两侧角色略白，前后缀连整体上形成深浅色相间的若干纵纹；腹面除腹鳞两外侧色稍深外，其余均白色无斑。头及体背灰黑色，头背棕褐鳞被，头腹及颌部浅黄色。

颊鳞2～4，个别一侧或两侧只有1枚颊鳞；眶前鳞1，有一较小的眶前下鳞；眶后鳞2，有一号标本的右侧为1，有一号标本的右侧为3；颞鳞2+2，海南有一号一侧的上唇鳞部分愈合，尚有部分鳞沟存在，该号标本的右侧后颞鳞为1枚；上唇鳞8（3-2-3式）；下唇鳞10，第1对在颏鳞后相接，前5枚接前颏片，第5及第6枚最大，海南一号标本两侧的第8、第9枚下唇鳞均愈合为一枚较大的鳞片，该鳞片中部上下缘还可看出极浅的凹痕（前后两枚鳞之间鳞沟的痕迹）；颏片2对，后对大于前对，后对的后部被1～3枚小鳞分隔，背鳞平滑，15-15（13）-11；腹鳞157～183；肛鳞二分；尾下鳞102～142对。上颌齿每侧24～27枚。半阴茎不分叉，圆柱形。

[生活习性] 栖息环境为平原、丘陵和山区，也见于灌丛、杂草地、路边、各种水域附近。多于白昼活动，栖于灌木上。卵生。

[地理分布] 茂兰地区分布于洞塘。国内分布于安徽、澳门、福建、广东、广西、贵州、海南、湖南、江西、台湾、香港、云南、浙江等地。国外分布于印度（阿萨姆邦）、缅甸、泰国、马来西亚、印度尼西亚。

[模式标本] 模式标本（USNM 42932）保存于美国国家自然历史博物馆

[模式产地] 印度尼西亚爪哇

[保护级别] 三有保护动物；贵州省重点保护野生动物

[IUCN濒危等级] 易危（VU）

[讨论] 赵尔宓（2006）指出颊鳞1枚以上是鼠蛇属的典型特征，但已发现海南有一定数量的灰鼠蛇标本也只有1枚颊鳞，因此需注意检查其他特征是否符合，不能仅依据颊鳞数目作出鉴定。

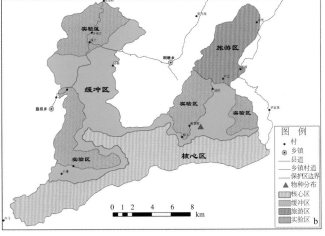

a 整体照（祝非摄）
b 在茂兰地区的分布

（39）黑线乌梢蛇 *Ptyas nigromarginata* (Blyth, 1854)

Coluber nigromarginatus Blyth, 1854, J. Asiatic Soc. Bengal, Calcutta, 23 (3): 287-302. Type locality: Vicinity of Darjeeling, India.

Zaocys nigromarginatus Boulenger, 1893: 376. Catalogue of the snakes in the British Museum (Nat. Hist.) I. London (Taylor & Francis): 448.

Zaocys nigromarginatus Wall, 1908: 325. J. Bombay Nat. Hist. Soc., 18: 312-337.

Zaocys nigromarginatus Smith, 1943: 165. Taylor & Francis, London: 583.

Zaocys nigromarginatus Swan and Leviton, 1962: 116. Proc. Cal. Acad. Sci., 32 (6): 103-147.
Ptyas nigromarginata David and Das, 2004, Hamadryad, 28 (1 & 2): 113-116.
Zaocys nigromarginatus Sharma, 2004. Handbook Indian Snakes. Akhil Books, New Delhi: 292.
Ptyas nigromarginata Whitaker and Captain, 2004. Snakes of India. Draco Books: 500.
Ptyas nigromarginata Pyron and Burbrink, 2013, Ecology Letters, 17 (1): 13-21.
Ptyas nigromarginata Wallach et al., 2014: 617. Taylor & Francis, CRC Press: 1237.

[同物异名] *Zaocys nigromarginatus*；*Coluber nigromarginatus*；*Ptyas nigromarginatus*

[俗名] 乌梢蛇

[鉴别特征] 颊鳞1；背中央4～6行鳞起棱，背面绿色或黄绿色。与相近种乌梢蛇的区别是，背脊两侧2黑纵纹在体前部时不显，在后部清晰可见，腹鳞两侧有1黑纵纹。

[形态描述] 依据茂兰地区标本GZNU320082描述 体全长/尾长为890.2mm/245.1mm。通体背面绿色或黄绿色，腹面浅黄绿色有2黑纵纹。头背黄绿色无斑，上唇缘及口角黄色；头腹面白色。颊鳞1；眶前鳞1，眶后鳞+眶下鳞3；上唇鳞以8（3-2-3式）；下唇鳞9，前4枚接前颏片；颏片2对，前对短于后对；前对左右相接，后对后1/2或1/3为小鳞相隔；背鳞17-17-14。

[生活习性] 多栖于丘陵或山区，尤其是林木繁盛茂密的地方。白天到耕作地或其附近活动觅食，有攀爬上树的习性。卵生。

[地理分布] 茂兰地区分布于永康。国内分布于贵州、四川（西南部）、西藏、云南等地。国外分布于尼泊尔、印度（阿萨姆邦、大吉岭、锡金）、缅甸（北部）、越南。

[模式标本] 正模标本（ZSI 7343）保存于爱尔兰动物学协会

[模式产地] 印度大吉岭附近

[保护级别] 三有保护动物；贵州省重点保护野生动物

[IUCN 濒危等级] 易危（VU）

成体量度

编号	产地	头体长（mm）	体全长（mm）	尾长（mm）	背鳞（枚）	腹鳞（枚）	肛鳞（枚）	上唇鳞（枚）	下唇鳞（枚）	颊鳞（枚）	眶前鳞（枚）	眶后鳞+眶下鳞（枚）	前颞鳞（枚）	后颞鳞（枚）
GZNU320082	荔波茂兰	645.1	890.2	245.1	17-17-14	171	2	8	9	1	1	3	2	3

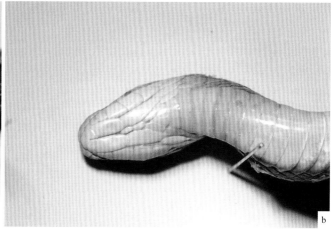

a 背面照
b 腹面照

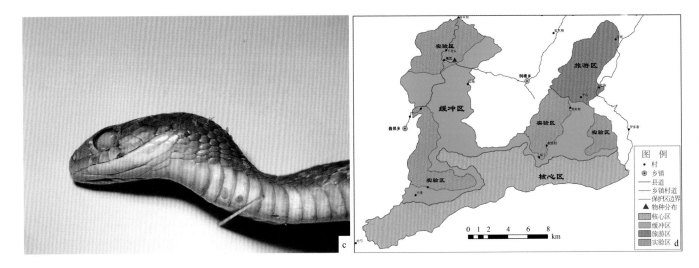

c 侧面照
d 在茂兰地区的分布

28）绿蛇属 *Rhadinophis* Vogt, 1922

Rhadinophis Vogt, 1922, Archiv. Naturg. 88, Abt. A. Heft, 10: 140. Type species: *Rhadinophis melli* Vogt, 1922. Type locality: India, Khasi Hills.

本种体型长，圆柱形；头较长，头、颈区分明显；体背面翠绿色，肛前鳞不超过15行，上颌齿14～24，大小几乎相等。卵生。

本属我国有2种，贵州有2种（绿蛇 *Rhynchophis prasina* 和灰腹绿蛇 *Rhynchophis frenatum*）。茂兰地区有1种，即灰腹绿蛇 *Rhadinophis frenatum*。

（40）灰腹绿蛇 *Rhadinophis frenatum* (Gray, 1853)

Herpetrodryas [sic] *frenatus* Gray, 1853, Ann. Mag. Nat. Hist., (2) 12: 386-392. Type locality: Khasi Hills, India.

Coluber frenatus Boulenger, 1894: 58. British Mus. (Nat. Hist.), London, 11: 382.

Rhadinophis melli Vogt, 1922, Archiv. Naturg. 88, Abt. A. Heft, 10: 140.

Gonyosoma caldwelli Schmidt, 1925, Amer. Mus. Novitates, No. 157: 4.

Coluber caldwelli Werner, 1929, Zool. Jahrb. Syst., 57: 90.

Gonyosoma melli Mell, 1931, Lingnan Sci. Jour., Canton, 8 [1929]: 199-219.

Elaphe frenata Pope, 1935, Rept. China.: 244.

Elaphe frenata Sdmith, 1943: 144. Taylor & Francis, London: 583.

Elaphe frenata Gumprecht, 2003, Reptilia (Münster), 8 (44): 37-41.

Elaphe frenata Mao et al., 2003, Coll. and Res., 16: 1-6.

Elaphe frenata Sharma, 2004. Akhil Books, New Delhi: 292.

Gonyosoma frenatum Utiger et al., 2005, Russ. J. Herpetol., 12 (1): 39-60.

Elaphe frenata Zhao, 2006. The snakes of China. Hefei, China, Anhui Sience & Technology Publ. House, Vol. I: 372.

Rhadinophis frenata Burbrink and Lawson, 2007, Molecular Phylogenetics and Evolution, 43 (1): 173-189.
Gonyosoma frenatum Sang et al., 2009, Chimaira, Frankfurt: 768.
Rhadinophis frenatum Pyron and Burbrink, 2013, Ecology Letters, 17 (1): 13-21.
Rhynchophis frenatus Wallach et al., 2014: 640. Taylor & Francis, CRC Press: 1237.
Gonyosoma frenatum Chen et al., 2014, Zootaxa, 3881 (6): 532-548.

[同物异名] *Herpetrodryas* [sic] *frenatus*；*Coluber frenatus*；*Elaphe frenata*；*Rhadinophis melli*；*Gonyosoma caldwelli*；*Gonyosoma melli*；*Gonyosoma frenatum*；*Rhadinophis frenata*；*Rhynchophis frenatus*；*Coluber caldwelli*

[俗名] 灰腹绿树锦蛇、灰腹绿瘦锦蛇

[鉴别特征] 全身翠绿色。吻部较长，无颊鳞，眼较大，瞳孔圆形，眼前后有1条黑色带纹，尾细而长。生活于山区森林中，喜树栖。

[形态描述] 依据茂兰地区标本GZNU20040716001描述 体细长，尾更细长。体全长为849.2mm。头、颈区分明显，吻部较长，全身背面翠绿色；腹面淡黄色，腹鳞两侧有侧棱，一直延伸到尾端。眼较大，瞳孔圆形，其侧有1黑纵线始自鼻孔，经眼到颌角，此黑线以下的上唇浅黄色。头腹浅黄白色。上下唇鳞及咽部呈灰白色。无颊鳞，前额鳞向头侧延伸并与上唇鳞相接；眶前鳞1，眶后鳞2；颞鳞2+3；上唇鳞9（3-3-3式），下唇鳞

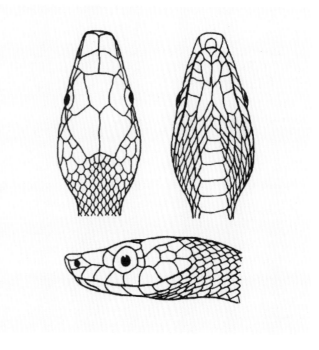

头部特征（引自《中国动物志 爬行纲 第三卷 有鳞目 蛇亚目》）

成体量度

编号	产地	体全长（mm）	头体长（mm）	尾长（mm）	背鳞（枚）	腹鳞（枚）	肛鳞（枚）	上唇鳞（枚）	下唇鳞（枚）	颊鳞（枚）	眶前鳞（枚）	眶后鳞（枚）	前颞鳞（枚）	后颞鳞（枚）
GZNU20040716001	荔波茂兰	849.2	597	252.2	19-19-15	215	2	9	11	0	1	2	2	3

a 整体照
b 在茂兰地区的分布

11，前3枚接前颏片；颏片2对，后对长于前对。背鳞19-19-15，微弱起棱，外侧1行平滑；腹鳞215；肛鳞二分。

[生活习性] 生活于山地、丘陵。常见于树木茂盛的林中，具攀援上树的能力，也见于山坡梯田田埂灌木上及溪畔灌木上。

[地理分布] 茂兰地区分布于翁昂。国内分布于贵州、安徽、福建、广东、广西、河南、湖南、四川、浙江等地。国外分布于印度、越南。

[模式标本] 正模标本（ZMB 27662 [*Rhadinophis melli*]）保存于德国柏林洪堡大学自然历史博物馆

[模式产地] 印度阿萨姆邦（Gray，1853）

[保护级别] 三有保护动物；贵州省重点保护野生动物

[IUCN濒危等级] 无危（LC）

29）链蛇属 *Lycodon* Boie, 1826

Lycodon Boie, 1826, Bull. Sci. Nat., 9: 238. Type species: *Lycodon aulicus* Linnaeus, 1758.

Dinadon Duméril, 1853, Mem. Acad. Sci., Paris, 23: 463. Type species: *Dinodon canccellcatum* Duméril (= *Lycodon rufozonatus* Cantor, 1842), by monotypy.

Lycodon Siler et al., 2013, Zoologica Scripta, 42: 262-277.

体型中等，全长1m左右。体型细长，头、颈略能区分；头较宽扁；眼较小，瞳孔直立椭圆形。头略大于颈部，吻端稍宽而略扁。通体背面黑褐色，有红色、黄色、粉红色或白色窄横纹；腹面黄白色，腹鳞略有侧棱。上颌齿分为3个齿群，具2个齿间隙，每侧上颌齿被两个齿间隙分隔为前、中、后3组，后组有3枚较大而无沟的普通牙齿。本属均为无毒蛇。

本属目前有50种左右，分布于印度（锡金）、越南、老挝、泰国、缅甸、马来西亚、新加坡、印度尼西亚、中国、日本、朝鲜、俄罗斯等地。我国有15种，茂兰地区分布4种，分种检索表。

茂兰地区种检索表

1. 全身有黑白相间的环纹，躯干部环纹20～46条 ················· 黑背链蛇 *Lycodon ruhstrati*
   体背无黑白环纹，具有粉红色、白色或黄色窄横斑 ················· 2
2. 体背具有60个以上白色的窄横斑 ················· 北链蛇 *Lycodon septentrionalis*
   体背具有60个以下粉红色或黄色的窄横斑 ················· 3
3. 背鳞平滑，或仅在体后端中央少数几行微起棱；具红色窄横纹 ················· 赤链蛇 *Lycodon rufozonatum*
   背鳞中央几行起棱；具黄色窄横纹 ················· 黄链蛇 *Lycodon flavozonatum*

（41）赤链蛇 *Lycodon rufozonatum* Cantor, 1842

Lycodon rufozonatus Cantor, 1842, Ann. Mag. Nat. Hist., (1) 9: 265-278. Type locality: Chusan, Zhejiang, China.

Dinodon cancellatum Duméril and Bibron, 1854, Paris, 7 (1): 447.

Coronella striata Hallowell, 1857, Proc. Acad. Nat. Sci. Philadelphia, 8 (4): 146-153.

Eumesodon striatus Cope, 1860, Part II. Proc. Acad. Nat. Sci. Philadelphia, 12: 241-266.

Lycodon rufozonatus Günther, 1864: 319. London (Taylor & Francis), 27: 452.

Dinodon rufozonatus Peters, 1881, Naturf. Freunde Berlin, (6): 87-91.

Dinodon rufozonatus var. *formosona* Boettger, 1885, Berichte Offenb. Ver. Naturk., 24/25: 115-170.

Lycodon rufozonatus Günther, 1888, Ann. Mag. Nat. Hist., (6) 1: 165-172.

Dinodon rufozonatus Boulenger, 1893: 361. London (Taylor & Francis): 448.

Dinodon rufozonatum Stejneger, 1907: 358. Bull. U.S. Natl. Mus., 58 (20): 577.

Lycodon rufozonatus walli Stejneger, 1907: 364. Bull. U.S. Natl. Mus., 58 (20): 577.

Dinodon rufozonatum walli Stejneger, 1907: 364. Bull. U.S. Natl. Mus., 58 (20): 577.

Dinodon rufozonatum williamsi Mell, 1931, Lingnan Sci. Jour., Canton, 8 [1929]: 199-219.

Dinodon rufozonatum yunnanense Mell, 1931, Lingnan Sci. Jour., Canton, 8 [1929]: 199-219.

Dinodon cf. *rufozonatum* Ziegler, 2002: 228. Die Natur und Tier Verlag (Münster): 342.

Dinodon rufozonatum Szczerbak, 2003. Krieger, Malabar, FL: 260.

Dinodon rufozonatum meridionale Orlov, 2004, Russ. J. Herpetol., 11 (3): 181-197.

Dinodon rufozonatus Goris and Maeda, 2004: 222. Krieger, Malabar: 285.

Lycodon rufozonatus Siler et al., 2013, Zoologica Scripta, 42: 262-277.

Dinodon rufozonatum Wallach et al., 2014: 236. Taylor & Francis, CRC Press: 1237.

[同物异名] *Dinodon rufozonatus*；*Dinodon cancellatum*；*Coronella striata*；*Dinodon rufozonatum williams*；*Eumesodon striatus*；*Dinodon rufozonatum yunnanense*；*Dinodon* cf. *rufozonatum*；*Lycodon rufozonatus walli*；*Dinodon rufozonatus* var. *formosona*；*Lycodon rufozonatus*；*Dinodon rufozonatum*；*Dinodon rufozonatum meridionale*

[俗名] 火赤链、红斑蛇、红百节蛇、红麻子

[鉴别特征] 体型中等偏大的无毒蛇。背鳞平滑无棱，体背黑褐色，有多个红色窄横斑；颊鳞常入眶。

[形态描述] 依据茂兰地区标本 GZNU20160727001 描述 体全长 1118.5mm 左右。头宽扁，头、颈略能区分；眼小，瞳孔直立椭圆形；背鳞平滑无棱，体背黑褐色，有多个红色窄横斑，头颈部有一"Λ"形红色斑，前端达顶鳞后部，后端延伸至颈侧；眼后自背部眶后鳞向后，有 1 红色线纹斜向口角。颊鳞 1，窄长，入眶；眶前鳞 2，眶后鳞 2；上唇鳞 8（3-2-3 式）；下唇鳞 10，前 5 枚切前颔片；颔片 2 对，约等长。背鳞 18-17-16；腹鳞 216；肛鳞完整。

[生活习性] 生活于平原、丘陵、山区的田野和村舍附近。多于傍晚或夜间活动觅食。在调查过程中曾见其捕食花臭蛙。

[地理分布] 茂兰地区分布于洞塘。国内分布于安徽、北京、重庆、福建、甘肃、广东、广西、贵州、海南、河北、河南、黑龙江、湖北、湖南、江苏、江西、吉林、辽宁、山东、山西、陕西、上海、四川、台湾、天津、云南等地。国外分布于俄罗斯（南部）、朝鲜、日本。

[模式标本] 正模标本（BMNH）保存于英国自然历史博物馆（伦敦），正模标本（USNM 34007 [*walli*]）保存于美国国家自然历史博物馆（华盛顿），正模标本（AMNH [*williamsi*]）保存于美国自然历史博物馆（纽约）

[模式产地] 浙江

[保护级别] 三有保护动物；贵州省重点保护野生动物

[IUCN 濒危等级] 无危（LC）

成体量度

编号	产地	头体长（mm）	体全长（mm）	尾长（mm）	背鳞（枚）	腹鳞（枚）	肛鳞（枚）	上唇鳞（枚）	下唇鳞（枚）	颊鳞（枚）	眶前鳞（枚）	眶后鳞（枚）	前颞鳞（枚）	后颞鳞（枚）
GZNU20160727001	荔波茂兰	848.5	1118.5	270	18-17-16	216	1	8	10	1	2	2	2	3

a 背面照
b 腹面照
c 侧面照
d 整体照
e 在茂兰地区的分布

（42）北链蛇 *Lycodon septentrionalis* (Günther, 1875)

Ophites septentrionalis Günther, 1875, Proc. Zool. Soc., London: 224-234. Type locality: Himalayas or Khasi Hills, North India.

Ophites septentrionalis Günther, 1888, Ann. Mag. Nat. Hist., (6) 1: 165-172.

Lycodon septentrionalis Boulenger, 1890. Reptilia and Batrachia. Taylor & Francis, London, 18: 541.

Dinodon septentrionalis Boulenger, 1893: 363. London (Taylor & Francis): 448.

Dinodon septentrionale chapaense Angel and Bourret, 1933, Bull. Soc. Zool. Fr., 58: 129-140.

Dinodon septentrionalis Smith, 1943: 270. Reptilia and Amphibia. 3 (Serpentes). Taylor & Francis, London: 583.

Dinodon septentrionale var. *chapaense* Deuve, 1961, Bull. Soc. Sci. Nat. Laos, 1: 5-32.
Dinodon septentrionalis Das, 1996: 56. Hamadryad, 37 (1-2): 127-131.
Dinodon septentrionale David et al., 2004, The Natural History Journal of Chulalongkorn University, 4 (1): 47-80.
Dinodon septentrionalis Stuart et al., 2006, Raffles Bull. Zool., 54 (1): 129-155.
Dinodon septentrionalis Sang et al., 2009, Chimaira, Frankfurt: 768.
Lycodon septentrionalis Siler et al., 2013, Zoologica Scripta, 42: 262-277.
Dinodon septentrionalis Wallach et al., 2014: 237. Taylor & Francis, CRC Press: 1237.
Dinodon septentrionalis Chan-ard et al., 2015: 200. A field guide to the reptiles of Thailand. Oxford University Press, N. Y.: 352.

[同物异名] *Dinodon septentrionalis*；*Ophites septentrionalis*；*Dinodon septentrionale*；*Dinodon septentrionale* var. *chapaense*；*Dinodon septentrionale chapaense*

[俗名] 白链蛇

[鉴别特征] 体型中等偏大的无毒蛇。体中段背鳞17-17-15，平滑无棱；背面有（29～36）+（10～27）个约占2枚鳞宽的白色横斑，横斑间距在体前段约为体后段的2倍。

[形态描述] 依据茂兰地区标本GZNU20160805010描述 雄性体全长/尾长为1245mm/173mm。头略大，吻端宽扁，与颈可以区分。眼小，瞳孔直立椭圆形；躯尾较长。头及体背黑褐色；颈部无白色倒"V"形斑纹；有约等距排列的白色横斑。色斑背面黑褐色，有约等距排列的白色横斑（29～36）+（10～27）个，横斑宽占2枚鳞长，在体侧D_5或D_6处分叉达腹鳞，横斑间距在体前部约为体后部的2倍；腹面黄白色。颊鳞1，窄长而不入眶；眶前鳞1，眶后鳞2；颞鳞2+5；上唇鳞8（2-3-3式或3-2-3式）；下唇鳞8，前5（或4）枚切前颏片。背鳞平滑，17-17-15；腹鳞211；肛鳞完整；尾下鳞双行，50～87对。上颌齿每侧13枚，由两个齿间隙分为3组，7+3+3，最后3枚最大。

[生活习性] 栖息于热带或亚热带山区阔叶林边缘的草丛或水域附近。傍晚或夜间活动。卵生。

[地理分布] 茂兰地区分布于翁昂等地。国内分布于贵州、云南。国外分布于印度、缅甸、泰国、老挝、越南。

[模式标本] 正模标本（BMNH 1946.1.14.96）保存于英国伦敦自然历史博物馆

[模式产地] 印度北部

[保护级别] 三有保护动物；贵州省重点保护野生动物

[IUCN 濒危等级] 无危（LC）

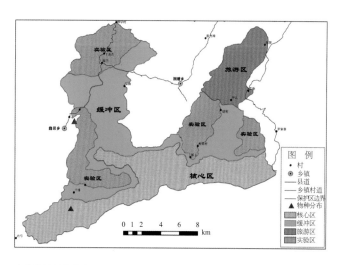

在茂兰地区的分布

成体量度

编号	产地	头体长（mm）	体全长（mm）	尾长（mm）	背鳞（枚）	腹鳞（枚）	肛鳞（枚）	上唇鳞（枚）	下唇鳞（枚）	颊鳞（枚）	眶前鳞（枚）	眶后鳞（枚）	前颞鳞（枚）	后颞鳞（枚）
GZNU20160805010	荔波茂兰	1072	1245	173	17-17-15	211	1	8	8	1	1	2	2	5

（43）黑背链蛇 *Lycodon ruhstrati* (Fischer, 1886)

Ophites ruhstrati Fischer, 1886, Abh. Geb. Naturw. Hamb., 9: 51-67. Type locality: Süd-Formosa [= South Taiwan], China.

Dinodon septentrionalis Boulenger, 1893: 363. London (Taylor & Francis): 448.

Dinodon septentrionalis var. *ruhstrati* Boulenger, 1899, Proc. Zool. Soc., London: 159-172.

Dinodon septentrionale ruhstrati Stejneger, 1907: 370. Bull. U.S. Natl. Mus., 58 (20): 1-577.

Ophites ruhstrati Zhao and Adler, 1993. Herpetology of China. SSAR, Oxford/Ohio: 522.

Lycodon ruhstrati Lanza, 1999, Tropical Zoology, 12: 89-104.

Lycodon cf. *ruhstrati* Ziegler et al., 2007, Zootaxa, 1493: 1-40.

Lycodon ruhstrati Wallach et al., 2014: 395. Taylor & Francis, CRC Press: 1237.

[同物异名] *Ophites ruhstrati*；*Dinodon septentrionalis*；*Dinodon septentrionalis* var. *ruhstrati*；*Dinodon septentrionale ruhstrati*；*Lycodon* cf. *ruhstrati*；*Lycodon ruhstrati abditus*

[俗名] 黑背白环蛇

[鉴别特征] 颊鳞和鼻间鳞不相邻接；眶前鳞不与额鳞邻接，全身有黑白相间环纹，在躯干部有20～46条，在尾部有11～22条，仅在尾部的环纹围绕周身。

[形态描述] 依据茂兰地区标本GZNU20160808033描述 体全长/尾长为463.3mm/114.2mm。头略大而稍扁，与颈略可区分。眼略小，瞳孔直立椭圆形；躯尾修长适度。背面黑褐色，全身有黑白相间环纹，在躯干部有20～46条，在尾部有11～22条，仅在尾部的环纹围绕周身，横纹中央散有浅褐色，在颈部的一节黑环纹最宽。颊鳞1，不入眶；颊鳞和鼻间鳞不相邻接；眶前鳞1，眶后鳞2；眶前鳞不与额鳞邻接；上唇鳞8（2-3-3式）；下唇鳞10，前后对大小相近。背鳞平滑，19-17-15，中央几行弱棱；腹鳞210；肛鳞完整。

[生活习性] 多生活于山地及山坡、山溪内石上、阴沟石缝等环境。

[地理分布] 茂兰地区分布于永康。国内分布于安徽、福建、甘肃、广东、广西、贵州、湖南、江苏、江西、陕西、四川、台湾、香港、浙江。国外未见报道。

成体量度

编号	产地	头体长（mm）	体全长（mm）	尾长（mm）	背鳞（枚）	腹鳞（枚）	肛鳞（枚）	上唇鳞（枚）	下唇鳞（枚）	颊鳞（枚）	眶前鳞（枚）	眶后鳞（枚）	前颞鳞（枚）	后颞鳞（枚）
GZNU20160808033	荔波茂兰	349.1	463.3	114.2	19-17-15	210	1	8	10	1	1	2	2	3

a 背面照
b 腹面照

[模式标本] 正模标本（ZFMK 86451）（基因序列号：EU999209）（成体雌性）保存于德国波恩·柯尼希动物研究博物馆（Zoologisches Forschungsinstitut und Museum Alexander Koenig）

[模式产地] 台湾南部

[保护级别] 三有保护动物；贵州省重点保护野生动物

[IUCN濒危等级] 无危（LC）

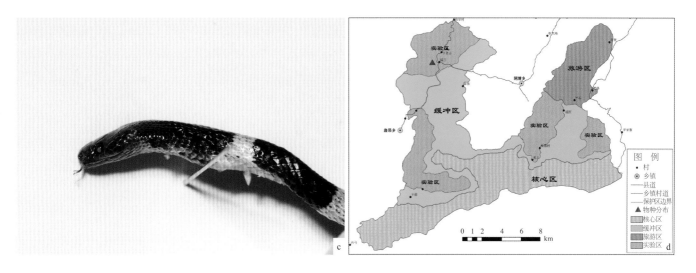

c 侧面照
d 在茂兰地区的分布

（44）黄链蛇 *Lycodon flavozonatus* (Pope, 1928)

Dinodon flavozonatum Pope, 1928, American Museum Novitates, 325: 1-4. Type locality: China, Fukien [= Fujian], Ch'unganhsien.
Dinodon rufozonatum meridionale Bourret, 1935 (fide Smith, 1943).
Dinodon flavozonatum Smith, 1943: 271. Reptilia and Amphibia. 3 (Serpentes). Taylor & Francis, London: 583.
Dinodon flavozonatum Orlov and Ryabov, 2004, Russ. J. Herpetol., 11 (3): 181-197.
Lycodon flavozonatus Siler et al., 2013, Zoologica Scripta., 42: 262-277.
Dinodon flavozonatum Wallach et al., 2014: 236. Taylor & Francis, CRC Press: 1237.

[同物异名] *Dinodon flavozonatum*；*Dinodon rufozonatum meridionale*；*Dinodon flavozonatus*；*Lycodon flavozonatus*

[俗名] 大牙蛇

[鉴别特征] 体中段背鳞15行，中央5～9行起棱，颊鳞不入眶，有（50～96）+（13～28）个约占半枚鳞长的黄色窄横斑。

[形态描述] 依据茂兰地区标本 GZNU20170513005 描述 体全长/尾长为980mm/185mm；头宽扁，头、颈略能区分。头略大，吻端宽扁，与颈可以区分；躯干较长。体中段背鳞15行，中央7行起棱。背面黑褐色，有约等距排列的多数黄色窄横斑55+14个，色斑体尾背面黑褐色，横斑宽占半枚鳞长。颊鳞1，窄长，不入眶；眶前鳞1，眶后鳞2；上唇鳞8（2-3-3式）；下唇鳞10，前4枚切前颏片。背鳞17-15-15，中段中央7行具弱棱；腹鳞237；肛鳞完整。

[生活习性] 多生活于丘陵、山区植被繁茂的水域附近。傍晚或夜间活动。曾在茂兰保护区板寨的翁根坡采集到一条，周围为灌木林和草丛。

[地理分布] 茂兰地区分布于洞塘，为茂兰保护区新纪录。国内分布于安徽、福建、广东、广西、贵州、海南、湖南、江西、四川、云南、浙江。国外分布于缅甸、越南。

[模式标本] 正模标本（AMNH 34371）保存于美国自然历史博物馆

[模式产地] 福建

[保护级别] 三有保护动物；贵州省重点保护野生动物

[IUCN濒危等级] 无危（LC）

成体量度

编号	产地	头体长(mm)	体全长(mm)	尾长(mm)	背鳞(枚)	腹鳞(枚)	肛鳞(枚)	上唇鳞(枚)	下唇鳞(枚)	颊鳞(枚)	眶前鳞(枚)	眶后鳞(枚)	前颞鳞(枚)	后颞鳞(枚)
GZNU20170513005	荔波茂兰	795	980	185	17-15-15	237	1	8	10	1	1	2	2	2

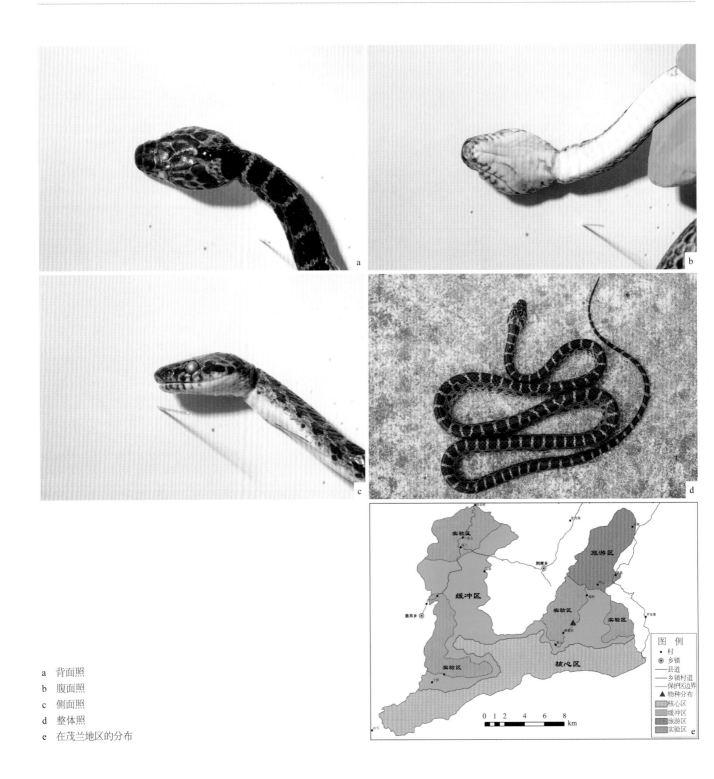

a 背面照
b 腹面照
c 侧面照
d 整体照
e 在茂兰地区的分布

30）玉斑蛇属 *Euprepiophis* Fitzinger, 1843

Euprepiophis Fitzinger, 1843, Vienna: Braumüller et Seidel: 106. Type species: *Euprepiophis mandarinus* Cantor, 1842.

体型中等偏大的无毒蛇，头、颈区分明显；体背面灰色或紫灰色，体及尾背有很多明显的黄色斑纹。

本属现有3种。我国分布2种，即玉斑蛇 *Euprepiophis mandarinus* 和横纹玉斑蛇 *Euprepiophis perlacea*。茂兰地区有1种——玉斑蛇。

（45）玉斑蛇 *Euprepiophis mandarinus* (Cantor, 1842)

Coluber mandarinus Cantor, 1842b, Ann. Mag. Nat. Hist., (1) 9: 481-493.

Coluber mandarinus Boulenger,1894: 42. Containing the Conclusion of the Colubridæ Aglyphæ. British Mus. (Nat. Hist.), London, Volume II: 382.

Elaphe mandarinus Fleck, 1985, Salamandra., 21 (2-3): 157-160.

Elaphe mandarina Schulz, 1996, Bushmaster, Berg (CH): 1-460.

Euprepiophis mandarinus Utiger et al., 2002, Russ. J. Herpetol., 9 (2): 105-124.

Euprepiophis mandarinus Gumprecht, 2003, Reptilia (Münster), 8 (44): 37-41.

Elaphe mandarina Zhao, 2006. The snakes of China. Hefei, China, Anhui Science & Technology Publ. House, Vol. I: 372.

Euprepiophis mandarinus Chen et al., 2013. Molecular Phylogenetics and Evolution.

Euprepiophis mandarina Pyron and Burbrink, 2013, Ecology Letters, 17 (1): 13-21.

Euprepiophis mandarinus Wallach et al., 2014: 292. Taylor & Francis, CRC Press: 1237.

［同物异名］ *Elaphe mandarina*；*Coluber mandarinus*；*Elaphe mandarinus*；*Elaphe takas ago*；*Eupriophis mandarina*

［俗名］ 美女蛇、玉斑丽蛇

［鉴别特征］ 体背灰色或紫灰色。背中央有1行黑色菱形大块斑镶着黄边及黄色中心；头背黄色，具有明显的黑斑；背鳞平滑。

［形态描述］ 体全长1000mm左右，中等偏大，最大体全长/尾长：雄性1425mm/235mm，雌性1240mm/210mm。头略大，与颈明显区分。眼大小适中，瞳孔圆形；躯尾修长适度。色斑体尾背面灰色或紫灰色，正背有1行大的黑色菱形斑，（18～31）+（6～11）个，菱形斑中心黄色，外侧亦镶以黄色边缘，体侧有紫红色斑；腹面黄白色，散以左右交错排列的黑色方斑。头背黄色，具有明显的3条黑斑；第1条横跨吻背；第2条横跨两眼，在眼下分2支分别达口缘；第3条为倒"V"形，其尖端始自额鳞，左右支分别斜经口角达喉部。本种色斑变异较多。颊鳞1（个别一侧无）；眶前鳞1，眶后鳞2（个别两侧或一侧1或3）；颞鳞2（1）+3（2）；上唇鳞7（2-2-3式），个别8（3-2-3式）或6（1-2-3、2-1-3或2-2-2式）；下唇鳞9（8～10），前4枚切前颔片；颔片2对。背鳞平滑，23（21～25）-23（21）-19（17～21）；腹鳞119～238，略有侧棱；肛鳞二分；尾下鳞49～76对。上颌齿每侧14～17枚。

［生活习性］ 多栖于平原、丘陵、山地、稻田、林中、溪流边、草丛、路边、居民点附近都有发现。卵生。每年的6～7月可产卵5～20枚，卵长椭圆形。

［地理分布］ 茂兰地区分布于永康。国内分布于江西、安徽、北京、重庆、福建、甘肃、广东、广西、贵州、河北、河南、湖北、湖南、江苏、辽宁、山西、陕西、上海、四川、台湾、天津、西藏、云南、浙江等地。国外分布于越南、缅甸。

［模式标本］ 正模标本（BMNH RR 60.3.19.6241）保存于英国伦敦自然历史博物馆

［模式产地］ 浙江舟山群岛

[保护级别]　三有保护动物；贵州省重点保护野生动物　　　[IUCN 濒危等级]　易危（VU）

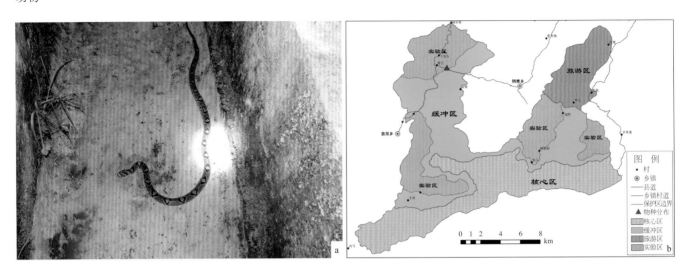

a　整体照
b　在茂兰地区的分布

31）紫灰蛇属 *Oreocryptophis* Cantor, 1839

Oreocryptophis Cantor, 1839, Proc. Zool. Soc., London, 1839: 31-34. Type species: *Oreocryptophis porphyraceus* Cantor, 1839. Type locality: India, Assam, Mishmi [Mishmee] Hills.

体型中等的无毒蛇，背面紫灰色或紫铜色，头背部有3条黑色纵纹，体尾背面有多条形如马鞍形的淡黑色横斑，背鳞平滑；腹部玉白色。卵生。

本属我国仅有1种。茂兰地区分布1种，即紫灰蛇 *Oreocryptophis porphyraceus*。

（46）紫灰蛇 *Oreocryptophis porphyraceus* (Cantor, 1839)

Coluber porphyraceus Cantor, 1839, Proc. Zool. Soc., London, 1839: 31-34. Type locality: India: Assam, Mishmi [Mishmee] Hills.

Coronella callicephalus Gray, 1853, Ann. Mag. Nat. Hist., (2) 12: 386-390.

Coluber porphyraceus Anderson, 1871, Proc. Zool. Soc., London: 149-211.

Simotes vaillanti Sauvage, 1877, Bull. Soc. Philom., (7) 1: 107-115.

Ablabes porphyraceus Boulenger, 1890. Taylor & Francis, London, 18: 1-541.

Coluber porphyraceus Boulenger, 1894: 34. Containing the Conclusion of the Colubridæ Aglyphæ. British Mus. (Nat. Hist.), London, Volume II, 11: 1-382.

Coluber porphyraceus Wall, 1908, J. Bombay Nat. Hist. Soc., 18: 312-337.

Liopeltis kawakamii Oshima, 1910, Annot. Zool. Japon., Tokyo, 7 (3): 185-207.

Holarchus vaillanti Mell, 1931, Lingnan Sci. Jour., Canton, 8 [1929]: 199-219.

Elaphe porphyracea Schulz, 1996: 196. Bushmaster, Berg (CH): 1-460.

Oreophis porphyraceus David et al., 2004. The Natural History Journal of Chulalongkorn University, 4 (1): 47-80.

Oreocryptophis prophyraceus Utiger et al., 2005, Russ. J. Herpetol., 12 (1): 39-60.

Oreocryptophis prophyraceus Ziegler et al., 2007, Zootaxa, 1493: 1-40.

Oreocryptophis porphyracea Pyron and Burbrink, 2013, Ecology Letters, 17 (1): 13-21.

[同物异名] *Elaphe porphyracea*；*Coluber porphyraceus*；*Coronella callicephalus*；*Ablabes porphyraceus*；*Oreophis porphyraceus*；*Liopeltis kawakamii*；*Simotes vaillanti*；*Holarchus vaillanti*；*Oreocryptophis porphyracea*

[俗名] 紫灰锦蛇

[鉴别特征] 本种体全长一般不超过1000mm，头体背紫灰色或紫铜色，头背有3条黑纵纹，体尾背有若干个马鞍形黑色横斑块，有2条黑纵纹纵贯全身或仅见于体后端；背鳞19-19-17，平滑无棱。

[形态描述] 依据茂兰地区标本描述 体型中等大小，体全长一般在1000mm之内，采集个体中最大体全长/尾长为：雄性921mm/131mm，雌性896mm/124mm。头略大，与颈明显区分；鼻孔侧位，略近鼻鳞前半；眼略小，瞳孔圆形；躯尾修长适度。头体背紫铜色，头背有黑色粗纵纹3条，正中1条沿前额鳞沟经额鳞再沿顶鳞沟，侧面2条分别起自眼后沿顶鳞外缘向后与体背两条细纵线连续，上唇、头腹、D_1（或包括D_2）及腹鳞玉白色无斑纹。体背2条黑纵纹在躯干沿D_6与D_7后延至尾末；体尾背有若干个马鞍形黑色横斑块，（7～17）+（1～6）个，斑中央浅紫褐色，边缘为暗紫褐细线；鞍形斑宽在背脊占3～7枚鳞，向两侧下延达D_1，鞍形斑尾后端几个小如点状斑。颊鳞1，前额鳞大，弯向头侧，故颊鳞低矮；眶前鳞1，眶后鳞2，颞鳞1+2，顶鳞宽大，其前端扩展到眶后鳞，故前颞鳞不与眶后鳞接触；上唇鳞8（3-2-3式），偶有9（4-2-3式），个别7（2-2-3式）；下唇鳞8～11，前4～5（个别有3或6）枚切前颏片；颏片2对。背鳞平滑，19（18，17）-19（17）-17（15）；腹鳞176～214；肛鳞二分；尾下鳞双行，49～77对。上颌齿每侧20～24枚。

[生活习性] 主要栖息于山区森林、茶山、农耕地、溪沟边、山路旁、秧田、村舍等附近。食物主要来源为鼠类等小型哺乳动物。卵生。7月产卵5～7枚，卵呈长椭圆形。

[地理分布] 茂兰地区分布于翁昂。国内广泛分布于安徽、重庆、福建、甘肃、广东、广西、贵州、海南、河南、湖北、湖南、江苏、江西、陕西、四川、台湾、香港、西藏、云南、浙江。国外分布于印度、缅甸、泰国、马来西亚、印度尼西亚。

[模式标本] 正模标本（BMNH [*nigrofasciatus*]）保存于英国伦敦自然历史博物馆；正模标本（AMNH 17705）保存于美国自然历史博物馆

[模式产地] 印度

[保护级别] 三有保护动物；贵州省重点保护野生动物

[IUCN濒危等级] 近危（NT）

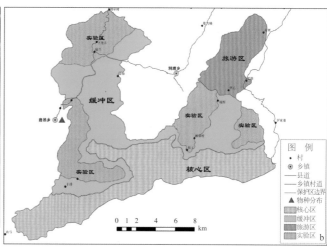

a 整体照
b 在茂兰地区的分布

32）晨蛇属 *Orthriophis* Utiger et al., 2002

Elaphe Fitzinger, 1832, in Wager (ed.), Descr. Icon. Amphib. Pt. 3, Pl: 27. Type species: *Elaphe parreysii* (= *quatuor-lineata*).
Orthriophis Utiger et al., 2002, Russ. J. Herpetol., 9 (2): 105-124.

体型中等偏大的日行性蛇类，头、颈区分明显；体背面有各种斑纹，背鳞起棱或弱棱，肛前鳞超过17行。卵生。

本属我国分布4种。茂兰地区有2种，即百花晨蛇 *Orthriophis moellendorffi* 和黑眉晨蛇 *Orthriophis taeniurus*。

茂兰地区种检索表

体型较小；头背红褐色；体背具有3行约30个近六角形的红褐色斑纹；尾具浅红色横斑·················百花晨蛇 *Orthriophis moellendorffi*
体型较大；头背黄绿色或棕灰色；体背呈明显的纵纹；头侧眼后有一黑色纵纹，尾部腹面两侧有两条黑色纵纹·········黑眉晨蛇 *Orthriophis taeniurus*

（47）百花晨蛇 *Orthriophis moellendorffi* (Boettger, 1886)

Cynophis moellendorffi Boettger, 1886, Zool. Anz., 9: 519-520. Type locality: China, Guangxi, Nanning.
Coluber moellendorffi Boulenger, 1894: 56. British Mus. (Nat. Hist.), London, 11: 1-382.
Elaphe möllendorffii Mell, 1931, Lingnan Sci. Jour., Canton, 8 [1929]: 199-219.
Elaphe moellendorffi Smith, 1943: 153. Reptilia and Amphibia. 3 (Serpentes). Taylor & Francis, London: 583.
Amblycephalus moellendorffii Deuve, 1961, Bull. Soc. Sci. Nat. Laos, 1: 5-32.
Elaphe moellendorffi Schulz, 1996, Bushmaster, Berg (CH): 1-460.
Orthriophis moellendorffi Utiger et al., 2002, Russ. J. Herpetol., 9 (2): 105-124.
Elaphe moellendorffi Zhao, 2006. The snakes of China. Hefei, China, Anhui Science & Technology Publ. House, Vol. I: 372.
Orthriophis moellendorffi Wallach et al., 2014: 512. Taylor & Francis, CRC Press: 1237.

[同物异名] *Elaphe moellendorffi*；*Amblycephalus moellendorffii*；*Coluber moellendorffi*；*Cynophis moellendorffi*；*Elaphe möllendorffii*

[俗名] 百花蛇、白花蛇、花蛇、菊花蛇、红头锦蛇。

[鉴别特征] 体型大，头部梨形；头背红褐色，体背具有3行约30个近六角形的红褐色斑纹。尾背具浅红色横斑；腹鳞多于250。

[形态描述] 依据茂兰地区标本GZNU20170513002描述 体全长/尾长为1165.3mm/147.2mm。头较大，呈梨形，与颈区分明显。眼大小适中，瞳孔呈圆形。躯干背面灰绿色，具有3行近六角形的红褐色斑纹，中央的1行斑块较大，其边缘蓝黑色，有28～32个，尾背黑色，有11个淡红色横斑，腹面白色，具有黑白相间的方格斑，头背红褐色，唇部灰色。颊鳞1；眶前鳞1，眶后鳞2；颞鳞2+3；上唇鳞8（4-2-2式）；下唇鳞10，第1对在颏鳞之后相接，前6枚切前颏片，第7枚最大，约为第6枚的3倍；颏片2对，前颏片长宽均大于后颏片，颏沟长而明显。背鳞21-27-19，具弱棱，仅外侧2行平滑；腹鳞295；肛鳞二分；尾下鳞92～102对。

[生活习性] 多生活于丘陵、低山、乱石堆、草丛、路边、沟渠、小河旁。但茂兰地区都是在洞穴洞口的上方发现，并且发现时都是在更换鳞片的时节。晚上蝙蝠飞出时，便张口捕捉。

[地理分布] 茂兰地区分布于翁昂、洞塘等地，数量较少。国内分布于贵州、广东、广西。国外分布于越南。

[模式标本] 模式标本（SMF 8056）保存于德国森根堡自然历史博物馆

[模式产地] 广西南宁

[保护级别] 三有保护动物；贵州省重点保护野生动物

[IUCN 濒危等级] 濒危（EN）

[讨论] 谢莉等于 2010~2012 年在贵州荔波的"中国南方喀斯特"世界自然遗产地进行脊椎动物物种多样性调查时，采集到 5 条蛇类标本，经鉴定为百花锦蛇，标本发现地分别为瑶山（25°13′N，107°46′E，海拔 395m）、板寨（25°13′N，108°01′E，海拔 696m）、洞塘（25°14′N，108°01′E，海拔 706m）、翁昂（25°17′N，108°04′E，海拔 746m）。该蛇栖息于荔波喀斯特地区的

成体量度

编号	产地	体全长（mm）	头体长（mm）	尾长（mm）	背鳞（枚）	腹鳞（枚）	肛鳞（枚）	上唇鳞（枚）	下唇鳞（枚）	颊鳞（枚）	眶前鳞（枚）	眶后鳞（枚）	前颞鳞（枚）	后颞鳞（枚）
GZNU20170513002	荔波茂兰	1165.3	1018.1	147.2	21-27-19	295	2	8	10	1	1	2	2	3

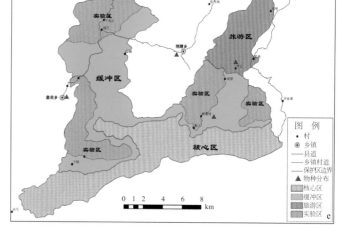

a 背面照
b 腹面照
c 侧面照
d 整体照
e 在茂兰地区的分布

岩洞洞口石隙处，其栖息的岩洞位于山体半山腰，岩洞中多有流水。采集标本的岩洞洞口温度为 18℃，湿度达 90%，阴凉潮湿，附近有灌丛和流水，采集时间为上午，阵雨，洞外温度为 26℃。调查发现岩洞中有白腹巨鼠（*Niviventer coninga*）、绿臭蛙（*Odorrana margaretae*）等小型脊椎动物，为百花晨蛇食物来源。在板寨捕获的 CNGZNU20120001 号标本腹内留存有未消化完全的白腹巨鼠。捕捉时蛇行动缓慢，性温顺，与其他常见的无毒蛇的快速逃离有差异。

采集的 2 条雌性标本（CNGZNU20120001 和 CNGZNU20120002）体全长分别为 1780mm 和 1326mm，尾下鳞分别为 82 对和 43 对，CNGZNU20120002 号标本尾背有 10 个淡红色横斑，尾下鳞和体全长均明显少于或小于 CNGZNU20120001 号标本及赵尔宓等（1998）、赵尔宓（2006）文献中的记载。CNGZNU20120002 号标本尾完整，出现上述情况可能是由个体发育阶段的差异造成的（谢莉等，2013）。

（48）黑眉晨蛇 *Orthriophis taeniurus* (Cope, 1861)

Elaphe taeniurus Cope, 1861, Proc. Acad. Nat. Sci. Philadelphia, 12 [1860]: 553-566. Type locality: Zhejiang, Ningbo, China.

Elaphis yunnanensis Anderson, 1879: 813. Volume II. Atlas. Bernard Quaritch, London "1878".

Elaphe taeniura grabowskyi Fischer, 1885, Archiv für Naturgeschichte, 51: 41-72.

Elaphis grabowskyi Fischer, 1885, Archiv für Naturgeschichte, 51: 41-72.

Elaphis taeniurus Boulenger, 1887, Ann. Mag. Nat. Hist., (5) 19: 169-170.

Coluber taeniurus Boulenger, 1890: 333. Taylor & Francis, London, 18: 541.

Elaphe taeniura schmackeri Boettger, 1895, Zool. Anz., 18: 266-270.

Coluber schmackeri Boettger, 1895, Zool. Anz., 18: 266-270.

Coluber schmackeri Boulenger, 1896: 627. London (Taylor & Francis), 14: 727.

Elaphe taeniura ridleyi Butler, 1899, J. Bombay Nat. Hist. Soc. (1898-1899), 12 (2): 425-426.

Coluber vaillanti Mocquard, 1905, Bull. Soc. Philomat., Paris, (9) 7: 317-322.

Elaphe taeniurus Stejneger, 1907, Bull. U.S. Natl. Mus., 58 (20): 319.

Elaphe schmackeri Stejneger, 1907, Bull. U.S. Natl. Mus., 58 (20): 322.

Coluber taeniurus var. *friesi* Werner, 1927: 245. Sitzungsb. Akad. Wiss. Wien, Math. Naturwiss. Kl., 143: 243-257 [1926].

Elaphe taeniura Smith, 1943: 150. Reptilia and Amphibia. 3 (Serpentes). Taylor & Francis, London: 583.

Elaphe taeniura grabowskyi Taylor, 1965, Univ. Kansas Sci. Bull., 45 (9): 609-1096.

Elaphe taenura [sic] *grabowskii* Manthey and Denzer, 1982, Herpetofauna, 4 (21): 11-19.

Elaphe taeniura Schulz, 1996: 253. Bushmaster, Berg (CH): 1-460.

Elaphe taeniura Schulz, 1996: 264. Bushmaster, Berg (CH): 1-460.

Elaphe taeniura mocquardi and *Elaphe taeniura yunnanensis* Schulz, 1996, Bushmaster, Berg (CH): 1-460.

Elaphe taeniura mocquardi Schulz, 1996: 253. Bushmaster, Berg (CH): 1-460.

Elaphe taeniura Manthey and Grossmann, 1997: 345. Natur und Tier Verlag (Münster): 512.

Elaphe taeniura yunnanensis Cox et al., 1998: 50. Ralph Curtis Publishing: 144.

Elaphe taeniura friesi Tiedemann and Grillitsch, 1999, Herpetozoa., 12 (3-4): 147-156.

Orthriophis taeniurus Utiger et al., 2002, Russ. J. Herpetol., 9 (2): 105-124.

Elaphe taeniura Ziegler, 2002: 233. Natur und Tier Verlag (Münster): 342.

Orthriophis taeniurus friesi Gumprecht, 2003, Reptilia (Münster), 8 (44): 37-41.

Orthriophis taeniurus ridleyi Grossmann and Tillack, 2004, Reptilia (Münster), 9 (50): 42-49.
Elaphe taeniura mocquardi Switak, 2006: 64. Natur und Tier Verlag (Münster): 364.
Elaphe taeniura Grismer et al. 2007, Hamadryad, 31 (2): 216-241.
Orthriophis taeniurus taeniurus Schulz, 2010, Sauria, 32 (2): 3-26.
Elaphe taeniura callicyanous Schulz, 2010, Sauria, 32 (2): 3-26.
Orthriophis taeniurus callicyanous Schulz, 2010, Sauria, 32 (2): 3-26.
Orthriophis taeniurus friesi Schulz, 2010, Sauria, 32 (2): 3-26.
Orthriophis taeniurus grabowskyi Schulz, 2010, Sauria, 32 (2): 3-26.
Elaphe taeniura helfenbergeri Schulz, 2010, Sauria, 32 (2): 3-26.
Orthriophis taeniurus helfenbergeri Schulz, 2010, Sauria, 32 (2): 3-26.
Orthriophis taeniurus mocquardi Schulz, 2010, Sauria, 32 (2): 3-26.
Orthriophis taeniurus ridleyi Schulz, 2010, Sauria, 32 (2): 3-26.
Orthriophis taeniurus schmackeri Schulz, 2010, Sauria, 32 (2): 3-26.
Orthriophis taeniurus Wallach et al., 2014: 512. Taylor & Francis, CRC Press: 1237.

[同物异名] *Elaphe taeniurus*；*Coluber nuthalli*；*Elaphis yunnanensis*；*Coluber taeniurus*；*Elaphe taeniura*；*Orthriophis taeniurus taeniurus*；*Orthriophis taeniurus callicyanous*；*Elaphe taeniura friesi*；*Coluber friesi*；*Coluber taeniurus* var. *friesi*；*Orthriophis taeniurus friesi*；*Elaphe taeniura grabowskyi*；*Elaphis grabowskyi*；*Elaphe taenura* [sic] *grabowskii*；*Orthriophis taeniurus grabowskyi*；*Elaphe taeniura helfenbergeri*；*Elaphe taeniura taeniura*；*Elaphe taeniura yunnanensis*；*Elaphe taeniuras*；*Elaphe taeniura ridley*；*Orthriophis taeniurus helfenbergeri*；*Elaphe taeniura mocquardi*；*Coluber vaillanti*；*Orthriophis taeniurus mocquardi*；*Elaphe taeniura schmackeri*；*Coluber schmackeri*；*Elaphe schmackeri*；*Orthriophis taeniurus schmackeri*；*Elaphe taeniura vaillanti*；*Orthriophis taeniurus yunnanensis*；*Elaphe taeniura callicyanous*；*Elaphis taeniurus*；*Orthriophis taeniurus ridleyi*

[俗名] 菜花蛇、黄颌蛇、枸皮蛇

[鉴别特征] 本种属于大型无毒蛇。头体背黄绿色或棕灰色，眼后有明显的粗黑纹。体前中段有黑色梯状或蝶状斑纹，至后段逐渐不显，从体中段开始，两侧有明显的黑纵带达尾端。背中央数行背鳞稍有起棱。

[形态描述] 依据茂兰地区标本GZNU20160909009描述 体大型。最大体全长/尾长为1127mm/244mm。头略大，与颈明显区分；眼大小适中，瞳孔圆形；头体背黄绿色或棕灰色；眼后有1明显的粗黑纹；体前中段有黑色梯状或蝶状斑纹，至后段逐渐不显，从体中段开始，两侧有明显的黑纵带达尾端。上、下唇鳞及下颌浅黄色。颊鳞1；眶前鳞1，眶后鳞2；上唇鳞9（4-2-3式），下唇鳞11，前5枚切前颔片；颔片2对，前对大于后对，前对左右相接，后对为数枚小鳞分隔。背鳞24-21-17；背中央数行背鳞稍有起棱。腹鳞107，略有侧棱；肛鳞二分。

[生活习性] 多生活于平原、丘陵和山区。路边、耕作地、竹林、针叶林也有发现。

[地理分布] 茂兰地区分布于永康。国内广泛分布于安徽、北京、重庆、福建、甘肃、广东、湖北、湖南、江苏、江西、辽宁、陕西、上海、广西、贵州、海南、河北、河南、台湾、天津、西藏、云南、浙江等地。国外分布于日本、俄罗斯（南部滨海地区）、泰国、越南、朝鲜、印度（大吉岭、阿萨姆邦）、缅甸。

[模式标本] 模式标本（ZFMK 86360 [*callicyanous*]）保存于德国亚利山大·柯尼希动物研究博物馆；模式标本（SMF 18411 [*schmackeri*]）保存于德国森根堡自然历史博物馆；模式标本（BMNH 1946.6.2.1 [*yunnanensis*]）保存于英国伦敦自然历史博物馆

[模式产地] 浙江宁波

[保护级别] 三有保护动物；贵州省重点保护野生动物

[IUCN 濒危等级] 濒危（EN）

成体量度

编号	产地	头体长（mm）	体全长（mm）	尾长（mm）	背鳞（枚）	腹鳞（枚）	肛鳞（枚）	上唇鳞（枚）	下唇鳞（枚）	颊鳞（枚）	眶前鳞（枚）	眶后鳞（枚）	前颞鳞（枚）	后颞鳞（枚）
GZNU20160909009	荔波茂兰	883	1127	244	24-21-17	107	2	9	11	1	1	2	2	3
GZNU20160909001	荔波茂兰	1220	1551	331	23-23-17	220	2	10	11	1	1	2	2	3

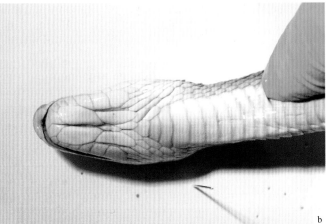

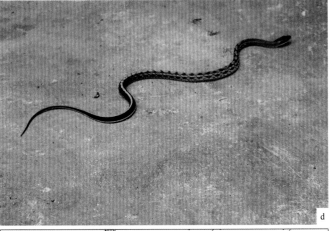
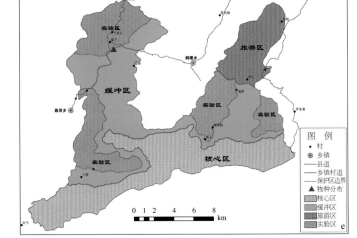

a 背面照
b 腹面照
c 侧面照
d 整体照
e 在茂兰地区的分布

33）腹链蛇属 *Amphiesma* Duméril, Bibron and Duméril, 1854

Amphiesma Duméril, Bibron and Duméril, 1854, Herpetol. Gen., Paris, 7 (1): 724. Type species: *Coluber solatus* Linnaeus, 1758, by subsequent designation (Steineger, 1907, Bull. U.S. Natl. Mus., Washington, 58: 264).

本属为体型较小的陆栖蛇类，头大小适中，与颈可以区分；体背面褐色，有 2 条浅色纵纹，纵纹间有多数黑色横斑；半阴茎，其顶端分叉成"丫"字形，满布细刺，自基部至顶端渐小，基部外侧还各有两枚较大的刺。卵生。

分子系统进化关系显示原腹链蛇属 *Amphiesma* 物种不形成单系，该类群至少包含 3 个种组（Guo et al., 2014），本书采用 Guo 等（2014）的结果，将原广义腹链蛇属分为 3 属（*Amphiesma*、*Hebius*、*Herpetoreas*），中文名不变。

本属分布于东亚、南亚和东南亚。我国分布 1 种。茂兰地区分布 1 种，即草腹链蛇 *Amphiesma stolatum*。

（49）草腹链蛇 *Amphiesma stolatum* (Linnaeus, 1758)

Coluber stolatus Linnaeus, 1758, Syst. Nat. Ed., 10: 219. Type locality: America (in error); Indien fide Kramer, 1977.

Elaps bilineatus Schneider, 1801: 299. Amphisbaenas et Caecilias. Frommanni, Jena: 374.

Natrix stolatus Merrem, 1820: 123. J. C. Kriegeri, Marburg: 191.

Tropidonotus stolatus Boie, 1827, Isis van Oken, 20: 508-566.

Tropidonotus stolatus Günther, 1864: 266. London (Taylor & Francis), 27: 452.

Tropidonotus ruficeps Peters, 1869, Monatsber. k. preuss. Akad. Wiss. Berlin, 1869: 432-447.

Tropidonotus stolatus Stoliczka, 1870: 191. J. Asiat. Soc. Bengal, 39: 134-228.

Tropidonotus stolatus Boulenger, 1890: 348. Taylor & Francis, London, 18: 541.

Natrix stolata Stejneger, 1907: 280. Bull. U.S. Natl. Mus., 58 (20): 577.

Tropidonotus stolatus Wall, 1908, J. Bombay Nat. Hist. Soc., 18: 312-337.

Amphiesma stolata and *Rhabdophis stolatus* Wall, 1921. Colombo Mus. (Cottle, govt. printer), Colombo., 22: 581.

Natrix stolata chinensis Mell, 1931, Lingnan Sci. Jour., Canton, 8 [1929]: 199-219.

Rhabdophis stolatus var. *chinensis* Deuve, 1961, Bull. Soc. Sci. Nat. Laos, 1: 5-32.

Amphiesma stolata Cox et al., 1998: 45. Ralph Curtis Publishing: 144.

Amphiesma stolatum David, Vogel and Pauwels, 1998, J. Taiwan Museum, 51 (2): 83-92.

Amphiesma stolata chinesis Chan-ard et al., 1999: 154. Bushmaster Publications, Würselen, Germany: 240 [book review in Russ. J. Herp., 7: 87].

Amphiesma stolata Sharma, 2004. Akhil Books, New Delhi: 292.

Amphiesma stolata Janzen et al., 2007, Sri Lankas Schlangenfauna. Draco, 7 (30): 56-64.

Amphiesma stolatus Guo et al., 2014, Zootaxa, 3873 (4): 425-440.

Amphiesma stolatum Wallach et al., 2014: 34. Taylor & Francis, CRC Press: 1237.

［同物异名］ *Coluber stolatus*；*Elaps bilineatus*；*Natrix stolatus*；*Tropidonotus stolatus*；*Tropidonotus stolbatus* [sic]；*Tropidonotus ruficeps*；*Natrix stolata*；*Rhabdophis stolatus*；*Amphiesma stolata*；*Natrix stolata chinensis*；*Amphiesma stolata chinesis*；*Amphiesma stolatus*；*Rhabdophis stolatus* var. *chinensis*

［俗名］ 黄头蛇、花浪蛇

［鉴别特征］ 背面褐色，有 2 条浅色纵纹，2 纵纹间有多数黑色横斑；头黄褐色，颌下黄色。

[形态描述] 依据茂兰地区标本 GZNU20170508036 描述 最大体全长/尾长为 380.4mm/66.2mm，头大小适中，与颈可以区分；鼻孔大而圆，靠近鼻鳞上部，几与鼻间鳞相切，鼻鳞下沟达第一上唇鳞；眼较大，瞳孔圆形。体背面褐色，有 2 条浅色纵纹，2 纵纹间有多数黑色横斑；头黄褐色，颌下黄色。颊鳞 1；眶前鳞 1；眶后鳞 3；颞鳞 1+2；上唇鳞 8（2-3-3 式）；下唇鳞 10，第 1 对在颏鳞后相接，前 5 枚接前颏片；颏片 2 对。背鳞 19-19-17，除两侧最外

成体量度

编号	产地	头体长（mm）	体全长（mm）	尾长（mm）	背鳞（枚）	腹鳞（枚）	肛鳞（枚）	上唇鳞（枚）	下唇鳞（枚）	颊鳞（枚）	眶前鳞（枚）	眶后鳞（枚）	前颞鳞（枚）	后颞鳞（枚）
GZNU20170508036	荔波茂兰	314.2	380.4	66.2	19-19-17	154	2	8	10	1	1	3	1	2

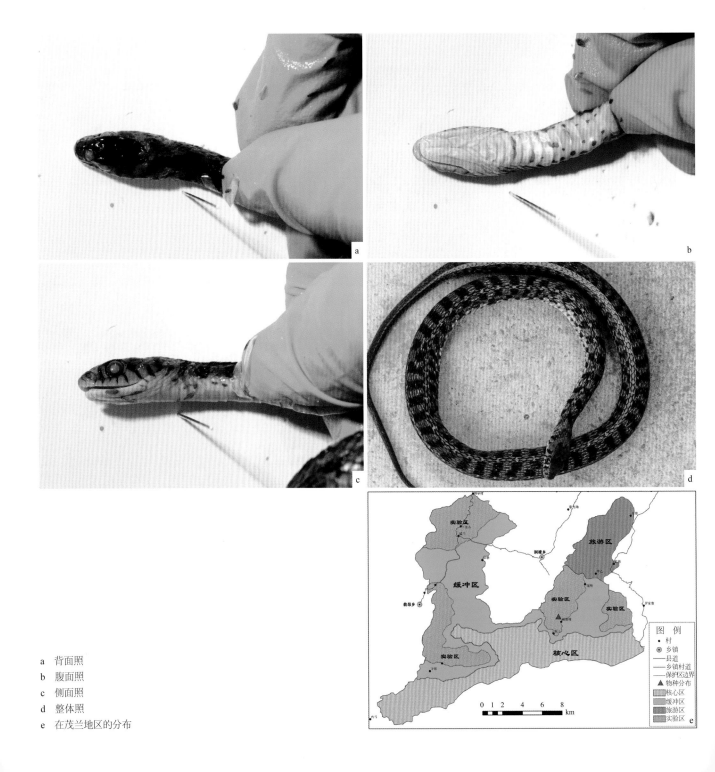

a 背面照
b 腹面照
c 侧面照
d 整体照
e 在茂兰地区的分布

一行平滑外，其余具棱；腹鳞154；肛鳞二分。

[**生活习性**] 常见于河边、山坡、路旁、耕地、谷草堆、院内、住屋附近，甚至树上。体色与周围环境相似，不易被发现。

[**地理分布**] 茂兰地区分布于洞塘。国内分布于安徽、澳门、福建、广东、广西、贵州、河南、海南、湖南、江西、台湾、香港、云南、浙江。国外分布于巴基斯坦、印度、斯里兰卡、尼泊尔、中南半岛。

[**模式标本**] 正模标本（NRM）保存于瑞典国家自然历史博物馆（斯德哥尔摩）

[**模式产地**] 印度

[**保护级别**] 三有保护动物；贵州省重点保护野生动物

[**IUCN 濒危等级**] 无危（LC）

34）东亚腹链蛇属 *Hebius* Thompson, 1913

Hebius Thompson, 1913, Proc. Zool. Soc., London, 1913: 414-426. Type species: *Tropidonotus vibakari* Boie, 1826: 207. Type locality: Japan.

陆栖或半水栖中小型蛇类。鼻间鳞前缘约与后缘等宽或略窄，鼻孔侧位或背侧位；正常有1枚颊鳞、1枚（个别2枚）眶前鳞和3枚眶后鳞。本属蛇类的腹鳞及尾下鳞两侧常各有1黑色点斑，前后点斑缀连成两条黑色"腹链纹"贯通体尾。本属雄性个体的半阴茎及精沟都不分叉；上颌齿呈连续的一列，由前到后逐渐增大。

本属蛇种原隶属于广义腹链蛇属 *Amphiesma*，本书采用 Guo 等（2014）研究结果，将原广义腹链蛇属分为3属（*Amphiesma*、*Hebius*、*Herpetoreas*），中文名不变。

本属世界现有42种，分布于东亚、南亚和东南亚，南到澳大利亚（东北部）。我国分布16种。茂兰地区有5种，分种检索表如下。

茂兰地区种检索表

1. 颞鳞无或在第5枚上唇鳞与顶鳞之间有1枚小鳞片 ·············· 无颞鳞腹链蛇 *Hebius atemporale*
   有颞鳞 ·············· 2
2. 背鳞通体17行；背面无纵纹或横纹 ·············· 棕黑腹链蛇 *Hebius sauteri*
   中段背鳞19行；背面呈纵纹或横纹 ·············· 3
3. 背面呈长短不一的黄色横纹，左右交叉或在背中线相连；绝不呈纵纹状 ·············· 丽纹腹链蛇 *Hebius optatum*
   背面呈点线或网纹；或虽有横纹，但同时具有明显或不明显的纵纹 ·············· 4
4. 腹鳞两侧构成腹链的黑褐色点斑较长，前后点斑几乎相连 ·············· 锈链腹链蛇 *Hebius craspedogaster*
   腹鳞两侧构成腹链的黑褐色点斑短，前后点斑相距较宽 ·············· 坡普腹链蛇 *Hebius popei*

（50）锈链腹链蛇 *Hebius craspedogaster* (Boulenger, 1899)

Tropidonotus craspedogaster Boulenger, 1899, Proc. Zool. Soc., London, 1899: 159-172. Type locality: Kuatun [= Guadun], Fukien [= Fujian], China, 3000-4000 ft elevation.

Natrix craspedogaster Stejneger, 1907: 265. Bull. U.S. Natl. Mus., 58 (20): 577.

Tropidonotus gastrotaenia Werner, 1922, Anz. Akad. Wissensch. Wien, ser. Mathemat.-Naturwissensch., Vienna, 59 (24-25): 220-222.

Natrix craspedogaster Mell, 1931, Lingnan Sci. Jour., Canton, 8 [1929]: 199-219.

Natrix chrysarga chekiangensis Mell, 1931, Lingnan Sci. Jour., Canton, 8 [1929]: 199-219.

Amphiesma craspedogaster David et al., 2007, Zootaxa, 1462: 41-60.

Amphiesma craspedogaster Wallach et al., 2014: 29. Taylor & Francis, CRC Press: 1237.

Hebius craspedogaster Guo et al., 2014, Zootaxa, 3873 (4): 425-440.

[同物异名] *Tropidonotus craspedogaster*；*Tropidonotus gastrotaenia*；*Natrix craspedogaster*；*Natrix chrysarga chekiangensis*；*Amphiesma craspedogaster*

[俗名] 锈链游蛇

[鉴别特征] 体型中等大小、半水栖、具腹链的无毒蛇，背面黑褐色，背侧有2行铁锈色纵纹，枕部两侧有1对椭圆形黄色枕斑，腹面黄色具有腹链纹。头、颈可以区分，瞳孔圆形。腹鳞135～159。

成体量度

编号	产地	头体长（mm）	体全长（mm）	尾长（mm）	背鳞（枚）	腹鳞（枚）	肛鳞（枚）	上唇鳞（枚）	下唇鳞（枚）	颊鳞（枚）	眶前鳞（枚）	眶后鳞（枚）	前颞鳞（枚）	后颞鳞（枚）
GZNU20170501001	荔波茂兰	395.4	431.5	36.1	19-19-17	141	2	8	10	1	2	3	2	2
GZNU20170423099	荔波茂兰	401.1	576.4	175.3	19-19-17	141	2	9	10	1	2	3	2	3

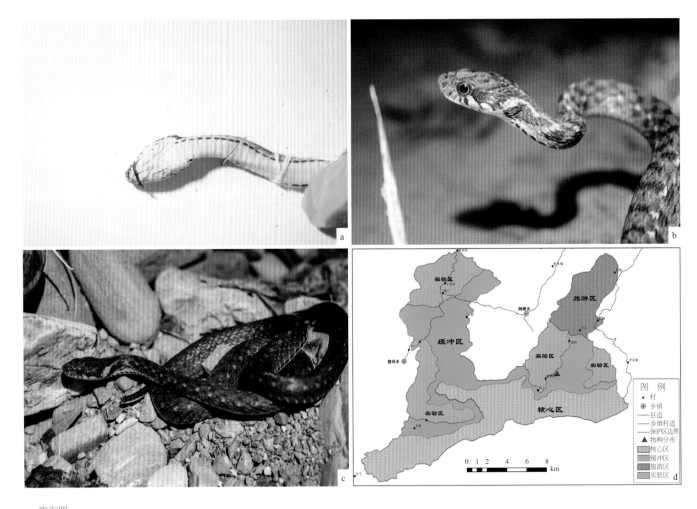

a 腹面照
b 侧面照
c 整体照
d 在茂兰地区的分布

［形态描述］ 依据茂兰地区标本 GZNU20170501001 描述 最大体全长/尾长为 431.5mm/36.1mm。头、颈可以区分，瞳孔圆形。头背黑褐色，背侧有 2 行铁锈色纵纹，枕侧各有 1 黄色枕斑，始自最后一枚上唇鳞，斜向颈背，略呈椭圆形。唇部淡黄色，唇鳞沟黑褐色。躯尾背面黑褐色，腹面淡黄色，有腹链纹，两侧位置各有 1 行浅黄色纵纹，沿此纵纹可看出 1 列铁锈色点斑。颊鳞 1；眶前鳞 2，眶后鳞 3；颞鳞 2+2；上唇鳞 8（2-3-3 式），下唇鳞 10；颏片 2 对。背鳞 19-19-17，均具棱；腹鳞 141；肛鳞二分。

［生活习性］ 多栖息于山区常绿阔叶林林带下，常见于各种水域，如冬水田、稻田、井边、水沟、河流附近，或路边、草丛、石砾、落叶丛中。白昼活动。食物来源主要为两栖动物成体及幼体或一些小型鱼类。

［地理分布］ 茂兰地区分布于洞塘。国内分布于山西、安徽、重庆、福建、甘肃、广东、广西、贵州、河南、湖北、湖南、江苏、江西、陕西、四川、浙江。国外未见报道。

［模式标本］ 模式标本（BMNH 1946.1.15.42）保存于英国伦敦自然历史博物馆

［模式产地］ 福建挂墩

［保护级别］ 三有保护动物；贵州省重点保护野生动物

［IUCN 濒危等级］ 无危（LC）

（51）丽纹腹链蛇 *Hebius optatum* (Hu and Zhao, 1966)

Natrix optata Hu and Zhao, 1966, Acta Zootaxon. Sinica, Peking, 3 (2): 158-164. Type locality: Liangfeng Gang, Mt. Omei, Sichuan, 700m elevation.

Amphiesma optatum David et al., 1998, J. Taiwan Museum, 51 (2): 83-92.

Amphiesma optatum David et al., 2007, Zootaxa, 1462: 41-60.

Amphiesma optatum Wallach et al., 2014: 32. Taylor & Francis, CRC Press: 1237.

Hebius optatum Guo et al., 2014, Zootaxa, 3873 (4): 425-440.

［同物异名］ *Natrix optata*；*Amphiesma optatum*

［俗名］ 无

［鉴别特征］ 小型至中型大小、具腹链、半水栖的无毒蛇类。头、颈可以区分，头背棕褐色，眼后有 1 白色线纹，体背及尾部黑褐色，具有较多的橘黄色横斑。背鳞 19-19-17。

［形态描述］ 依据茂兰地区标本 GZNU20170424001 描述 头、颈可以区分；体全长/尾长为 543.3mm/186.2mm。头背棕褐色，眼后有 1 白色线纹，体背及尾部黑褐色，两侧有黑色链纹，并具有较多的橘黄色横斑；头腹色白，上下唇缘散以黑褐点。颊鳞 1，不入眶；眶前鳞 1，眶后鳞 3；前颞鳞 1，后颞鳞 1；上唇鳞 8（3-2-3 式），下唇鳞 8，前 2 对在颏鳞后相切，第 4、第 5 枚最大；颏片 2 对。背鳞 19-19-17，中央 7～15 行起弱棱；腹鳞 152；肛鳞二分。

［生活习性］ 生活在山区，常见于小溪及沟坡草地中，也见于稻田、路边。以鱼类为主食。白天活动。

［地理分布］ 茂兰地区分布于翁昂、洞塘、永康等地。国内分布于山西、安徽、重庆、福建、甘肃、广东、广西、贵州、河南、湖北、湖南、江苏、江西、陕西、四川、浙江。国外未见报道。

［模式标本］ 正模标本（CIB Chuan 3624）保存于中国科学院成都生物研究所

［模式产地］ 四川峨眉山

［保护级别］ 三有保护动物；贵州省重点保护野生动物

［IUCN 濒危等级］ 近危（NT）

成体量度

编号	产地	头体长（mm）	体全长（mm）	尾长（mm）	背鳞（枚）	腹鳞（枚）	肛鳞（枚）	上唇鳞（枚）	下唇鳞（枚）	颊鳞（枚）	眶前鳞（枚）	眶后鳞（枚）	前颞鳞（枚）	后颞鳞（枚）
GZNU20170424001	荔波茂兰	357.1	543.3	186.2	19-19-17	152	2	8	8	1	1	3	1	1

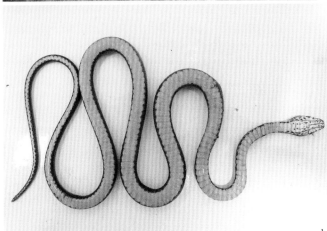
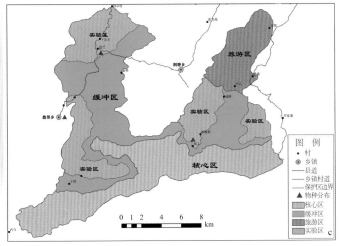

a 背面照
b 腹面照
c 在茂兰地区的分布

（52）坡普腹链蛇 *Hebius popei* (Schmidt, 1925)

Natrix popei Schmidt, 1925, American Museum Novitates, (157): 1-5. Type locality: Nodoa, Hainan, China.
Natrix vibakari popei Mell, 1931, Lingnan Sci. Jour., Canton, 8 [1929]: 199-219.
Natrix popei Glass, 1946, Copeia, (4): 249-252.
Amphiesma popei David et al., 2007, Zootaxa, 1462: 41-60.
Hebius popei Guo et al., 2014, Zootaxa, 3873 (4): 425-440.
Amphiesma popei Wallach et al., 2014: 32. Taylor & Francis, CRC Press: 1237.

[同物异名]　*Natrix popei*；*Natrix vibakari popei*；*Amphiesma popei*

[俗名]　无

[鉴别特征]　半水栖具腹链的小型无毒游蛇。头背土红色，近口角处有1浅色圆斑，枕侧各有1较大浅色椭圆形斑；躯干及尾背面灰褐色，背侧有2条由浅色短横斑缀成的点线纵贯全身。腹鳞142以下，背鳞最外行平滑。

[形态描述]　最大体全长/尾长：雄性518mm/156mm，雌性507mm/162mm。头略大，与颈区分明显，瞳孔圆形；头背土红色，近口角处有1浅色圆斑，枕侧各有1较大浅色椭圆形斑；躯干及尾背面灰褐色，背侧 D_{5-7} 每相距2～3鳞有镶黑边的浅色短横斑（或呈2个并列的小点斑），前后缀连成点线，形成2条由浅色短横斑缀成的点线纵贯全身。腹鳞及尾下鳞两外侧灰褐色，近外侧各有1黑点，前后缀连成腹链纹，左右腹链纹之间灰白色。顶鳞沟前1/3两侧有1对镶深色边的浅色小点，即顶斑；上唇鳞前6枚

白色，后缘黑褐色；下唇鳞白色，鳞沟黑褐色。颊鳞1；眶前鳞1（偶2），眶后鳞3（个别一侧2）；颞鳞2+1或1+2，少数1+1或2+2；上唇鳞以8（3-2-3式）为主，少数8（2-3-3式）；下唇鳞10（5）或9（4），个别8（4）或7（4）；颔片2对。背鳞19-19-17，除两侧最外行，均起棱；腹鳞131～142；肛鳞二分；尾下鳞66～88对，个别有数枚成单。上颌齿每侧24～28枚，由前到后渐次增大，或最后2枚骤然增大。

[生活习性] 多见于低山溪流或其他水域，偶见于稻田。卵生。

[地理分布] 茂兰地区分布于洞塘。国内分布于广东、广西、贵州、海南、湖南、云南。国外无报道。

[模式标本] 正模标本（AMNH 27763，成年雄性）保存于美国自然历史博物馆

[模式产地] 海南

[保护级别] 三有保护动物；贵州省重点保护野生动物

[IUCN濒危等级] 无危（LC）

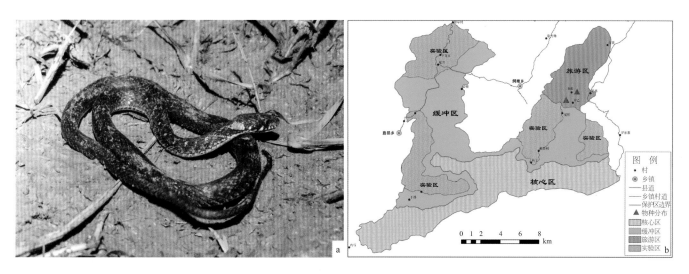

a 整体照
b 在茂兰地区的分布

（53）棕黑腹链蛇 *Hebius sauteri* Boulenger, 1909

Tropidonotus sauteri Boulenger, 1909: 495. Ann. Mag. Nat. Hist., (4) 8: 492-495. Type locality: Kosempo, Formosa [= Taiwan].
Natrix sauteri Mell, 1931, Lingnan Sci. Jour., Canton, 8 [1929]: 199-219.
Natrix sauteri Smith, 1943: 287. Taylor & Francis, London: 583.
Amphiesma sauteri Malnate, 1960, Proc. Acad. Nat. Sci. Philadelphia, 112 (3): 41-71.
Amphiesma sauteri David et al., 2007, Zootaxa, 1462: 41-60.
Hebius sauteri Guo et al., 2014, Zootaxa, 3873 (4): 425-440.
Amphiesma sauteri Wallach et al., 2014: 33. Taylor & Francis, CRC Press: 1237.

[同物异名] *Tropidonotus sauteri*；*Natrix sauteri*；*Amphiesma sauteri*

[俗名] 棕黑游蛇、梭德氏游蛇

[鉴别特征] 半水栖或潮湿山地型具腹链的小型游蛇，背鳞通体17行。颊鳞正常；下唇鳞8（6～9）。

[形态描述] 头略大，与颈可以区分，瞳孔圆形；背鳞通体17行；前4～5枚上唇鳞白色，其后缘黑褐色。头背黑褐色，没有顶斑或顶斑不明显，前数枚上唇鳞白色，鳞沟黑褐色，口角有或无镶黑边的浅色圆斑，枕斑大小不一；头腹灰白色，部分下唇鳞沟黑褐色。躯尾背面黑褐色，

隐约可见碎黑斑，D_{4-6}鳞行每隔 2～3 鳞有 1 镶黑边的浅色短横斑（或呈 2 小点），前后缀连成纵列点线；腹面有腹链纹，腹链纹外侧的腹鳞黑褐色，左右两链纹之间灰白色，无斑。颊鳞 1；眶前鳞 1（偶 2），眶后鳞 3，个别一侧 2；颞鳞 1+1 或 1+2，少数 2+1 或 2+2；上唇鳞 7（2-2-3 或 3-2-2 式）、6（2-2-2、1-2-3 或 2-1-3 式）、5（1-2-2 式），或个别一侧 8（3-2-3 式）；下唇鳞 8（6～9），前 4（5 或 3）枚接前颏片，第 1 对在颏鳞后相接；颏片 2 对。背鳞通体 17 行，除两外侧 1～2 行平滑或微棱外，其余均起棱；腹鳞 120～147；肛鳞二分；尾下鳞 65～92 对，个别有数枚成单。上颌齿每侧 22～26 枚，连续排列，由前到后渐次增大或最后 2 枚骤然增大。

[**生活习性**] 多生活于水域附近或山坡荒草丛中，偶见于水田周围。周围有泽陆蛙和锯腿原指树蛙等两栖动物。

[**地理分布**] 茂兰地区分布于洞塘。国内分布于安徽、重庆、福建、广东、广西、贵州、海南、湖南、江西、四川、台湾、云南。国外分布于越南（北部）。

[**模式标本**] 正模标本（CAS 18004）保存于中国科学院

[**模式产地**] 台湾

[**保护级别**] 三有保护动物；贵州省重点保护野生动物

[**IUCN 濒危等级**] 无危（LC）

a 整体照（祝非摄）
b 在茂兰地区的分布

（54）无颞鳞腹链蛇 *Hebius atemporale* (Bourret, 1934)

Natrix atemporalis Bourret, 1934, Bull. Gén. Instr. Publique, Hanoi, (4): 73-84. Type locality: Ton-king [= North Vietnam].

Natrix atemporalis Smith, 1943: 287. Taylor & Francis, London: 583.

Amphiesma atemporalis Malnate and Romer, 1969, Notulae Naturae, Acad. Nat. Sci. Philadelphia, 424: 1-8.

Amphiesma atemporalis Zhao, 2006. The snakes of China. Hefei, China, Anhui Science & Technology Publ. House, Vol. I: 372.

Amphiesma atemporale David et al., 2007, Zootaxa, 1462: 41-60.

Hebius atemporale Guo et al., 2014. Zootaxa, 3873 (4): 425-440.

Amphiesma atemporale Wallach et al., 2014: 28. Taylor & Francis, CRC Press: 1237.

[**同物异名**] *Natrix atemporalis*；*Amphiesma atemporalis*；*Hebius atemporale*；*Amphiesma atemporale*

[**俗名**] 无颞鳞游蛇

[**鉴别特征**] 小型具腹链的无毒蛇类。全长 400～500mm，尾长约占体全长的 1/3。体背棕褐色，体前段有不明显的白色小点斑，没有颞鳞或仅在第 5 枚上唇鳞与顶鳞

之间有 1 枚较小鳞片，背鳞通体 17 行，中央鳞片起棱。

[**形态描述**] 依据茂兰地区标本 GZNU20170423098 描述 体全长/尾长为 473.5mm/169.4mm，尾长约占体全长的 1/3。体背棕褐色，体前段有不明显的白色小点斑，没有颊鳞或仅在第 5 枚上唇鳞与顶鳞之间有 1 枚较小鳞片，背鳞通体 17 行，中央鳞片起棱，外侧 1～3 行平滑。颊鳞 1；眶前鳞 1，眶后鳞 3；颞鳞 0；上唇鳞 6（2-2-2 式），第 5 枚特别大，为黑褐色；下唇鳞 6，第 4 枚切前颔片；颔片两对，前对大于后对，后对后半分开呈"八"字形，其间介以小鳞。腹鳞 132；肛鳞二分。

[**生活习性**] 生活于山区，常见于小路边。白昼活动。

[**地理分布**] 茂兰地区分布于洞塘。国内分布于广东、广西、云南、贵州、香港。国外分布于越南。

[**模式标本**] 副模标本（MNHN-RA 1935.0077-0078、MNHN-RA 1935.0450）保存于法国国家自然历史博物馆

[**模式产地**] 越南（北部）

[**保护级别**] 三有保护动物；贵州省重点保护野生动物

[**IUCN 濒危等级**] 近危（NT）

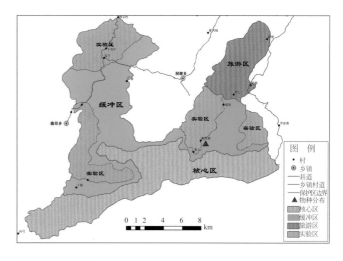

在茂兰地区的分布

成体量度

编号	产地	头体长（mm）	体全长（mm）	尾长（mm）	背鳞（枚）	腹鳞（枚）	肛鳞（枚）	上唇鳞（枚）	下唇鳞（枚）	颊鳞（枚）	眶前鳞（枚）	眶后鳞（枚）	前颞鳞（枚）	后颞鳞（枚）
GZNU20170423098	荔波茂兰	304.1	473.5	169.4	17-17-17	132	2	6	6	1	1	3	0	0

35）颈槽蛇属 *Rhabdophis* Fitzinger, 1843

Rhabdophis Fitzinger, 1843, Syst. Rept., Vienna, 1: 27. Type species: *Tropidonotus subminiatus* Schlegel, 1837, by original designation.

体型中等或略偏大的陆栖无毒蛇类。鼻间鳞前缘约与后缘等宽，鼻孔侧位。主要识别特征是颈背正中两行背鳞左右并列，其间形成 1 个浅的沟槽，故名"颈槽蛇"。本属个别蛇种颈槽的皮肤下有脊腺（或颈腺），分泌物有毒。本属物种的半阴茎分支，最后 2 枚上颌齿明显较大，向后弯曲，与前方的齿列形成齿间隙。

本属共 22 种，主要分布于俄罗斯、日本、中国、菲律宾、马来西亚、印度尼西亚、越南、老挝、泰国、缅甸、孟加拉国、印度及斯里兰卡等地。我国已知 10 种。茂兰地区有 2 种，即虎斑颈槽蛇和红脖颈槽蛇。

茂兰地区种检索表

颈及躯干前端鳞片间皮肤猩红色···红脖颈槽蛇 *Rhabdophis subminiatus*
颈及躯干前端两侧有红黑色相间的粗大斑纹···虎斑颈槽蛇 *Rhabdophis tigrinus*

（55）虎斑颈槽蛇 *Rhabdophis tigrinus* (Boie, 1826)

Tropidonotus tigrinus Boie, 1826: 205. Isis von Oken, Jena., 18-19: 203-216. Type locality: Japan; Dejima of Nagasaki City, eastern Kyushu, Japan (fide Zhao and Adler, 1993).

Amphiesma tigrinum Duméril and Bibron, 1854: 732. Vol. 7 (Partie 1), Paris, 16: 780.

Tropidonotus lateralis Berthold, 1859, Nachr. Georg-August-Univ. Königl. Ges. Wiss. Göttingen, (17): 179-181.

Amphiesma tigrinum Hallowell, 1861, John Rogers, U. S. N. Proc. Acad. Nat. Sci. Philadelphia, 12 [1860]: 480-510.

Tropidonotus orientalis Günther, 1861, Proc. Zool. Soc., London, 1861: 390-391 [1862].

Tropidonotus tigrinus Günther, 1888, Ann. Mag. Nat. Hist., (6) 1: 165-172.

Tropidonotus tigrinus Boulenger, 1893, London (Taylor & Francis): 249.

Tropidonotus tigrinus Boulenger, 1896, London (Taylor & Francis), 14: 607.

Natrix tigrina Stejneger, 1907: 272. Bull. U.S. Natl. Mus., 58 (20): 577.

Natrix tigrina lateralis Stejneger, 1907: 278. Bull. U.S. Natl. Mus., 58 (20): 577.

Natrix tigrina lateralis Mell, 1931, Lingnan Sci. Jour., Canton, 8 [1929]: 199-219.

Natrix tigrina lateralis ab. *caerulescens* Emelianov, 1936, Bull. Far E. Br. Acad. Sci. U.S.S.R., (19): 111-113.

Natrix tigrina lateralis Glass, 1946, Copeia, (4): 249-252.

Natrix tigrina Alexander and Diener, 1958, Copeia, (3): 218-219.

Rhabdophis tigrinus lateralis Szyndlar, 1984, Acta Zool. Cracov., 27 (1): 3-18.

Rhabdophis tigrinus formosanus Ota and Mori, 1985, Japanese Journal of Herpetology, 11: 41-45.

Rhabdophis tigrina Kharin, 2011, Biodiversity and Environment of Far East Reserves, Vladivostok, (1): 30-48.

Rhabdophis formosanus and *Rhabdophis lateralis* and *Rhabdophis tigrinus* Takeuchi et al., 2012, Biological Journal of the Linnean Society, 105: 395-408.

Rhabdophis tigrinus Wallach et al., 2014: 637. Taylor & Francis, CRC Press: 1237.

[同物异名] *Tropidonotus tigrinus*；*Amphiesma tigrinum*；*Tropidonotus lateralis*；*Tropidonotus orientalis*；*Natrix tigrina*；*Natrix tigrina lateralis*；*Natrix tigrina lateralis* ab. *caerulescens*；*Rhabdophis tigrina*；*Rhabdophis tigrinus lateralis*；*Rhabdophis lateralis*；*Rhabdophis tigrinus formosanus*；*Rhabdophis formosanus*

[俗名] 野鸡脖子、野鸡顶、竹竿青、鸡冠蛇、雉鸡蛇

[鉴别特征] 体型中等的无毒游蛇，体背面翠绿色或草绿色，体前段两侧有粗大的黑色与橘红色斑块相间排列，颈背有较明显的颈槽；枕部两侧有1对粗大的黑色"八"字形斑；躯干前段黑红色斑相间。

[形态描述] 依据茂兰地区标本GZNU20170423009描述 体型中等，雄性体全长/尾长为446.3mm/91.1mm。躯干及尾背面草绿色，躯干前段两侧有粗大的黑色与橘红色斑块相间排列，后段犹可见黑色斑块，橘红色则渐趋消失；颈背有较明显的颈槽；枕部两侧有1对粗大的黑色"八"字形斑；躯干前段黑红色斑相间；躯尾腹面黄绿色，腹鳞游离缘的颜色稍浅。头背绿色，上唇鳞灰白色，鳞沟色黑，眼正下方及斜后方各有1条黑纹最粗；头腹面白色。颊鳞1；眶前鳞2，眶后鳞3；颞鳞1+2；上唇鳞7（2-2-3式）；下唇鳞8，前4枚接前颔片；颔片2对。背鳞19-19-17，全部具棱或两侧最外行平滑；腹鳞148；肛鳞二分。

[生活习性] 多出没于有水草之处或农田、水沟、池塘等多蛙、蟾的地方，也见于远离水域但潮湿多草的山坡。多白天活动。

[地理分布] 茂兰地区分布于翁昂、洞塘。国内分布于安徽、北京、重庆、湖北、台湾、湖南、天津、福建、甘肃、广西、贵州、河北、河南、黑龙江、江苏、江西、吉林、辽宁、内蒙古、宁夏、青海、山东、山西、陕西、上海、四川、西藏、云南、浙江。国外分布于日本。

[模式标本] 正模标本（BMNH 1861.12.27.9）保存于英国伦敦自然历史博物馆

[模式产地] 日本长崎

[保护级别] 三有保护动物；贵州省重点保护野生动物　　　　　　[IUCN 濒危等级] 无危（LC）

成体量度

编号	产地	头体长（mm）	体全长（mm）	尾长（mm）	背鳞（枚）	腹鳞（枚）	肛鳞（枚）	上唇鳞（枚）	下唇鳞（枚）	颊鳞（枚）	眶前鳞（枚）	眶后鳞（枚）	前颞鳞（枚）	后颞鳞（枚）
GZNU20170423009	荔波茂兰	355.2	446.3	91.1	19-19-17	148	2	7	8	1	2	3	1	2

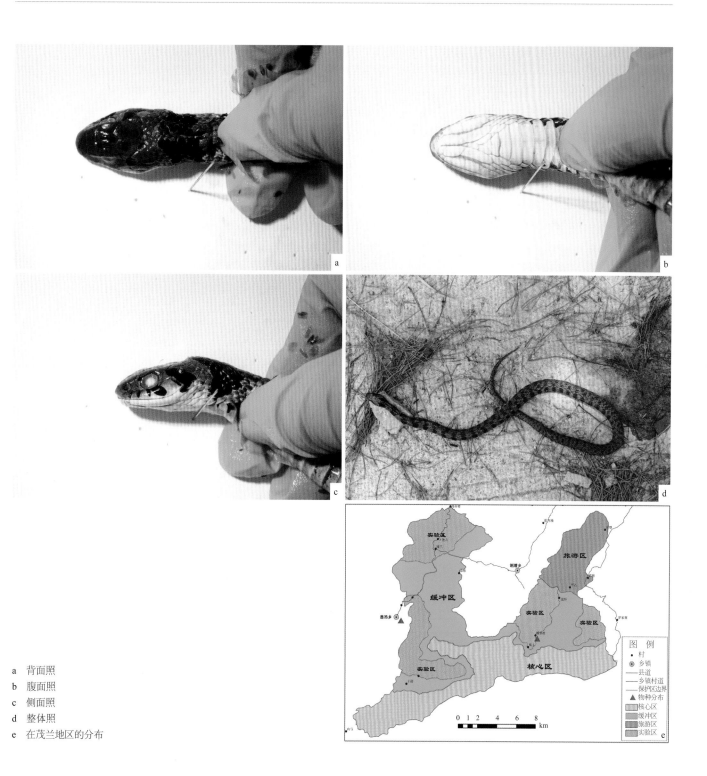

a　背面照
b　腹面照
c　侧面照
d　整体照
e　在茂兰地区的分布

（56）红脖颈槽蛇 *Rhabdophis subminiatus* (Schlegel, 1837)

Tropidonotus subminiatus Schlegel, 1837: 313. Partie Descriptive. La Haye (Kips and van Stockum), 16: 606. Type locality: Java.

Amphiesma subminiatum Duméril and Bibron, 1854: 734. Vol. 7 (Partie 1), Paris, 16: 780.

Tropidonotus subminiatus Boulenger, 1893: 256. London (Taylor & Francis): 448.

Pseudoxenodon intermedius Lönnberg, 1899, Zool. Anz., 22: 108-111.

Tropidonotus subminiatus Wall, 1908, J. Bombay Nat. Hist. Soc., 18: 312-337.

Natrix helleri Schmidt, 1925, American Museum Novitates, (157): 1-5.

Natrix subminiata hongkongensis and *Natrix subminiata subminiata* and *Natrix subminiata siamensis* Mell, 1931, Lingnan Sci. Jour., Canton, 8 [1929]: 199-219.

Natrix subminiata helleri Mell, 1931, Lingnan Sci. Jour., Canton, 8 [1929]: 199-219.

Rhabdophis subminiata Malnate, 1960, Proc. Acad. Nat. Sci. Philadelphia, 112 (3): 41-71.

Rhabdophis subminiatus siamensis Deuve, 1961, Bull. Soc. Sci. Nat. Laos, 1: 5-32.

Rhabdophis subminiatus var. *helleri* Deuve, 1961, Bull. Soc. Sci. Nat. Laos, 1: 5-32.

Rhabdophis subminiatus subminiatus and *Rhabdophis subminiata helleri* Taylor, 1965, Univ. Kansas Sci. Bull., 45 (9): 609-1096.

Rhabdophis subminatus Ferlan et al., 1983, Toxicon, 21 (4): 570-574.

Rhabdophis subminiatus helleri Zhao et al., 1986, Acta Herpetologica Sinica, 5 (2): 157.

Rhabdophis subminiatus Manthey and Grossmann, 1997: 389. Natur und Tier Verlag, Münster: 512.

Rhabdophis subminiatus Cox et al., 1998: 47. Ralph Curtis Publishing: 144.

Rhabdophis subminiata Sharma, 2004. Akhil Books, New Delhi: 292.

Rhabdophis subminiatus helleri Sang et al., 2009, Chimaira, Frankfurt: 768.

Rhabdophis subminiatus Wallach et al., 2014: 636. Taylor & Francis, CRC Press: 1237.

[同物异名] *Tropidonotus subminiatus*；*Natrix subminiatus*；*Amphiesma subminiatum*；*Pseudoxenodon intermedius*；*Natrix subminiata subminiata*；*Rhabdophis subminiata*；*Natrix helleri*；*Natrix subminiata hongkongensis*；*Natrix subminiata siamensis*；*Natrix subminiata helleri*；*Rhabdophis subminiatus* var. *helleri*；*Rhabdophis subminiatus siamensis*，*Rhabdophis subminiatus*，*Rhabdophis subminiatus subminiatus*，*Rhabdophis subminiata helleri*

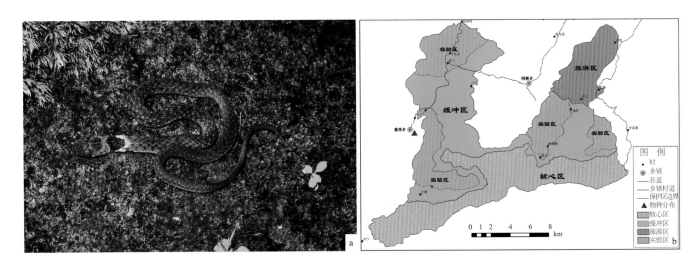

a 整体照（钟光辉摄）
b 在茂兰地区的分布

[俗名] 野鸡项、红脖游蛇、扁脖子

[鉴别特征] 本种通体草绿色，颈区及体前段鳞片间皮肤猩红色，颈背部的颈槽不显或无颈槽；枕部两侧无黑色"八"字形斑。

[形态描述] 通体草绿色，颈区及体前段鳞片间皮肤猩红色。受到惊扰时，体前段膨扁，鳞片张开露出皮肤，颈部及体前段显示猩红色，起到警戒威吓作用，故名"红脖颈槽蛇"。头、颈区分明显，颈及体前段背正中两行鳞片并列，鼻间鳞前端略窄。眼较大，瞳孔圆形。颈背部的颈槽不显或无颈槽；枕部两侧无黑色"八"字形斑。头部上唇鳞色稍浅，部分鳞沟黑色，头腹面灰白色，躯尾腹面黄白色。资料记载：最大体全长/尾长为雄性973mm/203mm、雌性1135mm/190mm。中等体型，颈背正中有1纵列浅凹槽。颊鳞1；眶前鳞1，眶后鳞3（4），个别为2；颞鳞2+3或2，个别一侧前颞鳞为1或3；上唇鳞8（2-3-3式，个别为3-2-3式），有的标本一侧为9（3-3-3式）；下唇鳞10，前5枚切前颏片，个别标本为9（4）；颏片2对。背鳞19-19-17，全部具棱或两侧最外行平滑；腹鳞144～184；肛鳞二分；尾下鳞56～97对。上颌齿每侧（21～24）+2，最后2枚骤然增大，与其前方齿列有1齿间隙。

[生活习性] 生活于山坡、路边、农耕区水沟附近或草丛中，也见于林间山地。卵生。无毒蛇。

[地理分布] 茂兰地区分布于翁昂。国内分布于福建、广东、广西、贵州、海南、四川、香港、云南。国外分布于东南亚各国。

[模式标本] 正模标本（AMNH 20149）保存于美国自然历史博物馆，副模（FMNH [*helleri*]）保存于芝加哥菲尔德自然历史博物馆

[模式产地] 印度尼西亚爪哇

[保护级别] 三有保护动物；贵州省重点保护野生动物

[IUCN 濒危等级] 无危（LC）

36）异色蛇属 *Xenochrophis* Günther, 1864

Xenochrophis Günther, 1864, Rept. Brit. India, London: 273. Type species: *Psammophis cerasogaster* Cantor, 1839. Type locality: India, by monotypy.

本属的特征是：半阴茎及精沟分支；上颌齿30枚呈现连续排列，后面的增大；鼻间鳞前端较窄，鼻孔背侧位；背鳞没有端窝或颈部的鳞片有不甚明显的端窝。本属多是半水栖物种。

本属已知7种，东起中国台湾，西到巴基斯坦的俾路支省及东南亚广泛分布。我国分布2种。茂兰地区有异色蛇 *Xenochrophis piscator* 和黄斑异色蛇 *Xenochrophis flavipunctatus* 2种。

茂兰地区种检索表

背面具有网状斑点或较大的黑点；背外侧无白色或黄色小斑点；颈部无黄色的"V"形斑··· 异色蛇 *Xenochrophis piscator*
背面具有黑色小斑点或纵纹；背外侧有白色或黄色组成的小斑点；颈部有一个黄色的"V"形斑··黄斑异色蛇 *Xenochrophis flavipunctatus*

（57）异色蛇 *Xenochrophis piscator* (Schneider, 1799)

Hydrus Piscator Schneider, 1799, Historiae Amphibiorum narturalis et literariae. Frommanni, Jena: 266. Type locality: East Indies.
Natrix piscator Merrem, 1820: 122. Versuch eines Systems der Amphibien I. J. C. Kriegeri, Marburg: 191.
Coluber bengalensis Gray, 1834, Illustrations of Indian Zoology, Vol. 2, London, (1833-1834): 263.
Tropidonotus quincunciatus Schlegel, 1837: 307. Essai sur la physionomie des serpens. Partie Descriptive, 16: 606.
Tropidonotus quincunciatus Günther, 1859, Ann. Mag. Nat. Hist., (3) 3: 221-237.

Amphiesma flavipunctatum Hallowell, 1861: 503. John Rogers, U. S. N. Proc. Acad. Nat. Sci. Philadelphia, 12 [1860]: 480-510.

Tropidonotus piscator Boulenger, 1893: 230. Catalogue of the snakes in the British Museum I. London (Taylor & Francis): 448.

Natrix piscator Stejneger, 1907: 288. Bull. U.S. Natl. Mus., 58 (20): 1-577.

Tropidonotus piscator Wall, 1908, J. Bombay Nat. Hist. Soc., 18: 312-337.

Nerodia (*Tropidonotus*) *piscator* Wall, 1921: 91. Ophidia Taprobanica or the Snakes of Ceylon. Colombo Mus., Colombo., 22: 581.

Natrix piscator Smith, 1943. Reptilia and Amphibia. 3 (Serpentes). Taylor & Francis, London: 583.

Natrix piscator Tweedie, 1954, Bull. Raffles Mus., No. 25: 107-117.

Xenochrophis piscator Cox et al., 1998: 43. Ralph Curtis Publishing: 144.

Xenochrophis piscator Chan-ard et al., 1999: 190. Amphibians and reptiles of peninsular Malaysia and Thailand-an illustrated checklist. Bushmaster Publications, Würselen, Germany: 240.

Xenochrophis piscator Wallach et al., 2014: 798. Snakes of the World: A catalogue of living and extinct species: 1273.

Xenochorphis piscator Fellows, 2015, Entomol Ornithol Herpetol., 4: 136.

[同物异名] *Hydrus piscator*；*Natrix piscator*；*Coluber bengalensis*；*Tropidonotus quincunciatus*；*Tropidonotus piscator*；*Nerodia*（*Tropidonotus*）*piscator*；*Amphiesma flavipunctatum*

[俗名] 渔游蛇、红糟蛇、草花蛇、千布花甲、小黄蛇、鱼蛇、水草蛇

[鉴别特征] 半水栖无毒蛇。腹鳞白色，每一腹鳞基部黑色，整个腹面看来呈黑白相间的横纹；头侧眼后下方有2条黑色细线纹；鼻间鳞前端较窄，鼻孔位于近背侧。

[形态描述] 依据茂兰地区标本GZNU20170423010描述 体型中等大小。体全长939.28mm，头体长/尾长为684.14mm/255.14mm。头、颈区分明显；瞳孔圆形；鼻间鳞前端较窄，鼻孔背侧位。眼后下方有2条黑色细线纹分别斜达上唇缘和口角；腹面白色，腹鳞基部黑色，形成整个腹面黑白相间的横纹。背面橄榄绿色，前段两侧隐约可见数行黑色横斑；体侧许多背鳞边缘黑色，在体侧也形成黑横纹。头背橄榄灰色，顶鳞沟及其后有1镶黑边的短白纵纹，颈背有"V"形黑纹。上唇鳞色白，头腹面灰白色。颊鳞1；眶前鳞1；眶后鳞+眶下鳞5；颞鳞为2+3；上唇鳞8（2-2-4式）为主，下唇鳞10，前5枚切前颏片；颏片2对，后对大于前对。背鳞17-17-15，腹鳞139；肛鳞二分。

[生活习性] 我国南部平原丘陵低海拔地区稻田或其他水域附近常见的一种中型游蛇。多见于潮湿多水草的地方，如田边、水塘、水沟，或路边草丛、山野房舍院内。食物来源广泛，包括鱼、蛙、蝌蚪、蛙卵、蜥蜴、小型兽类等。

[地理分布] 茂兰地区分布于永康。国内分布于西藏、安徽、澳门、福建、广东、江西、贵州、海南、湖北、湖南、江苏、江西、陕西、台湾、香港、云南、浙江等地。国外分布于阿富汗、巴基斯坦、印度、斯里兰卡、越南、老挝、柬埔寨、缅甸、泰国、新加坡、马来西亚和印度尼西亚等地。

[模式标本] 正模标本（BMNH 1904.7.27.32）保存于英国伦敦自然历史博物馆

[模式产地] 东印度群岛

[保护级别] 三有保护动物；贵州省重点保护野生动物

[IUCN 濒危等级] 无危（LC）

成体量度

编号	产地	头体长（mm）	体全长（mm）	尾长（mm）	背鳞（枚）	腹鳞（枚）	肛鳞（枚）	上唇鳞（枚）	下唇鳞（枚）	颊鳞（枚）	眶前鳞（枚）	眶后鳞+眶下鳞（枚）	前颞鳞（枚）	后颞鳞（枚）
GZNU20170423010	荔波	684.14	939.28	255.14	17-17-15	139	2	8	10	1	1	5	2	3

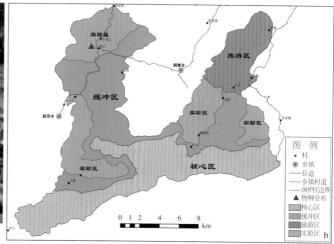

a 侧面照
b 在茂兰地区的分布

（58）黄斑异色蛇 *Xenochrophis flavipunctatus* (Hallowell, 1860)

Amphiesma flavipunctatum Hallowell, 1860, John Rogers, U. S. N. Proc. Acad. Nat. Sci. Philadelphia, 12 [1860]: 480-510. Type locality: Island of Hongkong [= Hong Kong], China.

Natrix piscator flavipunctata Smith, 1943. Reptilia and Amphibia. 3 (Serpentes). Taylor & Francis, London: 583.

Xenochrophis flavipunctatus Manthey and Grossmann, 1997: 397. Amphibien & Reptilien Südostasiens. Natur und Tier Verlag (Münster): 512.

Xenochrophis flavipunctatus Pauwels et al., 2000, Dumerilia, 4 (2): 123-154.

Xenochrophis flavipunctatus Vogel and David, 2006, Herpetologica Bonnensis, II: 241-246.

Xenochrophis flavipunctatus Wallach et al., 2014: 797. Snakes of the World. Taylor & Francis, CRC Press: 1237.

[同物异名] *Amphiesma flavipunctatum*; *Natrix piscator flavipunctata*

[俗名] 黄斑渔游蛇、渔游蛇

[鉴别特征] 头长椭圆形，与颈区分明显。体色变异较大，自颈后至尾有黑色网纹，网纹两侧有醒目的黑斑；头背灰绿色，眼下至唇边有1条短黑纹，眼后至口角有长黑纹。

[形态描述] 体型中等大小，体全长0.5～1m。体色变异较大，背面灰褐色、深灰色、灰棕色、橄榄绿色、暗绿色、黄褐色或橘黄色，头、颈区分明显，自颈后至尾有黑色网纹，网纹两侧有醒目的黑斑；头背灰绿色，眼下至唇边有1条短黑纹，眼后至口角有长黑纹。颈部有1个"V"形黑斑；上唇鳞色白，眼后下方有两条黑色细线纹分别斜达上唇缘和口角；腹面白色或黄白色或淡绿黄色，腹鳞基部黑色，使整个腹面呈现等距离的黑横纹。瞳孔圆形；鼻尖鳞前端比较窄，鼻孔背侧位。

[生活习性] 一般生活在山区丘陵、平原及田野的河湖水塘边。半水生，夜行性，能在水中潜游。性凶猛，常攻击捕蛇者。主要猎捕小鱼，兼食蛙、蟾蜍等。当受到惊吓时，它会抬起身体前部，采取攻击的姿势。每年5～7月产卵，每次产3～14枚，自然孵化，孵化期为一个多月。其胆可药用。无毒蛇。

[地理分布] 茂兰地区分布于洞塘。国内分布于香港、广东、贵州。国外分布于印度、泰国、缅甸、马来西亚（西部）、老挝、柬埔寨、越南、孟加拉国。

[模式标本] 正模标本（USNM 7387）保存于美国国家自然历史博物馆

[模式产地] 香港

[保护级别] 贵州省重点保护野生动物

[IUCN濒危等级] 无危（LC）

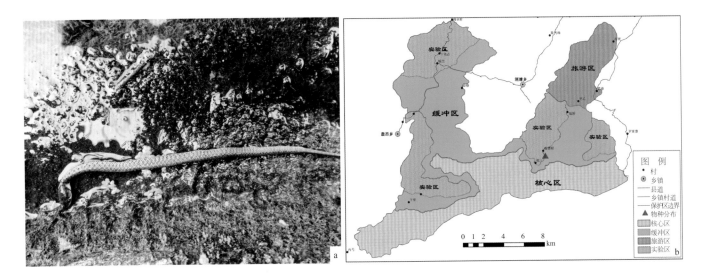

a 整体照
b 在茂兰地区的分布

37）后棱蛇属 *Opisthotropis* Günther, 1872

Opisthotropis Günther, 1872, Ann. Mag. Nat. Hist., London, (4) 9: 16. Type species: *Opisthotropis ater* Günther, 1872. Type locality: West Africa (in error), by monotype.

本属物种属体型中等大小的半水栖无毒蛇类。头较小而略扁，头与颈无明显区分；鼻间鳞较窄，鼻孔位于鼻鳞外侧上方；前额鳞单枚，吻宽大于长；上颌齿小，每侧20～40枚，最后2枚稍大。躯干圆柱形，背鳞15～19行，平滑或起棱，平滑者间隔若干行往往也有棱，因此称"后棱蛇"。尾长度适中，尾下鳞成对。卵生。

本属已知22种，分布于东南亚。我国有13种。茂兰地区分布1种，即山溪后棱蛇 *Opisthotropis latouchii*。

（59）山溪后棱蛇 *Opisthotropis latouchii* (Boulenger, 1899)

Tapinophis latouchii Boulenger, 1899, Proc. Zool. Soc., London, 1899: 159-172. Type locality: Kuatun, Fukien, China.
Cantonophis praefrontalis Werner, 1909, Jahreshefte Ver. Vaterl. Naturk. Württ., 65: 55-63.
Opisthotropis latouchii Pope, 1935, Amer. Mus. Nat. Hist., New York, Nat. Hist. Central Asia, 10 (52): 604.
Opisthotropis latouchii Brown and Leviton, 1961, Occasional Papers of the Natural History Museum of Stanford University, (8): 1-11.
Opisthotropis latouchii Li et al., 2010, Asian Herpetological Research, 1 (1): 57-60.
Opisthotropis latouchii Wallach et al., 2014: 510. Snakes of the World. Taylor & Francis, CRC Press: 1237.

[同物异名] *Tapinophis latouchii*；*Cantonophis praefrontalis*

[俗名] 福建颈斑蛇

[鉴别特征] 颊鳞1，入眶；上唇鳞8～9，第1枚直立，与鼻鳞的前半部邻接；背鳞通体17行；背部有黑黄相间的纵纹。

[形态描述] 依据茂兰地区标本GZNU201708描述 体全长/尾长为419.2mm/82.0mm。前额鳞单枚；鼻间鳞较窄，

鼻孔背侧位；背面棕黄色，背面呈多条黑黄相间的纵纹。体尾背面呈黑黄相间的多条纵纹，腹面浅黄白色。颊鳞1，入眶；眶前鳞1，眶后鳞2；颞鳞1+2；上唇鳞9（4-2-3式），第1枚直立，与鼻鳞的前半部邻接；下唇鳞9，前4枚接前颏片；颏片2对。背鳞17-17-17；腹鳞149；肛鳞二分，尾下鳞双列。

[生活习性] 多见于山溪水中或石下，活动于岩石、砂砾或腐烂植物间；也见于水沟或稻田附近。在调查过程

成体量度

编号	产地	头体长（mm）	体全长（mm）	尾长（mm）	背鳞（枚）	腹鳞（枚）	肛鳞（枚）	上唇鳞（枚）	下唇鳞（枚）	颊鳞（枚）	眶前鳞（枚）	眶后鳞（枚）	前颞鳞（枚）	后颞鳞（枚）
GZNU201708	荔波茂兰	337.2	419.2	82.0	17-17-17	149	2	9	9	1	1	2	1	2

a 背面照
b 腹面照
c 侧面照
d 整体照
e 在茂兰地区的分布

中见于竹林小路的岩石上，离水源较远。

[地理分布]　茂兰地区分布于洞塘。国内分布于安徽、重庆、福建、广东、广西、贵州、湖南、江西、浙江等地。国外无报道。

[模式标本]　正模标本（BMNH 1946.1.13.37）保存于英国伦敦自然历史博物馆

[模式产地]　福建

[保护级别]　三有保护动物；贵州省重点保护野生动物

[IUCN濒危等级]　无危（LC）

38）华游蛇属 *Sinonatrix* Rossman and Eberle，1977

Sinonatrix Rossman and Eberle, 1977, Herpetologica, Lawerence, 33: 42. Type species: *Tropidonotus annular* Hallowell, 1857, by original designation.

本属物种多是半水栖中型或偏大的无毒蛇类，鼻间鳞前端较窄，鼻孔背侧位。本属蛇种原隶属于广义的游蛇属 *Natrix sensu lato*，Rossman 和 Eberle（1977）以 *Tropidonotus annularis* Hallowell, 1856 为模式种，建立华游蛇属 *Sinonatrix*。由于本属蛇种主要分布于中国，故称为"华游蛇"。本属主要特征：染色体 $2n = 42$；中段背鳞19行；前颞鳞2；肛鳞二分；尾下鳞无棱；腭骨与上颌骨正常；半阴茎中等程度分叶，具单一的左旋精沟，顶部裸区不显著。

本属已知有4种，分布于东南亚。我国分布4种。茂兰地区分布2种，即乌华游蛇 *Sinonatrix percarinata* 和环纹华游蛇 *Sinonatrix aequifasciata*，分种检索表如下。

茂兰地区种检索表

体背棕褐色，背通体具镶黑色边的粗大环纹，环纹中央绿褐色；在体侧形成粗大"X"形斑·····················环纹华游蛇 *Sinonatrix aequifasciata*

体背暗橄榄绿色，通体具有黑褐色环纹，在体侧不形成"X"形斑··乌华游蛇 *Sinonatrix percarinata*

（60）乌华游蛇 *Sinonatrix percarinata* (Boulenger, 1899)

Tropidonotus percarinata Boulenger, 1899, Proc. Zool. Soc., London, 1899: 163. Type locality: Kuatun [= Guadun], Chongan Co., about 270 miles from Foochow, NW Fokien [= Fujian], China.

Sinonatrix percarinata percarinata Boulenger, 1899, Proc. Zool. Soc., London, 1899: 159-172.

Natrix annularis Barbour, 1912, Mem. Mus. Comp. Zool., XL: 130.

Natrix percarinatus Parker, 1925, Ann. Mag. Nat. Hist., (9) XV: 302-304.

Natrix suriki Maki, 1931, Monogr. Sn. Japan: 38-40.

Natrix percarinata Mell, 1931, Lingnan Sci. Jour., Canton, 8 [1929]: 199-219.

Natrix percarinata Smith, 1943. Reptilia and Amphibia. 3 (Serpentes). Taylor & Francis, London: 299.

Sinonatrix percarinata percarinata and *Sinonatrix percarinata suriki* Zhao and Jiang, 1986, Acta Herpetol. Sinica, Chengdu, 5: 240.

Sinonatrix percarinata Ziegler, 2002: 263. Natur und Tier Verlag (Münster): 342.

Sinonatrix percarinata percarinata and *Sinonatrix percarinata suriki* Sang et al., 2009. Herpetofauna of Vietnam. Chimaira, Frankfurt: 768.

Sinonatrix percarinata Wallach et al., 2014: 677. Snakes of the World. Taylor & Francis, CRC Press: 1237.

[同物异名] *Tropidonotus percarinata*；*Natrix annularis*；*Natrix percarinata*；*Natrix suriki*；*Sinonatrix percarinata percarinata*；*Sinonatrix percarinata suriki*；*Natrix percarinatus*

[曾用名] 华游蛇

[鉴别特征] 本种通体具多数环纹，腹面不呈橘红色或橙黄色，鼻间鳞前端极窄，鼻孔位于近背侧，2枚上唇鳞入眶。山区溪流或水田内的一种中型游蛇。

[形态描述] 依据茂兰地区标本GZNU320172描述 体全长/尾长为464.9mm/132.5mm。头、颈可以区分；鼻间鳞前端极窄，鼻孔位于近背侧；眼较小。体尾有几十个环纹，体侧清晰可见。躯尾背面暗橄榄绿色，体侧橘红色，有明显的黑褐色环纹，背面基本色调较深，环纹模糊不清，体侧则清晰可数，均呈"Y"形；腹面灰白色，尾下鳞边缘黑色，构成尾腹面双行网格及左右尾下鳞沟交错而成的中央黑色折线纹。头背橄榄灰色，上唇鳞色稍浅淡，鳞沟色较深；头腹面灰白色。颊鳞1；眶前鳞为1，眶后鳞3，颞鳞2+3，上唇鳞9（3-2-4式），第4对仅以后上角入眶；下唇鳞10，第1对在颏鳞后相接，前5枚以接前颏片为主；颏片2对，后对长于前对，后对左右之间局部有3枚小鳞以各种排列方式相隔。背鳞19-19-17，均具棱，背正中各行的棱特别强；腹鳞142；肛鳞二分。

[生活习性] 多栖息于平原、丘陵或山区，常出没于稻田、水沟、溪流、大河等各种水域及其附近。

[地理分布] 茂兰地区分布于翁昂、洞塘、永康等地。国内分布于陕西、安徽、重庆、福建、甘肃、广东、广西、贵州、海南、河南、湖北、湖南、江苏、江西、上海、四川、香港、云南、浙江等地。国外分布于缅甸（北部）、泰国、越南。

[模式标本] 未知（Sang et al., 2009）

[模式产地] 福建挂墩

[保护级别] 三有保护动物；贵州省重点保护野生动物

[IUCN 濒危等级] 易危（VU）

成体量度

编号	产地	头体长（mm）	体全长（mm）	尾长（mm）	背鳞（枚）	腹鳞（枚）	肛鳞（枚）	上唇鳞（枚）	下唇鳞（枚）	颊鳞（枚）	眶前鳞（枚）	眶后鳞（枚）	前颞鳞（枚）	后颞鳞（枚）
GZNU320172	荔波茂兰	332.4	464.9	132.5	19-19-17	142	2	9	10	1	1	3	2	3

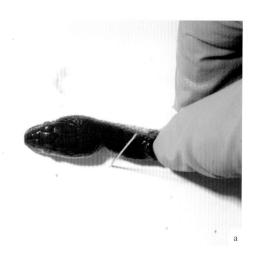
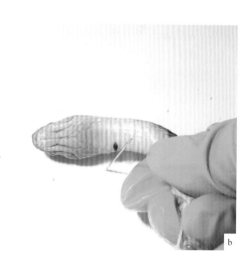

a 背面照
b 腹面照

c 侧面照
d 整体照
e 在茂兰地区的分布

（60a）乌华游蛇指名亚种 *Sinonatrix percarinata percarinata* (Boulenger, 1899)

[特征]　腹鳞 131～160，平均 137.8（$n = 228$）。

[分布]　除台湾以外，本种在我国广泛分布。国外分布于越南。

（60b）乌华游蛇台湾亚种 *Sinonatrix percarinata suriki* (Maki, 1931)

[特征]　腹鳞 142～153，平均 148（$n = 6$）。　　[分布]　台湾。

（61）环纹华游蛇 *Sinonatrix aequifasciata* (Barbour, 1908)

Natrix aequifasciata Barbour, 1908, Bull. Mus. Comp. Zool. Harvard, 51 (12): 317. Type locality: Mt. Wuchi, now Mt. Wuzhi, Hainan Island, People's Republic of China.

Natrix aequifasciata Mell, 1931, Lingnan Sci. Jour., Canton, 8 [1929]: 199-219.

Natrix aequifasciata Chang, 1932, Contr. Biol. Lab. Sci. Soc. China, (Zool. Series) 8: 33.

Sinonatrix aequifasciata Rossman and Eberle, 1977, Herpetologica, Lawerence, 33: 42.

Sinonatrix aequifasciata Vogel et al., 2004, Hamadryad, 29 (1): 110-114.

Sinonatrix aequifasciata Sang et al., 2009. Herpetofauna of Vietnam. Chimaira, Frankfurt: 768.
Sinonatrix aequifasciata Wallach et al., 2014: 677. Snakes of the World. Taylor & Francis, CRC Press: 1237.

[同物异名]　*Natrix aequifasciata*

[曾用名]　环纹游蛇

[鉴别特征]　体型粗大的水栖无毒蛇。开阔山区溪流中的一种大型游蛇。体型粗大，周身有粗大环纹，在体侧形成"X"形斑，鼻间鳞前端极窄，鼻孔位于近背侧，常有0～3枚眶下鳞，上唇鳞通常只有1或2枚入眶或全不入眶。

[形态描述]　体全长/尾长：雄性930mm/210mm，雌性1095mm/250mm。头、颈区分明显；鼻间鳞前端极窄，鼻孔背侧位；眼较小。雄性唇鳞及颏片疣粒显著。

躯干及尾背面棕褐色，体侧及腹面黄白色，通体有粗大环纹（17～25）+（10～13）个。环纹镶黑色或黑褐色边，中央绿褐色，在体侧每一环纹的两黑边相交，再分叉而达腹中线；从体侧看，每一环纹形成一个黑色的"X"形斑。头背灰褐色，腹面灰白色。亚成体环纹清晰，随年龄增长环纹颜色变淡，老年个体的背面环纹模糊，很难辨别，仅体侧可看出"X"形斑。颊鳞1；眶前鳞1或2，眶后鳞3（2，4），常有1枚较小的眶下鳞，眶周围鳞片数目较多，有的眶前鳞与眶后鳞相接，以致上唇鳞有的全不入眶；颞鳞2（1）+3（2），个别为3+2；上唇鳞9（3-4-2式，少数3-3-3或4-1-4式），个别一侧为8（4-1-3式）或10（3-3-4、5-1-4或4-2-4式）；下唇鳞10（8～11），前5（4～6，个别3）枚接前颏片；颏片2对，后对长于前对。背鳞19-19-17，中段中央13～17行起强棱，个别全部起棱；腹鳞144～164；肛鳞二分；尾下鳞双行，61～78对。上颌齿每侧25～28枚。

[生活习性]　多栖息于平原、丘陵和山区。多见于地形比较开阔的较大溪流，也见于河旁洞内，有时可攀援在岸边灌木上，如遇惊扰会潜入水下。主要以鱼类为食。卵生。

[地理分布]　茂兰地区分布于永康。国内分布于重庆、福建、广东、广西、贵州、海南、湖南、江西、香港、云南、浙江。国外未见报道。

[模式标本]　正模标本（MCZ 7101）保存于哈佛大学比较动物学博物馆（Museum of Comparative Zoology，MCZ）

[模式产地]　海南五指山

[保护级别]　三有保护动物；贵州省重点保护野生动物

[IUCN濒危等级]　易危（VU）

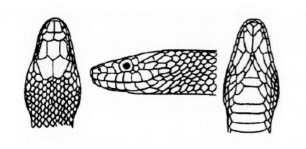

头部特征（引自《贵州爬行类志》，1985）

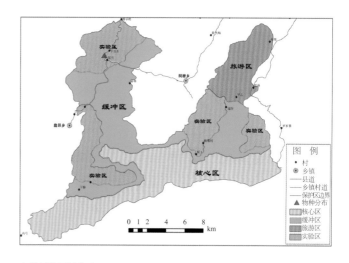

在茂兰地区的分布

39）颈棱蛇属 *Macropisthodon* Boulenger, 1893

Macropisthodon Boulenger, 1893, Cat. Sn. Brit. Mus., I: 265. Type species: *Macropisthodon flaviceps* Duméril, Bibron and Duméril, 1854.

本属物种身体粗大，呈圆柱形，能膨扁；头与颈区分明显，眼大，瞳孔圆形；背鳞具强棱，尾下鳞成对。无毒。

本属共有4种，分布于中国（南部）、印度、马来群岛。我国仅分布1种。茂兰地区有1种，即颈棱蛇 *Macropisthodon rudis*。

（62）颈棱蛇 *Macropisthodon rudis* Boulenger, 1906

Macropisthodon rudis Boulenger, 1906, An. Mag. Nat. Hist., 17 (7): 568. Type locality: Dongchuan, Yunnan.

Pseudagkistrodon carinatus van Denburgh, 1909, Proc. Cal. Acad. Sci., 3 (4): 51.

Natrix namiei Oshima, 1910, Annot. Zool. Japon., Tokyo, 7 (3): 185-207.

Macropisthodon carinatus Stejneger, 1910, Proc. U.S. Natl. Mus., 38: 91-114.

Macropisthodon rudis carinatus Maki, 1931, Monogr. Snakes Japan, 55: P1, 17.

Pseudagkistrodon rudis Wallach et al., 2014. Snakes of the World: A catalogue of living and extinct species. Taylor & Francis, CRC Press: 594.

[同物异名] *Pseudagkistrodon carinatus*；*Natrix namiei*；*Macropisthodon rudis carinatus*；*Pseudagkistrodon rudis*；*Macropisthodon carinatus*

[俗名] 伪蝮蛇、老憨蛇

[鉴别特征] 颊鳞具棱，无颊窝；身体花纹近似蝮蛇。在受惊扰或威胁时常采取缩扁头部及身体的伪装策略，外观与山烙铁头和蝮蛇相似，但山烙铁头和蝮蛇瞳孔垂直，有颊窝，可作为辨别二者的依据。

[形态描述] 身体粗大，体全长1m左右，无毒，尾短，背面呈棕褐色，有两行粗大的深棕色斑块，腹面褐色，前部具有白色点斑，头部略呈三角形，外形像蝮蛇或蝰蛇，有"伪蝮蛇"之称。上唇鳞一般为7，不入眶，下唇鳞9或10，前4～5枚切前颏片。前颏片短于后颏片。颊鳞2或1，不入眶。眶前鳞3；眶后鳞3或4；前颞鳞2或3，后颞鳞4或3，前、后颞鳞均起棱，在前颞鳞与上唇鳞之间还有一枚较小的鳞片，其前缘与眶后鳞相接，而后缘不与眶后鳞相接。背鳞23-23-19，起棱明显，腹鳞123～158；肛鳞2；尾下鳞37～61对。上颌齿11～18，间隔一个短的无齿区后有2枚特别大的向后弯曲的牙齿，形似后沟牙。

[生活习性] 常见于灌丛、草丛或树林中。受惊时头体能变扁平。大多出现在天然阔叶林底层，性温驯，无毒，无攻击性。以蟾蜍和蛙类为食。卵胎生。

[地理分布] 茂兰地区分布于洞塘。国内分布于福建、云南、贵州、湖南、四川、江西、浙江、广西、广东。国外无报道。

[模式标本] 正模标本（ZMB 38069）保存于德国柏林洪堡大学自然历史博物馆；正模标本（CIB 6515143）保存于中国科学院成都生物研究所

[模式产地] 云南东川

[保护级别] 三有保护动物；贵州省重点保护野生动物

[IUCN 濒危等级] 无危（LC）

a 背面照
b 腹面照
c 在茂兰地区的分布

主要参考文献

蔡波, 王跃招, 陈跃英, 等. 2015. 中国爬行纲动物分类厘定. 生物多样性, 23 (3): 365-382.
陈晓虹, 周开亚, 郑光美. 2010a. 中国臭蛙属一新种. 动物分类学报, 35 (1): 206-211.
陈晓虹, 周开亚, 郑光美. 2010b. 中国臭蛙类一新种. 北京师范大学学报(自然科学版), 46 (5): 606-609.
陈晓虹, 朱命炜, 侯名根, 等. 2006. 河南游蛇科一新记录——平鳞钝头蛇. 四川动物, 25 (2): 269-269.
费梁, 胡淑琴, 叶昌媛, 等. 2006. 中国动物志 两栖纲(上卷 总论) 蚓螈目 有尾目. 北京: 科学出版社: 1-471.
费梁, 胡淑琴, 叶昌媛, 等. 2009a. 中国动物志 两栖纲(中卷) 无尾目. 北京: 科学出版社: 1-958.
费梁, 胡淑琴, 叶昌媛, 等. 2009b. 中国动物志 两栖纲(下卷) 无尾目 蛙科. 北京: 科学出版社: 959-1847.
费梁, 叶昌媛, 黄永昭. 1990. 中国两栖动物检索. 重庆: 科学技术文献出版社重庆分社.
费梁, 叶昌媛, 黄永昭, 等. 2005. 中国两栖动物检索及图解. 成都: 四川科学技术出版社: 1-340.
费梁, 叶昌媛, 江建平. 2010. 中国两栖动物彩色图鉴. 成都: 四川科学技术出版社: 1-519.
费梁, 叶昌媛, 江建平. 2012. 中国两栖动物及其分布彩色图鉴. 成都: 四川科学技术出版社: 1-619.
费梁, 叶昌媛, 谢锋, 等. 2007. 中国四川省蛙科一新种——南江臭蛙(两栖纲: 无尾目). 动物学研究, 28 (5): 551-555.
费梁, 叶昌媛, 杨戎生. 1984. 疣螈属一新种和一新亚种(蝾螈目, 蝾螈科). 动物学报, (1): 85-91.
贵州省林业厅. 1987. 茂兰喀斯特森林科学考察集. 贵阳: 贵州人民出版社: 1-386.
郭克疾, 邓学建. 2006. 湖南省蛇类一新纪录——平鳞钝头蛇. 四川动物, 25 (2): 270.
江建平, 周开亚. 2001. 从12S rRNA基因序列研究中国蛙科24种的进化关系. 动物学报, 47 (1): 38-44.
李成, 叶昌媛. 1999. 我国臭蛙属的分类进展(Anura: Ranidae)//中国动物学会. 中国动物科学研究——中国动物学会第十四届会员代表大会及中国动物学会65周年年会论文集.
李永民. 2015. 基于线粒体基因组的中国蛙科系统关系及花臭蛙的系统地理学研究. 安徽师范大学博士学位论文.
刘承钊, 胡淑琴. 1961. 中国无尾两栖类. 北京: 科学出版社: 1-364.
乔梁. 2011. 广义花臭蛙的形态量度和遗传分化研究. 河南师范大学硕士学位论文.
庆宁, 马天峰, 梁晓旭, 等. 2011. 华南地区黑眶蟾蜍的遗传变异和地理分化. 动物分类学报, 36 (2): 356-367.
史静耸, 杨登为, 张武元, 等. 2016. 西伯利亚蝮-中介蝮复合种在中国的分布及其种下分类(蛇亚目: 蝮亚科). 动物学杂志, 51 (5): 777-798.
陶娟. 2010. 基于线粒体基因序列探讨广义花臭蛙种群遗传结构. 河南师范大学硕士学位论文.
田婉淑, 江耀明. 1986. 中国两栖爬行动物鉴定手册. 北京: 科学出版社: 1-164.
魏刚, 陈服官, 李德俊. 1989. 贵州两栖动物区系及地理区划的初步研究. 10 (3): 241-249.
伍律, 董谦, 须润华. 1986. 贵州两栖类志. 贵阳: 贵州人民出版社: 1-192.
伍律, 李德俊, 刘积琛. 1985. 贵州爬行类志. 贵阳: 贵州人民出版社: 1-436.
谢莉, 宋先华, 蒋鸿, 等. 2013. 贵州省荔波发现百花锦蛇. 动物学杂志, 48 (3): 487-489.
熊洪林, 刘燕, 覃龙江, 等. 2010. 贵州茂兰泽陆蛙繁殖生态观察. 四川动物, 29 (5): 535-539.
杨大同, 饶定齐. 2008. 云南两栖爬行动物. 昆明: 云南科技出版社: 1-411.
叶昌媛, 费梁. 2001. 我国臭蛙属(两栖纲: 蛙科)的系统发育. 动物学报, (5): 528-534.
叶昌媛, 费梁, 胡淑琴. 1993. 中国珍稀及经济两栖动物. 成都: 四川科学技术出版社: 1-412.
疣宁, 马天峰, 梁晓旭, 等. 2011. 华南地区黑眶蟾蜍的遗传变异和地理分化. 动物分类学报, 36 (2): 356-367.
玉屏, 兰洪波, 冉景丞, 等. 2011. 茂兰自然保护区生物多样性现状及保护对策. 现代农业科技, (15): 233-236.
翟晓飞. 2015. 基于线粒体基因的中国大绿臭蛙复合体遗传分化研究. 河南师范大学硕士学位论文.

张汾, 李友邦. 2017. 我国睑虎属的分类和保护研究进展. 安徽农业科学, 45 (01): 1-3+7.

张洁, 庆宁, 易祖盛, 等. 2014. *Goniurosaurus* indet.的有效性及中国睑虎属(Squamata: Sauria: Eublepharidae)种间的亲缘关系. 华南师范大学学报(自然科学版), 46 (2): 92-98.

张永宏, 龚大洁, 闫礼, 等. 2012. 贵州省从江县太阳山两栖爬行动物研究. 安徽农业科学, 40 (1): 194-195, 202.

赵尔宓. 1990. 我国有尾类分类学中的几个问题及其名录//赵尔宓. 从水到陆——刘承钊教授诞辰九十周年纪念文集. 北京: 中国林业出版社: 1-454.

赵尔宓. 2006. 中国蛇类(上册). 合肥: 安徽科学技术出版社: 1-372.

赵尔宓, 胡其雄. 1984. 中国有尾两栖类的研究. 成都: 四川科学技术出版社: 1-68.

赵尔宓, 黄美华, 宗愉, 等. 1998. 中国动物志 爬行纲 第三卷 有鳞目 蛇亚目. 北京: 科学出版社: 152-154.

郑建州, 周江. 2000. 佛顶山自然保护区两栖动物物种组成及区系分析//贵州省林业厅. 贵州佛顶山自然保护区科学考察集. 北京: 中国林业出版社: 244-247.

中国野生动物保护协会. 2002. 中国爬行动物图鉴. 郑州: 河南科学技术出版社: 1-161.

朱艳军. 2016. 花臭蛙复合体遗传分化研究. 河南师范大学硕士学位论文.

Aeberhard R. 2014. Bemerkungen zur Naturbrut der Chinesischen Nasenotter, *Deinagkistrodon acutus* (Günther, 1888). Ophidia, 8 (1): 23-32.

Alexander P R, Burbrink F T. 2013. Early origin of viviparity and multiple reversions to oviparity in squamate reptiles. Ecology Letters, 17 (1): 13-21.

Amphibia China. 2017. The database of Chinese amphibians. http://www. amphibiachina.org/. Kunming Institute of Zoology (CAS), Kunming, Yunnan, China [2018-2-14].

Amphibia Web. 2016. Amphibia Web: Information on amphibian biology and conservation. University of California, Berkeley, CA. http://amphibiaweb.org/. [2016-10-9].

Amphibia Web. 2017. Amphibia Web: Information on amphibian biology and conservation. University of California, Berkeley, CA. http://amphibiaweb.org/. [2017-10-9].

Ananjeva N B, Munkhbayar K, Orlov N L, et al. 1997. Amphibians and reptiles of Mongolia. Reptiles of Mongolia. Vertebrates of Mongolia. Moscow, Russia: KMK Press: 328-330 [in Russian with English summary].

Anderson J. 1871. On some Indian reptiles. Proc. Zool. Soc., London, 1871: 149-211.

Anderson J. 1879. Reptilia & Amphibian. Anatomical and zoological researches: comprising an account of the zoological results of the two expeditions to western Yunnan in 1868 & 1875. London: Quaritch.

Angel M F. 1920. Liste de reptiles récémment déterminés et entrés dans les collections et description d'une nouvelle espèce du genre *Amblycephalus*. Bull. Mus. Nation Hist. Nat., Paris, 1920: 111-114.

Barbour T. 1908. Some new reptiles and amphibians. Bull. Mus. Comp. Zool. Harvard, 51 (12): 315-325.

Blyth E. 1855. Notices and descriptions of various reptiles, new or little known [part 2]. J. Asiatic Soc. Bengal, Calcutta, 23 (3): 287-302.

Boettger O. 1886. Diagnoses reptilium novorum ab ill. viris O. Herz et Consule Dr. O. Fr. de Moellendorff in Sina meridionali repertorum. Zool. Anz., 9: 519-520.

Boie F. 1827. Bemerkungen über Merrem's Versuch eines Systems der Amphibien, 1. Lieferung: Ophidier. Isis van Oken, 20: 508-566.

Boie H. 1826. Merkmale einiger japanischer Lurche. Isis von Oken, 18-19: 203-216.

Boulenger G A. 1890a. List of the reptiles, batrachians, and freshwater fishes collected by Professor Moesch and Mr. Iversein in the district of Deli, Sumatra. Proc. Zool. Soc., London, 1890: 31-40.

Boulenger G A. 1890b. The Fauna of British India, Including Ceylon and Burma. Reptilia and Batrachia. London: Taylor & Francis: 541.

Boulenger G A. 1894. Catalogue of the Snakes in the British Museum (Natural History). Volume Ⅱ. Containing the Conclusion of the Colubridæ Aglyphæ. British Mus. (Nat. Hist.), London, 6: 382.

Boulenger G A. 1896. Catalogue of the Snakes in the British Museum. Vol. 3. London: Taylor & Francis: 727.

Boulenger G A. 1899. On a collection of reptiles and batrachians made by Mr. J. D. La Touche in N. W. Fokien, China. Proc. Zool. Soc., London, 1899: 159-172.

Boulenger G A. 1914. Descriptions of new species of snakes in the collection of the British Museum. Ann. Mag. Nat. Hist. Ser., 8 (14): 481-485.

Boulenger G A. 1908. A revision of the Oriental pelobatid batrachians (genus *Megalophrys*). Proceedings of the Zoological Society of London, 1908: 407-430.

Boulenger G A. 1918. XXVII. —On the races and variation of the edible frog, *Rana esculenta* L. Journal of Natural History, 2 (10): 241-257.

Burger W L. 1971. Genera of *Pitvipers*. Ph.D. Thesis, University of Kansas, Lawrence, USA.

Cantor T. 1842a. General features of Chusan, with remarks on the flora and fauna of that island [part 1]. Ann. Mag. Nat. Hist., 9 (1): 265-278.

Cantor T. 1842b. General features of Chusan, with remarks on the flora and fauna of that island [part 3]. Ann. Mag. Nat. Hist., 9 (1): 481-493.

Cantor T E. 1839. Spicilegium serpentium indicorum [part 1]. Proc. Zool. Soc., London, 1839: 31-34.

Chanard T, Parr J W K, Nabhitabhata J. 2015. A Field Guide to the Reptiles of Thailand. Oxford: Oxford University Press: 352.

Cannatella D C. 1985. A phylogeny of primitive frogs (Archaeobatrachians). Ph.D. Dissertation, The University of Kansas, Lawrence.

Che J, Hu J S, Zhou W W, et al. 2009. Phylogeny of the Asian spiny frog tribe Paini (Family Dicroglossidae) *sensu* Dubois. Molecular Phylogenetics & Evolution, 50 (1): 59-73.

Chen J M, Zhou W W, Poyarkov N A, et al. 2016. A novel multilocus phylogenetic estimation reveals unrecognized diversity in Asian horned toads, genus *Megophrys sensu lato* (Anura: Megophryidae). Molecular Phylogenetics and Evolution, 106: 28-43.

Chen L, Murphy R W, Lathrop A, et al. 2005. Taxonomic chaos in Asian ranid frogs: an initial phylogenetic resolution. Herpetological Journal, 15 (4): 231-243.

Chen X, Jiang K, Guo P, et al. 2014. Assessing species boundaries and the phylogenetic position of the rare Szechwan ratsnake, *Euprepiophis perlaceus* (Serpentes: Colubridae), using coalescent-based methods. Molecular Phylogenetics and Evolution, 70: 130-136.

Chen X, Lemmon A R, Lemmon E M, et al. 2017. Using phylogenomics to understand the link between biogeographic origins and regional diversification in ratsnakes. Molecular Phylogenetics and Evolution, 111: 206-218.

Chen X H, Chen Z, Jiang J P, et al. 2013. Molecular phylogeny and diversification of the genus *Odorrana* (Amphibia, Anura, Ranidae) inferred from two mitochondrial genes. Molecular Phylogenetics and Evolution, 69 (3): 1196-1202.

Chen Y, Zhang D, Jiang K, et al. 1992. Evaluation of snake venoms among *Agkistrodon* species in China. Asiatic Herpetological Research, 4: 58-61.

Conrad J L. 2008. Phylogeny and systematics of Squamata (Reptilia) based on morphology. Bulletin of the American Museum of Natural History: 1-182.

Cope E D. 1861. Catalogue of the Colubrids in the museum of the Academy of Natural Sciences of Philadelphia. Part 3. Proc. Acad. Nat. Sci. Philadelphia, 12 [1860]: 553-566.

Cox M J, van Dijk P P, Jarujin N, et al. 1998. A photographic guide to snakes and other Reptiles of Peninsular Malaysia, Singapore and Thailand. Copeia, 1999 (1): 239.

Currin C. 2016. Recent reptiles records from Kaeng Krachan National Park, Thailand. Seavr, 2016: 117-120.

Das A, Saikia U, Murthy B, et al. 2009. A herpetofaunal inventory of Barail Wildlife Sanctuary and adjacent regions, Assam, north-eastern India. Hamadryad, 34 (1): 117-134.

Das I. 1999. Biogeography of the amphibians and reptiles of the Andaman and Nicobar Islands, India// Ota H. Tropical Island Herpetofauna. Amsterdam: Elsevier: 43-77.

Das I. 2012. A naturalist's guide to the snakes of South-east Asia: Malaysia, Singapore, Thailand, Myanmar, Borneo, Sumatra, Java and Bali. Oxford: Oxford John Beaufoy Publishing.

David A. 1871. Rapport adressé à MM. les professeurs-administrateurs du Muséum d'Histoire Naturelle. Nouvelles archives du Muséum d'Histoire Naturelle de Paris, 7: 75-100.

David P, Bain R H, Truong N Q, et al. 2007. A new species of the natricine snake genus *Amphiesma* from the Indochinese Region (Squamata: Colubridae: Natricinae). Zootaxa, 1462 (1): 41-60.

David P, Merel J, Coxolivier S G, et al. 2004. Book review-when a bookreview is not sufficient to say all: an in-depth analysis of a recent book on the snakes of Thailand, with an updated checklist of the snakes of the Kingdom. The Natural History Journal of Chulalongkorn University, 4 (1): 47-80.

Dowling H G, Jenner J V. 1988. Snakes of Burma: checklist of reported species and bibliography. Smithsonian Herp. Inf. Serv., (76): 19.

Dubois A. 1975. Un nouveau sous-genre (*Paa*) et trois nouvelles espèces du genre *Rana*. Remarques sur la phylogénie des Ranidés (Amphibiens, Anoures). Bull. Mus. Nat. Hist. Nat. Zool., 231: 1091-1115.

Dubois A. 1987. Miscellanea taxinomicabatrchologica (1). Alytes, 5: 7-95.

Dubois A. 1992. Notes sur la classification des Ranidae (Amphibians Anoures). Bull. Mens. Soc. Linn. Lyon., 61 (10): 305-352.

Eimermacher T. 2016. Die Chinesische Nasenotter (*Deinagkistrodon acutus*): ein herpetologisches und kulturelles Phänomen. Reptilia, 21 (120): 44-49.

Faivovich J, Haddad C F B, Garcia P C A, et al. 2005. Systematic review of the frog family Hylidae, with special reference to hylinae: phylogentic analysis and taxonomic revision. Bulletin of the American Museum of Natural History, 294 (294): 1-240.

Fei L, Ye C Y. 2016. Amphibians of China. Vol. 1. Beijing: Science Press.

Fei L, Ye C Y, Yang R S. 1984. A new species and subspecies of the genus *Tylototriton* (Caudata: Salamandridae). Acta Zoologica Sinica, 30: 85-91.

Fellows S. 2015. Species diversity of snakes in Pachmarhi Biosphere Reserve. Entomol. Ornithol. Herpetol., 4: 136.

Fischer J G. 1886. Herpetologische Notizen. Abh. Geb. Naturw. Hamb., 9: 51-67.

Fleck J. 1987. Erst-Nachzucht von *Agkistrodon acutus* (Günther 1888) (Serpentes: Crotalidae). Salamandra, 23 (4): 193-203.

Frost D R. 2016. Amphibian species of the world: an online reference. Version 6.0 (Date of access). American Museum of Natural History, New York, USA. http://research. amnh. org/herpetology/amphibia/index. html [2016-11-12].

Frost D R. 2017. Amphibian species of the world: an online reference. Version 6.0 (Date of access). American Museum of Natural History, New York, USA. http://researchamnhorg/herpetology/amphibia/index. html [2017-11-12].

Frost D R, Grant T, Faivovich J, et al. 2006. The amphibian tree of life. Bulletin of the American Museum of Natural History, 297: 1-291.

Gloyd H K. 1979. A new generic name for the hundred-pace viper. Proc. Biol. Soc., Washington, 91: 963-964.

Gray J E. 1853. Descriptions of some undescribed species of reptiles collected by Dr. Joseph Hooker in the Khassia Mountains, East Bengal, and Sikkim Himalaya. Ann. Mag. Nat. Hist., 12 (2): 386-392.

Grismer L L, McGuire J A, Sosa R, et al. 2002. Revised checklist and comments on the terrestrial herpetofauna of Pulau Tioman, Peninsular Malaysia. Herpetological Review, 33 (1): 26-29.

Grismer L L, Neang T, Chav T, et al. 2008a. Additional amphibians and reptiles from the Phnom Samkos Wildlife Sanctuary in Northwestern Cardamom Mountains, Cambodia, with comments on their taxonomy and the discovery of three new species. The Raffles Bulletin of Zoology, 56 (1): 161-175.

Grismer L L, Neang T, Chav T, et al. 2008b. Checklist of the amphibians and reptiles of the Cardamom region of Southwestern Cambodia. Cambodian Journal of Natural History, (1): 12-28.

Grossmann W. 1996. Das Portrait: *Elaphe porphyracea* (Cantor). Sauria, 18 (4): 1-2.

Grossmann W, Klaus D S. 2000. *Elaphe porphyracea laticincta* Schulz & Helfenberger. Sauria, 22 (2): 2.

Grossmann W, Tillack F. 2004. Angaben zur Haltung und Vermehrung der Breitband-Porphyrnatter *Oreophis porphyraceus laticincta* (Schulz & Helfenberger, 1998) (Squamata: Colubridae). Sauria. Berlin, 26 (1): 37-44.

Gu X, Chen R, Tian Y, et al. 2012. A new species of *Paramesotriton* (Caudata: Salamandridae) from Guizhou Province, China. Zootaxa, 3510 (1): 41-52.

Gu X M, Wang H, Chen G R, et al. 2012. The phylogenetic relationships of *Paramesotriton* (Caudata: Salamandridae) based on partial mitochondrial DNA gene sequences. Zootaxa, 3150: 59-68.

Gumprecht A. 2003. Anmerkungen zu den chinesischen Kletternattern der Gattung *Elaphe* (*sensu lato*) Fitzinger, 1833. Reptilia (Münster), 8 (6): 37-41.

Gumprecht A, Tillack F, Orlov N L, et al. 2004. Asian Pitvipers. Berlin: Geitje Books: 368.

Günther A. 1858. Catalogue of Colubrine snakes of the British Museum. British Museum (Natural History), London, I - XVI: 1-281.

Günther A. 1864. The Reptiles of British India. London (Taylor & Francis), xxvii: 452.

Günther A. 1872. Seventh account of new species of snakes in the collection of the British Museum. Ann. Mag. Nat. Hist., (4) 9: 13-37.

Günther A. 1875. Second report on collections of Indian Reptiles obtained by the British Museum. Proc. Zool. Soc., London, 1875: 224-234.

Günther A. 1888. On a collection of reptiles from China. Ann. Mag. Nat. Hist., 1 (6): 165-172.

Günther A. 1889. Third contribution to our knowledge of reptiles and fishes from the Upper Yangtsze-Kiang. Ann. Mag. Nat. Hist., 4 (21): 218-229.

Guo K J, Deng X J. 2009. A new species of *Pareas* (Serpentes: Colubridae: Pareatinae) from the Gaoligong Mountains, southwestern China. Zootaxa, 2008: 53-60.

Guo P, Ermi Z. 2004. *Pareas stanleyi*—A record new to Sichuan, China and a Key to the Chinese Species. Asiatic Herpetological Research, 10: 280-281.

Guo P, Jadin R C, Malhotra A, et al. 2009. An investigation of the cranial evolution of Asian pitvipers (Serpentes: Crotalinae), with comments on the phylogenetic position of *Peltopelor macrolepis*. Acta Zoologica, 91: 401-407.

Guo P, Zhang F J, Chen Y Y. 1999. The hemipenes of Chinese species of *Deinagkistrodon* and Gloydius (Serpentes: Crotalinae). Asiatic Herpetological Research, 8: 38-42.

Guo P, Zhu F, Liu Q, et al. 2014. A taxonomic revision of the Asian keelback snakes, genus *Amphiesma* (Serpentes: Colubridae: Natricinae), with description of a new species. Zootaxa, 3873 (4): 425-440.

Hallermann J, Bohme W. 2000. A review of the genus *Pseudocalotes* (Squamata: Agamidae), with description of a new species from West Malaysia. Amphibia-Reptilia, 21 (2):193-210.

Hallowell E. 1860. Report upon the reptilia of the North Pacific exploring expedition, under command of Capt. John Rogers, U. S. N. Proc. Acad. Nat. Sci. Philadelphia, 12 (1861): 480-510.

Heiko W. 2013. Im Schlangenland—Ein Besuch bei Yvonne und Thomas Klesius. Terraria Elaphe, (2): 88-93.

Hoge A R, Sarwl R H. 1978/1979. Poisonous Snakes of the World: Cheek list of the Pit Vipers, Viperidae, Crotalinae. Butantan: Mem. Inst.: 179-284.

Hornig M. 2015. Thailändische Bambusnattern (*Oreocryptophis porphyraceus coxi*) in Terrarium. Reptilia (Münster), 20 (115): 41-45.

Hoser R T. 2012. A taxonomic revision of the Colubrinae genera *Zamenis* and *Orthriophis* with the creation of two new genera (Serpentes: Colubridae). Australasian Journal of Herpetology, 11: 59-64.

Huang S, He S, Peng Z, et al. 2007. Molecular phylogeography of endangered sharp-snouted pitviper (*Deinagkistrodon acutus*; Reptilia, Viperidae) in Mainland China. Molecular Phylogenetics and Evolution, 44: 942-952.

Huang Y, Hu J, Wang B, et al. 2016. Integrative taxonomy helps to reveal the mask of the genus *Gynandropaa* (Amphibia: Anura: Dicroglossidae). Integrative Zoology, 11 (2): 134-150.

Hua X, Fu C Z, Li J T, et al. 2009. A revised phylogeny of Holarctic treefrogs (genus *Hyla*) based on nuclear and mitochondrial DNA sequences. Herpetologica, 65: 246-259.

Jaureguy F, Landreau L, Passet V, et al. 2008. Evolution of the mitochondrial genome in snakes: gene rearrangements and phylogenetic relationships. BMC Genomics, 9: 560.

Jiang J P, Zhou K Y. 2005. Phylogenetic relationships among Chinese ranids inferred from sequence data set of 12S and 16S rDNA. The Herpetological Journal, 15 (1): 1-8.

Kaiser H, Crother B I, Kelly C M R, et al. 2013. Best practices: in the 21st century, taxonomic decisions in herpetology are acceptable only when supported by a body of evidence and published via peer-review. Herpetological Review, 44 (1): 8-23.

Kästle W, Rai K, Schleich H H. 2013. Field guide to Amphibians and Reptiles of Nepal. ARCO-Nepale, V: 625.

Kiew B H. 1987. An annotated checklist of the herpetofauna of Ulu Endau Johore Malaysia. Malayan Nature J., 41 (1-3): 413-423.

Knietsch V. 2005. Die Terrarienhaltung der Thailändischen Roten Bambusnatter (*Oreophis porphyraceus coxi*) sowie eine seltene Zwillingsgeburt. Reptilia (Münster), 10 (56): 61-66.

Kosuch J, Vences M, Dubois A, et al. 2001. Out of Asia: mitochondrial DNA evidence for an Oriental origin of tiger frogs, genus *Hoplobatrachus*. Molecular Phylogenetics & Evolution, 21 (3): 398-407.

Kunz K. 2015. Dauerbrenner in Terrarium: Nattern. Reptilia (Münster), 20 (115): 18-27.

Laita M. 2013. Serpentine. Auckland, New Zealand: Abrams and PQ Blackwell: 200.

Lathrop A. 1997. Taxonomic review of the megophryid frogs (Anura: Pelobatoidear). Asian Herpetological Research, 7: 68-79.

Lenk P, Joger U, Wink M. 2001. Phylogenetic relationships among European ratsnakes of the genus Elaphe Fitzinger based on mitochondrial DNA sequence comparisons. Amphibia-Reptilia, 22 (3): 329-339.

Lenz N, Gues M, Klump B. 2012. Von Schmetterlingen und Donnerdrachen-Natur und Kultur in Bhutan (Karlsruher Naturhefte) Taschenbuch-14. Verlag: Staatliches Museum f. Naturkde Karlsruhe: 124.

Leviton A E, Guinevere O U W, Michelle S K, et al. 2003. The dangerously venomous snakes of myanmar illustrated checklist with keys. Proc. Cal. Acad. Sci., 54 (24): 407-462.

Li D. 1989. A survey of reptiles in Leigongshan area [269-275]// Matsui M, Hikida T, Goris R C. Current Herpetology in East Asia. Proceedings of the Second Japan-China Herpetological Symposium Kyoto, July 1988. Herpetological Society of Japan, Kyoto Japan: i-ix, 1-521.

Li J T, Li Y, Klaus S, et al. 2013. Diversification of rhacophorid frogs provides evidence for accelerated faunal exchange between India and Eurasia during the Oligocene. Proceedings of the National Academy of Sciences of the United States of America, 110 (9):3441.

Lillywhite H B. 2014. How snakes work: structure, function and behavior of the world's snakes. New York: Oxford University Press: 256.

Lim K K P, Ng H H. 1999. The terrestrial herpetofauna of Pulau Tioman Peninsular Malaysia. Raffles Bull. Zool., 6: 131-155.

Linnaeus C. 1758. Systema naturæ per regna tria naturæ, secundum classes, ordines, genera, species, cum characteribus, differentiis, synonymis, locis. Tomus Ⅰ. Editio decima, reformata. Laurentii Salvii, Holmiæ. 10th Edition: 824.

Love B. 2010. Die China-connection. Reptilia (Münster), 15 (82): 14-15.

Lyu Z T, Zeng Z C, Wang J, et al. 2017. Resurrection of genus *Nidirana* (Anura: Ranidae) and synonymizing *N. caldwelli* with *N. adenopleura*, with description of a new species from China. Amphibia-Reptilia, 38 (4): 483-502.

Mahony S. 2010. Systematic and taxomonic revaluation of four little known Asian agamid species, *Calotes kingdonwardi* Smith, 1935, *Japalura kaulbacki* Smith, 1937, *Salea kakhienensis* Anderson, 1879 and the monotypic genus *Mictopholis* Smith, 1935 (Reptilia: Agamidae). Zootaxa, 2514 (1): 1-23.

Mahony S, Foley N M, Biju S D, et al. 2017. Evolutionary history of the Asian Horned Frogs (Megophryinae): integrative approaches to timetree dating in the absence of a fossil record. Molecular Biology and Evolution, 34 (3): 744-771.

Manthey U, Grossmann W. 1997. Amphibien & Reptilien Südostasiens. Münster: Natur und Tier Verlag: 512.

Marx H. 1958. Catalogue of type specimens of reptiles and amphibians in Chicago Natural History Museum. Fieldiana Zool., 36: 407-496.

Matsui M, Shimada T, Ota H, et al. 2005. Multiple invasions of the Ryukyu Archipelago by Oriental frogs of the subgenus *Odorrana* with phylogenetic reassessment of the related subgenera of the genus *Rana*. Molecular Phylogenetics & Evolution, 37 (3): 733-742.

Mattison C. 2007. The New Encyclopedia of Snakes. State of New Jersey: Princeton University Press.

McDiarmid R W, Campbell J A, Touré T A. 1999. Snake species of the world. Herpetologists' League, 1: 511.

Mell R. 1931. List of Chinese snakes. Lingnan Sci. Jour. Canton, 8 (1929): 199-219.

Meyen F J F. 1834. Reise um die Erde ausgeführt auf dem Königlich Preussischen Seehandlungs Schiffe Prinzess Louise, commandirt von Capitain W. Wendt, in den Jahren 1830, 1831 und 1832. 3 vols. Berlin: Sander'sche buchhandlung.

Ngo A, Murphy R W, Liu W, et al. 2006. The phylogenetic relationships of the Chinese and Vietnamese waterfall frogs of the genus *Amolops*. Amphibia Reptilia, 1 (27): 81-92.

Nishikawa K, Matsui M, Nguyen T T. 2013. A new species of *Tylototriton* from Northern Vietnam (Amphibia: Urodela: Salamandridae). Current Herpetology, 32 (1): 34-49.

Oliver L A, Prendini E, Kraus F, et al. 2015. Systematics and biogeography of the *Hylarana* frog (Anura: Ranidae) radiation across tropical Australasia, Southeast Asia, and Africa. Molecular Phylogenetics & Evolution, 90: 176-192.

Osbeck P. 1765. Reise nach Ostindien und China: Nebst O. Toreens Reise nach Suratte und CG Ekebergs Nachricht von der landwirthschaft der chineser. Aus dem schwedischen übersetzt von J G Georgi. Rostock: Johann Christian Koppe: 1-552.

Oshima M. 1910. An annotated list of Formosan snakes, with descriptions of four new species and one new subspecies. Annot. Zool. Japon Tokyo, 7 (3): 185-207.

Pallas P S. 1776. Reise durch verschiedene Provinzendes Russischen Reichs, Gedruckt bey der Kayserlichen. St. Petersburg: Imper. Acad. Sci.: 457-760.

Parkinson C L. 1999. Molecular systematics and biogeographical history of pitvipers as determined by mitochondrial ribosomal DNA sequences. Copeia, (3): 576-586.

Peloso P L V, Frost D R, Richards S J, et al. 2015. The impact of anchored phylogenomics and taxon sampling on phylogenetic inference in narrow-mouthed frogs (Anura, Microhylidae). Cladistics-the International Journal of the Willi Hennig Society, 32 (2): 113-140.

Phimmachak S, Aowphol A, Stuart B L, et al. 2015. Morphological and molecular variation in *Tylototriton* (Caudata: Salamandridae) in Laos, with description of a new species. Zootaxa, 4006 (2): 285-310.

Pope C H. 1929. Four new snakes and a new lizard from South China. American Museum Novitates, 325: 1-4.

Pyron R A, Burbrink F T. 2014. Early origin of viviparity and multiple reversions to oviparity in squamate reptiles. Ecology. Letters, 17 (1): 13-21.

Pyron R A, Burbrink F T, Colli G R, et al. 2011. The phylogeny of advanced snakes (Colubroidea), with discovery of a new subfamily and comparison of support methods for likelihood trees. Molecular phylogenetics and evolution, 58 (2): 329-342.

Pyron R A, Burbrink F T, Wiens J J. 2013. A phylogeny and revised classification of Squamata, including 4161 species of lizards and snakes. Bmc. Evolutionary Biology, 13 (1): 93.

Pyron R A, Wiens J J. 2011. A large-scale phylogeny of Amphibia including over 2800 species, and a revised classification of extant frogs, salamanders, and caecilians. Molecular Phylogenetics and Evolution, 61 (2): 543-583.

Reinhardt J T. 1844. Description of a new species of venomous snake, *Elaps* Macclellandi. Calcutta J. Nat. Hist., 4: 532-534.

Roelants K, Jiang J, Bossuyt F. 2004. Endemic ranid (Amphibia: Anura) genera in southern mountain ranges of the Indian subcontinent represent ancient frog lineages: evidence from molecular data. Molecular Phylogenetics and Evolution, 31 (2): 730-740.

Rossman D A, Eberle W G. 1977. Partition of the genus *Natrix*, with preliminary observations on evolutionary trends in natricine snakes. Herpetologica, 33 (1): 34-43.

Russell P. 1796. An account of Indian serpents collected on the coast of Coromandel, containing descriptions and drawings of each species, together with experiments and remarks on their several poisons. George Nicol., London, 8: 90.

Sang, N V, Ho T C, Nguyen Q T. 2009. Herpetofauna of Vietnam. Frankfurt: Chimaira: 768.

Schlegel H. 1837. Essai sur la physionomie des serpens. Partie Descriptive. La Haye (J. Kips, J. HZ. et W. P. van Stockum), 16: 606.

Schmidt K P. 1925. New reptiles and a new salamander from China. American Museum Novitates, (157): 1-5.

Schneider J G. 1799. Historiae Amphibiorum naturalis et literariae. Fasciculus primus, continens Ranas. Calamitas, Bufones, Salamandras et Hydros. Jena: Frommanni: 266.

Schulz K D. 1989. Die hinterasiatischen Kletternattern der Gattung Elaphe, Teil 17 *Elaphe porphyracea* (Cantor 1839). Sauria, 11 (4): 21-24.

Schulz K D. 1996. Eine Monographie der Schlangengattung *Elaphe* Fitzinger. Berg (CH): Bushmaster: 1-460.

Schulz K D. 2000. Haltung und Zucht von *Elaphe porphyracea coxi* Schulz & Helfenberger, 1998 und *Elaphe porphyracea vaillanti* (Sauvage, 1876). Sauria, 22 (3): 11-16.

Schulz K D, Helfenberger N. 1998. Eine Revision des Unterarten-Komplexes der Roten Bambusnatter *Elaphe porphyracea* (Cantor, 1839). Sauria, 20 (1): 25-45.

Schulz K D, Helfenberger N, Chettri B. 2010. Psammophis nigrofasciatus, ein Juniorsynonym von Oreophis porphyraceus porphyraceus. Sauria, 32 (3): 31-38.

Sharma R C. 2004. Handbook Indian Snakes. New Delhi: Akhil Books: 292.

Siebert B J, Ansermet M. 2006. Captive breeding of the broad-banded red bamboo Ratsnake *Oreophis porphyraceus laticinctus*. Reptilia, (46): 47-49.

Singh A, Tripathi M K, Singh R. 2016. Melatonin modulates leukocytes immune responses in freshwater snakes, natrix piscator. Journal of Herpetology, 50 (1): 130-137.

Slowinski J B, Boundy J, Lawson R. 2001. The phylogenetic relationships of Asian coral snakes (Elapidae: Calliophis and Maticora) based on morphological and molecular characters. Herpetologica, 57: 233-245.

Smedley N. 1932. Amphibians and reptiles from the Cameron Highlands, Malay Peninsula. Bull Raffl Mus Singapore, 6: 105-123.

Smith M A. 1943. The Fauna of British India, Ceylon and Burma, Including the Whole of the Indo-Chinese Sub-Region. Reptilia and Amphibia. 3 (Serpentes). London: Taylor & Francis: 583.

Song M. 1987. Survey of the reptiles of southern Shaanxi. Acta Herpetologica Sinica, 6 (1): 59-64.

Stejneger L. 1910. The batrachians and reptiles of Formosa. Proc. U.S. Natl. Mus., 38: 91-114.

Taylor E H. 1965. The serpents of Thailand and adjacent waters. Univ. Kansas Sci. Bull., 45 (9): 609-1096.

Teynié A, David P, Ohler A. 2010. Note on a collection of Amphibians and Reptiles from Western Sumatra (Indonesia), with the description of a new species of the genus *Bufo*. Zootaxa, 2416: 1-43.

Theobald W. 1868. Catalogue of reptiles in the Museum of the Asiatic Society of Bengal. J. Asiat. Soc. Bengal, 37: 84.

Tillack F, Shah R, Hermann R. 1997. Zur Verbreitung der roten Bambusnatter *Elaphe porphyracea porphyracea* in Nepal. Sauria, 19 (2): 31-33.

Townsend T M, Larson A, Louis E, et al. 2004. Molecular phylogenetics of Squamata: the position of snakes, amphisbaenians, and dibamids, and the root of the squamate tree. Systematic Biology, 53 (5): 735-757.

Trapp B. 2012. Terrarientiere fotografieren. Einsatz von Licht und Blitz bei Spiegelreflexkameras. Reptilia (Münster), 17 (94): 28-38.

Trutnau L. 2001. Zur Kenntnis der Chinesischen Nasenotter *Deinagkistrodon acutus* (Günther, 1888). Herpetofauna, 23 (135): 25-34.

Unterstein W. 1930. Beiträge zur Lurchund Kriechtierfauna Kwangsi's. 2. Schwanzlurche. Sitzungsberichte der Gesellschaft Naturforschender Freunde zu Berlin, 1930: 313-315.

Utiger U, Helfenberger N, Schätti B, et al. 2002. Molecular systematics and phylogeny of Old and New World ratsnakes, *Elaphe* Auct, and related genera (Reptilia, Squamata, Colubridae). Russian Journal of Herpetology, 9 (2): 105-124.

Vidal N, Delmas A S, David P, et al. 2007. The phylogeny and classification of caenophidian snakes inferred from seven nuclear protein-coding genes. C R Biol., 330 (2): 182-187.

Vogt T. 1922. Zur Reptilien-und Amphibienfauna Südchinas. Archiv für Naturgeschichte, 88A (10): 135-146.

Weisrock D W, Papenfuss T J, Macey J R, et al. 2006. A molecular assessment of phylogenetic relationships and lineage accumulation rates within the family Salamandridae (Amphibia, Caudata). Molecular Phylogenetics and Evolution, 41 (2): 368-383.

Werner F J M. 1903. Über Reptilien und Batrachier aus Guatemala und China in der zoologisch. Staats Sammlung in München nebst einem Anhang über seltene Formen aus anderen Gegenden. Abhandlungen der Mathematisch Physikalischen Klasse der königlich Bayerischen Akademie der Wissenschaften, 22 [1903-1904] (2): 341-384.

Wiens J J, Fetzner J W, Parkinson C L, et al. 2005. Hylid frog phylogeny and sampling strategies for speciose clades.

Systematic Biology, 54 (5): 778-807.

Wiens J J, Kuczynski C A, Smith S A, et al. 2008. Branch lengths, support, and congruence: testing the phylogenomic approach with 20 nuclear loci in snakes. Systematic Biology, 57 (3): 420-431.

Wiens J J, Kuczynski C A, Xia H, et al. 2010. An expanded phylogeny of treefrogs (Hylidae) based on nuclear and mitochondrial sequence data. Molecular Phylogenetics & Evolution, 55 (3): 871-882.

Wogan G O, Stuart B L, Iskandar D T, et al. 2016. Deep genetic structure and ecological divergence in a widespread human commensal toad. Biology Letters, 12 (1): 20150807. doi: 10.1098/rsbl.2015.0807.

Wu S P, Huang C C, Tsai C L, et al. 2016. Systematic revision of the Taiwanese genus *Kurixalus* members with a description of two new endemic species (Anura, Rhacophoridae). Zookeys, (557): 121.

Yan F, Zhou W, Zhao H, et al. 2013. Geological events play a larger role than Pleistocene climatic fluctuations in driving the genetic structure of *Quasipaa boulengeri* (Anura: Dicroglossidae). Molecular Ecology, 22 (4): 1120.

Yang J H, Orlov N L, Wang Y Y. 2011. A new species of pitviper of the genus *Protobothrops* from China (Squamata: Viperidae). Zootaxa, 2936: 59-68.

Yu G H, Rao D Q, Zhang M W, Yang J X. 2009. Re-examination of the phylogeny of Rhacophoridae (Anura) based on mitochondrial and nuclear DNA. Molecular Phylogenetics and Evolution, 50: 571-579.

Yu G H, Zhang M W, Yang J X. 2010. A species boundary within the Chinese Kurixalus odontotarsus species group (Anura: Rhacophoridae): New insights from molecular evidence. Molecular Phylogenetics and Evolution, 56: 942-950.

Yu G H, Zhang M W, Yang J X. 2013. Molecular evidence for taxonomy of *Rhacophorus appendiculatus*, and *Kurixalus*, species from northern Vietnam, with comments on systematics of *Kurixalus* and *Gracixalus*, (Anura: Rhacophoridae). Biochemical Systematics & Ecology, 47 (47): 31-37.

Yuan Z Y, Jiang K, Lü S Q, et al. 2011. A phylogeny of the *Tylototriton asperrimus* group (Caudata: Salamandridae) based on a mitochondrial study: suggestions for a taxonomic revision. Dongwuxue Yanjiu, 6 (32): 577-584.

Zaher H, Grazziotin F G, Cadle J E, et al. 2009. Molecular phylogeny of advanced snakes (Serpentes, Caenophidia) with an emphasis on South American Xenodontines: a revised classification and descriptions of new taxa. Papéis Avulsos de Zoologia (São Paulo), 49 (11): 115-153.

Zhang H, Yan J, Zhang G, et al. 2008. Phylogeography and demographic history of Chinese black-spotted frog populations (*Pelophylax nigromaculata*): evidence for independent refugia expansion and secondary contact. BMC Evolutionary Biology, 8 (1): 21.

Zhang Y, Li G, Xiao N, et al. 2017. A new species of the genus *Xenophrys* (Amphibia: Anura: Megophryidae) from Libo County, Guizhou, China. Asian Herpetological Research, 8: 75-85.

Zhao E M, Adler K A. 1993. Herpetology of China. Oxford (Ohio): Society for the Study of Amphibians and Reptiles.

中文名索引

B

白唇竹叶青蛇　159
百花晨蛇　208
斑腿泛树蛙　97
北草蜥　135
北链蛇　200
壁虎属　124
鳖属　119

C

草腹链蛇　213
草蜥属　134
侧褶蛙属　90
蟾蜍属　32
晨蛇属　208
赤链蛇　198
臭蛙属　80
粗皮姬蛙　48
脆蛇蜥　139
脆蛇蜥属　139
翠青蛇　190
翠青蛇属　189

D

大绿臭蛙　82
大树蛙　103
大眼斜鳞蛇　173
滇蛙　73
东亚腹链蛇属　215
短尾蝮　161
钝头蛇属　148
钝尾两头蛇　171
多疣壁虎　125

E

峨眉树蛙　102

F

繁花林蛇　179
泛树蛙属　97
福建钝头蛇　149
福建绿蝮　157
腹链蛇属　213

H

黑斑侧褶蛙　91
黑背链蛇　202
黑眶蟾蜍　38
黑眉晨蛇　210
黑头剑蛇　177
黑线乌梢蛇　194
横纹斜鳞蛇　175
红脖颈槽蛇　224
后棱蛇属　228
虎斑颈槽蛇　222
虎纹蛙　58
虎纹蛙属　58
花臭蛙　85
花姬蛙　53
华西雨蛙　43
华西雨蛙武陵亚种　43
华游蛇属　230
滑鼠蛇　192
环蛇属　168
环纹华游蛇　232
黄斑异色蛇　227
黄链蛇　203
灰腹绿蛇　196
灰鼠蛇　193

J

姬蛙属　47
棘侧蛙　68
棘腹蛙　65

棘蜥属　141
棘胸蛙　69
棘胸蛙属　65
尖吻蝮　155
尖吻蝮属　155
睑虎属　121
剑蛇属　177
绞花林蛇　181
颈槽蛇属　221
颈棱蛇　234
颈棱蛇属　234
锯腿原指树蛙　94

K

阔褶水蛙　77

L

蓝尾石龙子　130
里氏睑虎　121
丽棘蜥　141
丽纹腹链蛇　217
荔波壁虎　126
荔波睑虎　122
荔波异角蟾　30
链蛇属　198
两头蛇属　171
林蛇属　179
陆蛙属　55
瘰螈属　22
绿臭蛙　84
绿蝮属　157
绿蛇属　196

M

蟒蛇　146
茂兰瘰螈　23
茂兰原矛头蝮　153

孟加拉眼镜蛇 166

N

南草蜥 137
拟树蜥属 143

P

平鳞钝头蛇 148
坡普腹链蛇 218

Q

琴蛙属 72

R

蚺属 146

S

三港雨蛙 45
三索蛇 188
三索蛇属 188
山溪后棱蛇 228
石龙子属 129
饰纹姬蛙 50
鼠蛇属 191
树蛙属 101
树蛙未定种 105
双团棘胸蛙 61
双团棘胸蛙属 61

水蛙属 75

T

台湾小头蛇 183
蜓蜥属 127
铜蜓蜥 128
头棱蟾属 38

W

文县疣螈 28
乌华游蛇 230
乌华游蛇台湾亚种 232
乌华游蛇指名亚种 232
无颞鳞腹链蛇 220
无声囊泛树蛙 99
务川臭蛙 88

X

细鳞拟树蜥 144
细痣疣螈 25
小弧斑姬蛙 51
小头蛇属 182
斜鳞蛇属 173
锈链腹链蛇 215

Y

亚洲蝮属 161
眼镜蛇属 166

异角蟾属 30
异色蛇 225
异色蛇属 225
银环蛇 168
疣螈属 24
雨蛙属 41
玉斑蛇 205
玉斑蛇属 205
原矛头蝮 152
原矛头蝮属 151
原指树蛙属 94

Z

泽陆蛙 56
沼水蛙 75
中国小头蛇 184
中华鳖 119
中华蟾蜍 33
中华蟾蜍华西亚种 34
中华蟾蜍指名亚种 36
中华珊瑚蛇 164
中华珊瑚蛇属 163
中华石龙子 132
竹叶青 157
竹叶青属 159
紫灰蛇 206
紫灰蛇属 206
紫棕小头蛇 186
棕黑腹链蛇 219

拉丁名索引

A

Acanthosaura　141
Acanthosaura lepidogaster　141
Amphiesma　213
Amphiesma stolatum　213

B

Boiga　179
Boiga kraepelini　181
Boiga multomaculata　179
Bufo　32
Bufo gargarizans andrewsi　34
Bufo gargarizans gargarizans　36
Bufo gargarizans　33
Bungarus　168
Bungarus multicinctus　168

C

Calamaria　171
Calamaria septentrionalis　171
Coelognathus　188
Coelognathus radiatus　188
Cyclophiops　189
Cyclophiops major　190

D

Deinagkistrodon　155
Deinagkistrodon acutus　155
Dopasia　139
Dopasia harti　139
Duttaphrynus　38
Duttaphrynus melanostictus　38

E

Euprepiophis　205
Euprepiophis mandarinus　205

F

Fejervarya　55
Fejervarya multistriata　56

G

Gekko　124
Gekko japonicus　125
Gekko liboensis　126
Gloydius　161
Gloydius brevicaudus　161
Goniurosaurus　121
Goniurosaurus liboensis　122
Goniurosaurus lichtenfelderi　121
Gynandropaa　61
Gynandropaa phrynoides　61

H

Hebius　215
Hebius atemporale　220
Hebius craspedogaster　215
Hebius optatum　217
Hebius popei　218
Hebius sauteri　219
Hoplobatrachus　58
Hoplobatrachus chinensis　58
Hyla　41
Hyla gongshanensis wulingensis　43
Hyla gongshanensis　43
Hyla sanchiangensis　45
Hylarana　75
Hylarana guentheri　75
Hylarana latouchii　77

K

Kurixalus　94
Kurixalus odontotarsus　94

L

Lycodon　198
Lycodon flavozonatus　203
Lycodon rufozonatum　198
Lycodon ruhstrati　202
Lycodon septentrionalis　200

M

Macropisthodon　234
Macropisthodon rudis　234
Microhyla　47
Microhyla butleri　48
Microhyla fissipes　50
Microhyla heymonsi　51
Microhyla pulchra　53

N

Naja　166
Naja kaouthia　166
Nidirana　72
Nidirana pleuraden　73

O

Odorrana　80
Odorrana graminea　82
Odorrana margaretae　84
Odorrana schmackeri　85
Odorrana wuchuanensis　88
Oligodon　182
Oligodon chinensis　184
Oligodon cinereus　186
Oligodon formosanus　183
Opisthotropis　228
Opisthotropis latouchii　228
Oreocryptophis　206
Oreocryptophis porphyraceus　206

Orthriophis　208
Orthriophis moellendorffi　208
Orthriophis taeniurus　210

P

Paramesotriton　22
Paramesotriton maolanensis　23
Pareas　148
Pareas boulengeri　148
Pareas stanleyi　149
Pelodiscus　119
Pelodiscus sinensis　119
Pelophylax　90
Pelophylax nigromaculatus　91
Plestiodon　129
Plestiodon chinensis　132
Plestiodon elegans　130
Polypedates　97
Polypedates megacephalus　97
Polypedates mutus　99
Protobothrops　151
Protobothrops maolanensis　153
Protobothrops mucrosquamatus　152
Pseudocalotes　143
Pseudocalotes microlepis　144
Pseudoxenodon　173
Pseudoxenodon bambusicola　175
Pseudoxenodon macrops　173
Ptyas　191

Ptyas korros　193
Ptyas mucosa　192
Ptyas nigromarginata　194
Python　146
Python bivittatus　146

Q

Quasipaa　65
Quasipaa boulengeri　65
Quasipaa shini　68
Quasipaa spinosa　69

R

Rhabdophis　221
Rhabdophis subminiatus　224
Rhabdophis tigrinus　222
Rhacophorus　101
Rhacophorus dennysi　103
Rhacophorus omeimontis　102
Rhacophorus sp.　105
Rhadinophis　196
Rhadinophis frenatum　196

S

Sibynophis　177
Sibynophis chinensis　177
Sinomicrurus　163
Sinomicrurus macclellandi　164

Sinonatrix　230
Sinonatrix aequifasciata　232
Sinonatrix percarinata percarinata　232
Sinonatrix percarinata suriki　232
Sinonatrix percarinata　230
Sphenomorphus　127
Sphenomorphus indicus　128

T

Takydromus　134
Takydromus septentrionalis　135
Takydromus sexlineatus　137
Trimeresurus　159
Trimeresurus albolabris　159
Tylototriton　24
Tylototriton asperrimus　25
Tylototriton wenxianensis　28

V

Viridovipera　157
Viridovipera stejnegeri　157

X

Xenochrophis　225
Xenochrophis flavipunctatus　227
Xenochrophis piscator　225
Xenophrys　30
Xenophrys liboensis　30